Biodiversity in HORTICULTURAL CROPS

Biodiversity in HORTICULTURAL CROPS

— Volume 3 —

Editor
Professor K.V. Peter

2023
Daya Publishing House®
A Division of
Astral International Pvt. Ltd.
New Delhi – 110 002

First Published, 2011
Reprinted, 2023
ISBN 978-81-7035-672-1

Published by : **Daya Publishing House®**
A Division of
Astral International Pvt. Ltd.
– ISO 9001:2015 Certified Company –
4736/23, Ansari Road, Darya Ganj
New Delhi-110 002
Ph. 011-43549197, 23278134
E-mail: info@astralint.com
Website: www.astralint.com

Printed at : **Replika Press Pvt Ltd**

Devotion

The 3rd volume of ' Biodiversity in Horticultural Crops' is devoted to Prof. P.I. Peter, Chairman, Noni Bio Tech, Chennai who stood as a pillar of strength in my academic adventures. He provided the fullest academic freedom to express my talents in science journalism.

Acknowledgement

I acknowledge Dr. Kirti Singh, Chairperson, World Noni Research Foundation, Chennai for encouragement and Prof. P.I.Peter, Chairman, Noni Bio Tech for physical facilities. I express gratitudes to Dr. S. Ayyappan, Director General, Indian Council of Agricultural Research, New Delhi and Secretary, DARE, Government of India for writing the Foreword to the 3rd volume of the book. My wife Vimala, sons Anvar and Ajay and daughters in law Reeba and Cynara spared me to undertake serious Science Journalism to be of use to present and future.

Preface

Research and development in Horticultural Crops have ushered in a Golden Revolution in India. The vegetable production has touched 100 million tonnes, fruits 75 million tonnes and spices 5 million tonnes. India leads the whole world in productivity in rubber (one tonne/ha). India has the capability and plant genetic resources to rank above China in vegetable production and Brazil in Fruit production. Hi-tech Horticulture in Florist crops is giving dividents in states like Maharashtra, Karnataka and Gujarat. In spices especially black pepper and cardamom, India has distinct advantages. In plantation crops like tea, coffee, cocoa, cashew, cardamom, oilpalm, cinchona and rubber, India is in a position to lead the whole world. Despite above positive advantages and merits, there are challenges facing Indian Horticulture. Dwindling land area under horticultural crops, diminishing availability of irrigation water and migration of human labour to urban areas are emerging threats. Encroaching forests and hotspots of biodiversity by human greed, conversion of the best horticultural land for industrial purposes, challenges of globalization and free trade and above all climate change are posing newer challenges. These challenges are to be converted to opportunities by making use of new crops, hitech horticulture and competitive production and marketing.

The United Nations is celebrating 2010 as International Year of Biodiversity. The main theme is "Biodiversity is life, Biodiversity is our life". Health and wellness are essential pre-requisites to happy living. The Chennai based World Noni Research Foundation(WNRF) is holding the First World Noni Congress at Chennai during 1-3 October, 2010. The present book Biodiversity of Horticultural Crops edited by Dr K V Peter deals with biodiversity of Noni (*Morinda citrifolia* L.) in its totality. Biodiversity of ashgourd, satputia (ridge gourd), watermelon, cranberry, jamun, litchi, mango, pomegranate, heliconia, marygold, zinnia, bydagi chilli and bombara groundnut are also dealt with by the best experts in the respective crops. The book will be of much academic and practical use to students, scientists, policy makers and nature lovers. I congratulate the Editor and authors of various chapters for bringing out this very useful volume.

Kirti Singh

Chairperson,
World Noni Research Foundation,
Chennai – 96
Former Chairman,
ASRB, ICAR, New Delhi – 12

Foreword

Plant biodiversity is vital for sustainability of agriculture, nutritional and food security. Depletion of biodiversity due to human interference and the likely impacts of climate change are matters of concern. Realizing the need for conservation and bioprospecting of biodiversity, the United Nations has declared 2010 as the International Year of Biodiversity with the global theme of 'Biodiversity is life. Biodiversity is our life'. The Convention on Biological Diversity (CBD) which came into force on December 29, 1993 reaffirmed the sovereign rights of states over such own biological resources and requested states to evolve their own national laws to promote conservation of their own biodiversity and its sustainable use in an environment friendly manner. India is also a party to CBD since 1994. India also enacted the Biological Diversity Act, 2002 and Biological Diversity rules, 2004. With 2.4 per cent of the world's land area, India accounts for 7–8 per cent of the recorded species of the world including 45,968 plant species and is recognized as one of the global mega-centres of biodiversity. Horticultural plants like vegetables, fruits, ornamentals, spices, medicinal and aromatic plants, and plantation crops are major sources for nutritional and livelihood security.

Documenting existing variability in vegetables -ash gourd, satputia, watermelon, fruits–cranberry, jamun, litchi, mango, noni, pomegranate; ornamentals–heliconia, marigold, zinnia; and uncommon medicinal plants–bombara groundnut, is done in the present volume III–Biodiversity of Horticultural Crops. Each of the 14 chapters is contributed by the available best authority on the crop. The entire set of volumes are edited by Dr. K.V. Peter former Vice-Chancellor, Kerala Agricultural University. It is good that publication of 3rd volume is taking place during International Year of Biodiversity-2010.

S. Ayyappan

Secretary, Department of Agricultural Research & Education and
Director General, Indian Council of Agricultural Research,
Ministry of Agriculture, Krishi Bhavan, New Delhi – 110 114

Contents

List of Contributors

Chapter 1

I. Sreelathakumary
Professor, Kerala Agricultural University, P.O.Vellayani, Trivandrum
E-mail: sreelathakumary@rediffmail.com

Chapter 2

B.R. Choudhary and P.K. Singh*
Indian Institute of Vegetable Research, Sargatia, P.O. Seorahi, Kushinagar – 274 406, U.P.
*E-mail: choudharyiivr@gmail.com

Chapter 3

Dr. B.R. Choudhary and P.K. Singh*
Indian Institute of Vegetable Research, Sargatia, P.O. Seorahi, Kushinagar – 274 406, U.P.
*E-mail: choudharyiivr@gmail.com

Chapter 4

Hilary A. Sandler and Carolyn J. DeMoranville
UMass Cranberry Station, College of Natural Sciences, University of Massachusetts Amherst,
1 State Bog Road, P.O. Box 569, E.Wareham, MA 02538 USA

*Yu Zhou, Robert R. Boulanger, Jr and Bal Ram Singh**
Department of Chemistry and Biochemistry, University of Massachusetts Dartmouth,
285 Old Westport Road, N. Dartmouth, MA 02747 USA
*E-mail:bsingh@umassd.edu

Chapter 5

A.K. Singh* and Anju Bajpai
Central Institute of Subtropical Horticulture, Rehmankhera, P.O. Kakori, Lucknow – 227 107, U.P.
*E-mail: aksinghcish@rediffmail.com

Chapter 6

Rajesh Kumar
National Research Centre for Litchi, Muzaffarpur – 842 002, Bihar
E-mail: rajeshkr_5@yahoo.com

Chapter 7

S.P. Singh
ICAR Research Complex for Goa, Ela Old Goa, Goa – 403 402
E-mail: spsinghchandel@vsnl.com

Chapter 8

D.R. Singh
Central Agricultural Research Institute, Port Blair Andaman and Nicobar Islands – 744 101
E-mail: drsingh1966@yahoo.com

Chapter 9

V.T. Ram Chandra, P. Kumar Jadhav and K. Dhinesh Babu*
National Research Centre on Pomegranate, NH-9 By Pass Road, Shelgi, Solapur – 413 006, Maharashtra
*E-mail: dh_babu@yahoo.co.in

Chapter 10

T. Janakiram
Division of Floriculture, Indian Agricultural Research Institute, New Delhi – 110 012
E-mail: tolety@iihr.ernet.in

Anuradha Sane
Division of Plant Genetic Resources, Indian Institute of Horticultural Research, Hessaraghatta Lake, Bangalore – 560 089

Chapter 11

Namita*, Kanwar P. Singh and V.S. Raju Dantuluri
Division of Floriculture and Landscaping, Indian Agricultural Research Institute, New Delhi – 110 012
*E-mail: namitabanyal@yahoo.co.in

Chapter 12

R. Crofton Sloan
North Mississippi Research and Extension Center, Mississippi State University, Verona, Mississippi, USA
E-mail: rcsloan@ra.msstate.edu

Chapter 13

H.D. Mohan Kumar, M. Shivaprasad, M.H. Tatanaga and R.K. Mesta
Horticulture Research Station, Devihosur, Haveri, University of Horticultural Sciences, Bagalkot
E-mail: ars_devihosur@rediffmail.com

Chapter 14

S.L. Venter and E van den Heever*
Agricultural Research Council, Vegetable and Ornamental Plant Institute (ARC-VOPI),
Private Bag X293, Pretoria, South Africa, 0001
*E-mail: evdheever@arc.agric.za

C.J. Swanevelder
Agricultural Research Council, Grains Institute, Post Bag X1251, Potchefstroom, South Africa, 2520

Introduction

" Biodiversity is life; Biodiversity is our life" The convention on Biological Diversity (CBD) was negotiated by the heads of all nations met at the UNCED Earth Summit at Rio de Janeiro in Brazil on June 5, 1992. The CBD reaffirmed the sovereign rights of states over their own biological resources like flora and fauna including indigenous traditional knowledge (ITK). The states are encouraged to make their own laws of the land to promote conservation of their own biodiversity, its sustainable uses in an environment friendly manner and to allow access to the natural resources with pre- informed consent vis-a vis equitable sharing of the benefits resulting from the use of biodiversity and associated traditional knowledge. The CBD came into force on December 29, 1993. There are now 193 parties to this convention. Government of India enacted the Biological Diversity Act, 2002 and Biological Diversity Rules in 2004. The principles on which the benefits accrue form the bases of a benefit sharing agreement. Decisions on who has the right to provide access and benefit sharing (ABS) are left to countries to define. Bonn-Guidelines on ABS in 2002 during the 6th Conference of Parties (CoP) are expected to guide and assist countries in implementing ABS strategies/regimes. An International Regime (IR) can ensure that boidiversity–rich developing countries obtain a fair and equitable share of benefits of the utilization of genetic resources.

The previous volumes I and II dealt with variability in Horticultural crops–vegetables, fruits, plantation crops, spices, aromatic and medicinal plants and tuber crops -. The present volume III deals with diversity in 14 crops including three vegetables, six fruits, three ornamentals, one spice crop and one medicinal plant–Bombara Groundnut–uncommon to Asian readers but common in South Africa, where Mohandas Karamchand Gandhi started non violent satyagraha against the oppressors and apartheid.

Ashgourd, a native of Indo–Malaysia, finds mention in Chinese records of the fifth and sixth centuries. Fruits are used in confectionary, ayurvedic rasayanams and primarily as a vegetable in mixes and sambars. The ripe fruits are pealed, cut into pieces and candied to prepare the well known petha. The fruit is tonic, nutritive and diuretic. Genetic resources of ashgourd are maintained in Germany, India and USA. Breeding methods like mass selection, pedigree methods and bulk population are employed to increase productivity, quality and bring resistance to pests and diseases. Descriptors are available to catalogue the variability available in the crop. Monton (1971) described three cultivar types in wax gourd–fruits nearly round and essentially hairless, fruits nearly round and hairy and fruits oblong and hairy–.Resmi (2004) catalouged 25 land races of ash gourd from Kerala, Tamil Nadu and Karnataka, based on the description of National Bureau of Plant Genetic Resources (NBPGR) New Delhi. Changlin (1998) reported resistance to viral mosaic in six land races. Genetic divergence in ash gourd was studied by Lovely (2001) and Resmi and Sreelathakumary(2008). Mandal and Sirohi (2003) reported heterosis in ash gourd. Molecular characterization of ash gourd germplasm was carried out using PCR based DNA markers and RAPD techniques. The debilitating disease Diabetes mellitus Type II escalating and scorching, ash gourd is to receive intensive attention for preventive management.

Satputia is a curcurbit fruit vegetable related to ridge gourd and smooth gourd, both being monoecious and satputia bearing hermaphrodite flowers. It is described as a hermaphrodite form of ridge gourd. *Luffa*, to which satputia belongs, is an old world genus comprising nine species out of which seven are native to India. VRS-1 is the only line identified at Indian Institute of Vegetable Research, Kushinagar (UP) for early fruiting (44.0-46.2days) coupled with high fruit yield (5.2.6.3kg / plant). Ram *et al.* (2006) studied genetic variability of satputia in 24 genotypes and reported SP -59 to give maximum fruit yield (10.8t/ha), Choudhary *et al.* (2009) evaluated 35 genotypes and reported VRS–1, VRS–7 and VRS–11 as promising. Ram *et al* (2004) reported the highest heterosis in satputia for fruit yield / plant in VRST–2 x VRST–16 (121.74 per cent) followed by VRST–60 x VRST -16 (108.20 per cent) and VRST- 57 x VRST–53 (86.30 per cent) in a set of 15 hybrids. Satputia fruits on controlled drying possess ornamental value and health uses as a scrub. The seeds yield edible oil and cake a nutritious food to cattle.

Water melon is one of the most widely cultivated crops belonging to family Cucurbitaceae. It is a native of Tropical Africa especially the Kalahari Desert (the current nations of Namibia and Botswana) where wild forms are still found. Watermelon is domesticated at least 4000 years ago and now grown world wide where long and hot summers prevail. Whitaker (1933) considered wild water melon (*Citrullus colocynthis*) as possible ancestor of watermelon. Morphologically the *colocynthis* has similar characters with *vulgaris* but fruits are bitter and seeds small. Water melon differs from other economically important cucurbits by having pinnatifid leaves. It is monoecious, flowers unisexual and plant classified as a naturally cross pollinated crop. Water melon fruits are produced in different sizes (tea box, small, medium large or giant), different shapes (round, oval, blocky or elongate), different rind patterns (gray, narrow strip, medium strip, wide strip, light solid or dark solid), different flesh colours (white, yellow, orange or red) and different types (seeded, seedless). Sugar Baby,' 'Asahi Yamato,' 'Improved Shipper,' 'Arka Jyothi,' ' Arka Manik,' Pusa Badana, 'Durgapura Meetha,' and ' Lal' are a few popular varieties.

Heterosis was exploited in watermelon in Japan in 1949. Subsequently many F1 hybrids are released (Madhu, Milan, Mohini MHW-4,MHW -15, NS -295, Surchi, Samtrupti, Amruth, Century -2 and Nath–102). There are also watermelon varieties named after the localities grown-Farrukhabadi, Morada badi and Faizabadi. Watermelon fruits can be shaped to imaginable forms using molds at fruits development stage. There are square, round,octagonal and cylindrical melons. Watermelon juices are now available in tetrapacks and the juices quench thirst in the most pleasing way.

Cranberries are natives to North America and are used as a food crop, either as fresh produce or processed into sauces, juices or sweetened dried fruit. They are also used in nutraceutical products, primarily in powered form and in cosmetic products. Historically, cranberry was used as dyes and as a food preservative. As a medicinal plant, cranberry was used by Native Americans to treat bladder and kidney diseases. There are popular varieties like Early Black, House Ben Lear, McFarlin, Stevens, Pilgrim and Rutgers Hybrids. Vorsa and Novy (1995) used Random Amplified Polymorphic DNA (RAPD) to formally characterize varieties from field and green houses. Being an uncommon crop in India, detailed agronomic, soil, water and environment requirements of the crop are given in the chapter. Cranberry fruit worm, cranberry girdler, cranberry weevil and black - headed fire worm are serious pests. Fruit rot is the most serious disease. Cranberry contains a number of flavonoids and anthocyanins. Chemistry of cranberry is lucidly explained. Anthocyanins are extracted from the cranberry cell cultures *in vitro*.

The genus *Syzygium* to which jamun belongs, comprises about 1000 species, the most of them are tropical in origin. Jamun is found in Western Ghats and very extensively in the tropical region. Diversity in Jamun is due to heterozygocity and seed multiplication. Origin is traced to Indonesia and India and now grown in southern Asia. There are about 400 to 500 species of which a few provide edible fruits. Jamun exhibits considerable diversity in a wide range of ecological situations under tropics, sub tropics and temperates. The native diversity occurs largely in small orchards, backyards, roadside avenues and marginal lands, arisen from seedlings. Jamun (*Syzigium cumini* L) has related species of economic importance. They are watery rose apple (*E. aquem*), clove (*S. aromaticum*) wild jamun (*S. fruticosum*) rose apple (*S. jamboe*) and ornamental jamun (*S. mappaceum*). Jamun possesses incalcitrant seeds which are difficult to conserve. Jamun is propagated vegetatively–budding, soft wood grafting, inarching, veneer grafting, air layering, cuttage, top working-and through micro propagation. The Central Institute of Sub- Tropical Horticulture, Lucknow has developed varieties CISHJ–37 and CISHJ -42. Jamun GJ40 and Jamun GJ2 are selections from Gujarat. The information on the curative property of Jamun fruits against Diabetes mellitus type 2 has boosted its use and demand.

Litchies are cherished for the white arils covering the seeds. They have a long history of acceptance in many parts of South East Asia. The litchi growing countries are China, Brazil, India, Malaysia, Thailand, Vietnam, Myanmar, Mauritius, South Africa, Australia, New Zealand, Madegascar and Taiwan. The canned aril is a delicacy in non-producing developed countries. In China, dried fruits are common. Known as "Queen" of fruits, litchi originated in the sub- tropical elevations of southern China, first in the north tropical rain forests and mountain forests of southern China where it is a dominant tree species. Arora *et al.* (1996)concluded that the *Litchi chinensis* originated in northern Vietnam, parts of Malaysia and southern tropics of South China. In China, numerous wild litchi trees are found in moist forest in Hainan Island from low elevation to 600 to 1000 m and below 500 m in hilly areas of the Leizhou Peninsular, West Guangdong and East Guangxi. Litchi now reached Mexico, Israel, California (USA) and Madagascar. Singh (1990) attempted to identify genotypes based on plant traits, fruit yield, fruit quality, seedlessness and resistance to physiological disorders. When distinguishing the cultivars, shapes of skin segments and protuberances are the reliable and stable genetic characters. Recommended cultivars of litchi in India, China, Thailand, Philippines, Queensland, Florida and Hawaii, South Africa, Israel, Canary island, Pakistan and Bangladesh are listed in the chapter 4. Litchi flavour is much cherished and as such litchi flavour based confectionaries are emerging.

Mango the ' King of fruits', originated in India and now grown commercially in Egypt, Israel, Florida (USA), Australia, Thailand and India. Majority of countries grow cultivars from Florida like

Hadem, Tommy Atkins and Keith. India produces 18 million tonnes of mango/ year with the variety Alphoso accounting 30-40 per cent of the Rs. 750 million mango export market. The major markets are Japan, Australia, New Zealand and the Middle East. There are specific preferences and requirements for the above markets. The genus *Mangifera* to which mango belongs, has 41 species. The cultivated mango is *Mangi fera indica* and more than 1100 varieties exist in India alone. More than 100 countries in tropics and sub tropics grow mango. Mango, being the most ancient fruit of India, is intimately connected with the Hindu religion. In Hindu mythology, it is highly valued as the fabled transformation of prajapati (God of Creation). Sanskrit poetry is full of praise for its bloom and fruit. In one of the famous Sanskrit plays, Shakuntala, the mango flower is invoked as one of the five arrows of Kamadeva. Even today, the Pancha Pallacva or aggregate of five springs from mango trees, is used by the Hindus at ceremonies. Mango blossoms are indispensable for the workship of Saraswati (Goddess of Wisdom and of the Fine Arts) in early spring and of Shiva (one of the Hindu Trinity) which follows usually a month later. Festoons of mango leaves are used as decoration at all most all Hindu ceremonies and festivals. Their use is wide spread in Hindu rituals of divine blessing called pujas. A ' purna kumbha' or clay pot filled with water is tapped with fresh mango leaves and one coconut. Mango cultivation is as old as Indian civilization. The first country to which mango was introduced happens to be Malaya. The Malaya name for Mango, Mangga, being the same as the Tamil word for mangoes, mangga, in South India is evidence of introduction of superior races of this fruit from India to Malaya. In India, about 30 commercial varieties are grown on largo scale. Dashehari, Langra, Bombaygreen, Chausa and Fazil are popular in North India. Banganapalli, Totapuri, Neelam, Prairi, Mulgoa, Kalepad and Rumani are popular in Southern India. There are monoembryonic and polyembryonic types in mango. Mango is propagated by stone grafting, epicotyl grafting, and protocol for micropropagation is available. Polyembryonic seedlings are used as root stock. Mango fruit pulp is a valuable base for preparing puree, squash and ready to serve(RTS) drinks. Mango juice is a laxative and an instant restorer of sugar levels in case of hypoglycaemia.

Noni (*Morinda citrifolia,L.*) is a wonder plant whose nutritional, health and wellness values, understood by Polynesians 4000 BC. It is a fruit of present and future. The genetic diversity of the genus *Morinda* and the evolution of a large number of species having varieties are attributed to its wide adaptation and world wide distribution. Noni originated in India and South East Asia and to be precise Andaman and Nicobar Islands. From here, it was naturalized in many parts of Asia, Austrialia, South America, Caribbean and Polynesian Islands through sea and ocean currents. The Nicobares, one of the aboriginal tribes of Andaman and Nicobar Islands, is the healthy race with medium height and yellowish complexion and they use Noni as a regular vegetable. Noni called Lurange is very popular among them. The Nicobares have long association with Noni plant and they used different parts as medicine to cure ailments of different human organ systems and of various etiologies. The genus *Morinda* is distributed world wide and about 172 species are reported. *Morinda citrifolia* L, the perennial bush or small tree, 3 to 12m tall with medium stem diameter and abundant wide elliptical leaves, is the most economic crop species containing 160+ nutraceuticals. Fruits or flowers sprout from alternate nodes. Only one fruit is found in every alternate node. Genetic diversity among species and cultivars are studied using RAPD PCR Profiling. The phenotypic diversity of fruit shape and size among accessions of *Morinda citrifolia* collected from different locations of Andaman and Nicobar Islands was significant. *Morinda citrifolia var citrifolia* is gaining ever increasing global demand for its wide range of properties and uses. Noni is propagated through seeds, cuttings and tissue culture. Noni is projected as a wonder plant to manage new life style diseases like diabetes, obesity, hypertension and cardio vascular diseases.

Pomegranate is a favourite dessert fruit of tropical and sub tropical regions. It grows in a variety of climates from tropical to temperate and as a result, a large number of diverse forms have evolved over the years in nature. It is extensively grown in South China and South East Asia. Pomegranate was introduced by Spanish colonists into the Caribbean Islands and Latin America. The fruit has wider consumer preference for its attractive juicy, sweet, acidic and refreshing arils. There is growing demand for good quality fruits both for fresh use and processing into juices, syrup and wine. Time immemorial, it is being used as a symbol of fertility because of innumerable seeds. It is a food medicine of great importance.The seeds from wild pomegranate are used as spice.

Ornamental plants are pleasing to the eyes and smoothen human feelings. Growing of ornamental plants is a therapy against loneliness, stress and depression. In addition, it is an economic activity. The segment of cut flowers has gained attention for its various uses as in bouquet, garlands, vase plants etc. Heliconia is special for its diversity in colour and form. A stalk of heliconia can be displayed for 20-25 days in rooms and verandas. The whole plant is ideally suited for beds, shielding dirty places and above all providing green and coloured canvas to display other ornamentals. Heliconias are adapted to all major agro-climatic zones of India and can be cultivated at a height of 3000 -4000 metres above sea level.A native of tropical America, genus *Heliconia* has more than 250 species mostly native to New World where they are found in open clearings. Two main types are erect heliconias and pendent heliconias. Heliconias are propagated through seed and suckers. Micropropagation techniques like embryo culture, shoot tip culture and callus culture are successfully used. Mutation breeding is used to generate variability and isolate uncommon and unique plants. The flowers are harvested with peduncles of 70cm or more. The large heliconias are usually cut to fit a 150cm long box and often packed in a bunch of 10 and sleeved in plastic film or open weave netting. For its long vase life, uniqueness and rarity, heliconias have great potential. Moreover heliconias come up well where bananas are grown. It has wider adaptability and cultivation for cut flowers highly remunerative.

Marigold is a versatile ornamental with golden harvest grown all over the world. It is a native of Mexico. Marigold decorates beds in a garden and is ideal for animation. The plant type varies from miniatures to tall herbs and is also grown for extraction of natural yellow pigments used in feeds and pharmaceuticals. The generic name *Tagetes* was derived from tagtes, the name of an Estrusch God, known for his beauty. Marigold flowers are grown for loose flower production. They are used in making garlands, for beautification of mandaps and decoration in marriages and religious offerings. Marigold is one of the largest members of plants which holds promise as a bioresource for a variety of useful phytochemicals. Marigold belongs to genus *Tagetes* and there are 50 species recorded. Notable species are *Tagetes erecta* (African manigold), *T. patula* (French marigold), Signet marigold *(T. tenuifolia)* and sweet scented marigold (*T. lucida*).F1 hybrids are available in marigold. RAPD marker techniques are used to characterize marigold.

Zinnias are grown around the world as cut flowers and as annuals in the gardens. *Zinnia violaceae* is the predominant ornamental species used by the gardeners. Zinnias are members of the Heliantheae tribe within the Asteraceae family. Stimart and Boyle (2006) recognized 17 species of *Zinnia* which arose in North America. Bailey (1917) divided commercial zinnias into three classes- tall, medium and dwarf-. There are single and double petalled zinnias.Double zinnias carry a large number of ray florets and a low number of disk florets on the capitulam.Numerous cultivars of zinnia are available for home gardeners.Diversity in height,shape of flowers and duration exist.

In spices, chillies occupy an important place. Byadgi chillies are low pungent but highly red coloured cultivars grown in North Karnataka especially in Haveri,Dharwad,Gadag,parts of Belgaum,Kappal,Shimoga,Chickmangalur,Davangare and Chitradurga districts of Karnataka.The byadgi chillies possess a unique characteristic flavour and aroma,high colour,oleoresin and very less or no pungency.This is an ideal variety for oleoresin extraction and chilli powder industries.Quality wise,it is comparable to paprika varieties growing in European countries. There are two versions of byadgi chilies - kaddi and dabbi–kaddi fruits are very thin, 15-18cm long and dabbi fruits are broader at the shoulder and narrow down towards the tip. The University of Agricultural Sciences, Dharwad maintains 90 accessions of byadgi chillies at Horticulture Research Station, Haveri,Karnataka. Byadgi chilli is projected as the paprika of 21st century.

There is increasing awareness throughout the world that additional plants should be developed into crop plants to avoid future problems in terms of food shortage. Studies conducted in South Africa emphasized the roles of indigenous plants in providing many of the basic needs of people- shelter, fuel, medicines, dietary supplements. One crop that has the potential to address some of the above needs is the Bambara groundnut. It is an indigenous tropical African crop. The name Bambara groundnut, appears derived from the District of Bambara, near Timbuktu in Northern Nigeria. Bambara groundnut is eaten in various forms, either immature or fully ripe. The immature seeds are normally eaten fresh, boiled or grilled while the mature ones are often pounded into flower and then mixed with oil or butter to form porridge. Sometimes they are roasted in oil.

References related to 14 crops under deliberation are given at the end of each chapter.

2011, Biodiversity in Horticultural Crops Vol. 3 *Pages* **1–15**
Editor: **Professor K.V. Peter**
Published by: **DAYA PUBLISHING HOUSE, NEW DELHI**

Chapter 1

Ashgourd

I. Sreelathakumary
Professor, Kerala Agricultural University, P.O.Vellayani, Trivandrum
E-mail: sreelathakumary@rediffmail.com

Ash gourd, *Benincasa hispida* (Thunb.) Cogn. (syn. *B. cerifera* Savi.) is known by several names: wax gourd, winter melon, hairy melon, ash pumpkin, white pumpkin, white gourd and Chinese preserving melon. Benincasa is named for Count Benincasa, an Italian patron of Botany. The name wax gourd refers to the thick, waxy cuticle that typically develops on mature fruits. High temperature favours formation of the waxy bloom. The specific epithet previously used, '*cerifera*', means wax bearing. The specific epithet now recognized, '*hispida*' refers to the hirsute pubescence on the foliage and immature fruits (Robinson and Decker, 1997). It has a Sanskrit equivalent 'Kooshmanda' as well as Persian 'pazadaba' which signify its antiquity.

Origin and Distribution

Benincasa is unknown in the wild. It is probably a native of Indo- Malaysia. It is grown throughout the plains of India, Myanmar and Sri Lanka up to an altitude of 1,500 m. Cultivated forms may have originated in Southeast Asia. Simmonds (1976) reported that ash gourd is indigenous to Asian tropics and distributed from Japan to India. Ashgourd is reported as introduced to India from Japan and Java by foreign navigators and missionaries (Peter, 1998). De Candolle found it wild on the seashore of Java and it has spread northwards to Japan and also to Central America and West Indies. Chinese records of the fifth and sixth centuries AD mention ash gourd. Ash gourd is grown throughout the old world tropics and less commonly in the new world. In North India, it is mainly grown in Uttar Pradesh and Delhi.

Importance and Uses

Ashgourd is an important warm season cucurbit vegetable. The fruit has very high moisture content and is low in calories and carbohydrates (Morton, 1971). It contains 0.4 per cent protein, 1.9

per cent carbohydrates, 0.3 per cent minerals and traces of vitamin A, B and C/100g edible portion (Saimbhi, 1993). In addition, the fruit also contains Ca, P, Na, Mg, Fe, K, S and starch in minute quantities. Fruits are used in confectionary and ayurvedic medicinal preparations. The ripe fruits are peeled and cut into pieces and candied to make the well known sweet, *petha*. Mature and immature fruits are used as vegetable and are ingredients in curries. The pulp is used as adulterant or as substitute in tomato ketchup in place of tomatoes. Young leaves, vine tips and flower buds are boiled and eaten as greens. Seeds are fried and consumed.

Ash gourd is considered good for people suffering from nervousness and debility (Nadkarni, 1927). The small fruited type with thick rind, waxy coat and long keeping quality commonly referred to as *neykumbalam* is the medicinally important one. The shreddings of the hard pulp of *neykumbalam* are used for the preparation of ayurvedic medicines for diabetes like *kushmandalehyam*, *kushmandarasayanam* and *kushmandakhritham* and so this variety is locally known as *vaidhya kumbhalom* meaning thereby, medicinal ash gourd (Gopalakrishnan, 1957).

The fruit is considered as tonic, nutritive and diuretic (Dymock *et al.*, 1976; Chopra *et al.*, 1996). Decoction of fruit is laxative, styptic and given for respiratory troubles and internal haemorrhages (CSIR, 1986). Ripe fruits are useful for *tridoshas* and insanity and also improve intellect. The alkaline fruit juice is useful to reduce hyperacidity (Santhakumar, 1991). 'Kushmanda' is a rejuvenating drug capable of improving intellect and physical strength. The paste made of ash gourd fruit can be used in case of burns, guts and minor wounds (Mathewkutty, 1996).Tender stems are good in liver troubles and muscle pain. Seed powder is used for appendicitis and is antihelmintic (Agarwal, 1997). It is an effective antidote against mercurial and alcoholic poisoning (Chatterjee and Prakashi, 1997).

Fruit powder helps in increasing blood count and is used in treating anaemia (Supreeja, 2002). The dry fruit rind serves as containers and as an elegant serving bowl for soup. The fruit wax is used to make candles and as a vehicle for carrying poison. The *petha* wastes in the form of pith, pulp, seeds and peeling are subjected to vermicomposting (Kumari and Bhadauria, 2002).

Classification/Species/Cultivar

Benincasa is a monospecific genus belonging to the tribe *Benincaseae* Ser., subfamily Cucurbitoideae. It is distributed in South and South-East Asia. Two botanical forms in ash gourd –the typical and emarginated - are recognized in Japan. The emarginate form lacks the velvet testa and marginal band around the seeds which characterize the typical form. The waxy coat of the fruit ensures outstanding storage quality.

Certain cultivar groups are dominant in different geographical regions of the world. In order to classify inter specific variation in *Benincasa hispida*, four major categories are proposed as horticultural or cultivar groups (Walters and Walters, 1989).

1. *Unridged winter melon group*: seed unridged, fruit large (0.5–1 m long), cylindrical, rind dark–green with little or no waxy bloom, mature leaves with shallow to moderate lobes.
2. *Ridged winter melon group*: very similar to the unridged winter melon group, except that seeds have ridged margins and leaves have shallow lobes.
3. *Fuzzy gourd group*: seeds ridged, fruits small (20–25 cm in length), narrowly cylindrical, rind green, covered with white, soft hairs without waxy bloom, mature leaves shallowly lobed.
4. *Wax gourd group*: seed ridged, fruits small to large (10–60 cm in diameter), round and flattened at both ends to oblong, rind light-green, covered with hairs or hairless with a white, waxy bloom; mature leaves deeply lobed. In this group, three cultivar types are described based on

fruit characters (Morton, 1971). (*a*) fruits nearly round and essentially hairless, (*b*) fruits nearly round and hairy, and (*c*) fruits oblong and hairy.

Cytogenetics

Benincasa hispida is a diploid with 2n = 24 (x = 12). Varghese, (1973) observed secondary association of chromosomes which suggested that *Benincasa hispida* though apparently a diploid is in fact a stable tetraploid derived from an ancestor with x= 6. Meiotic preparations from root tip cells had 24 chromosomes. Analysis of the karyotype revealed that chromosome length was 2.0-3.0 μ. The complement consists of eight pairs of median and four pairs of sub-terminal chromosomes. According to their comparative length and arm ratio, the twelve pairs of chromosomes could be arranged into six groups. Total length of the complement and mean length of chromosomes are estimated to be 57.36 μ and 2.39 μ respectively. All stages of meiosis were normal. Pollen fertility was estimated to be 97.8 per cent.

Growth Habit

Ash gourd is an extensive trailing or climbing annual herb. The hairy vines of ash gourd are large and spreading. Each tendril is 2 or 3 branched, occurring opposite the stipule at the leaf axis. The large heavy leaves are lobed and have long petioles; they emit an unpleasant odour when bruised. Ash gourd is monoecious, with solitary male and female flowers. The five parted yellow flowers which are solitary in the leaf axils are 6.14 cm in diameter and 3.5 cm in length including the ovary. Staminate flowers have long peduncles; peduncles of the carpellate flowers are much shorter. Sepal lobes are foliaceous. The ratio of staminate to pistillate flowers is 34: 1. Anthesis takes place at 4.30–7.30 a.m., with anther dehiscence at 3-5 a.m. The stigma is receptive from 8 hours before anthesis until 18 hours thereafter. Fruit set after hand pollination of flowers is 65 to 75 per cent. Pollen fertility is maximum (95 per cent) at anthesis.

Germplasm Resources

Organizations which maintain germplasm collections of ash gourd are:

1. The Zentralinstitut fur Genetik und Kulturpflanzenforschung, Gatersleben, Germany;
2. National Bureau of Plant Genetic Resources,Regional station, Vellanikkara, India;
3. Kerala Agricultural University, Vellanikkara, India;
4. Southern Regional Plant Introduction Station, Georgia, USA;
5. Imperial Valley Conservation Research Center, California, USA,
6. Robinson's, Geneva, New York, USA.

Breeding Methods

Breeding methods like mass selection, pedigree method and bulk population can be used.

Cultivar/Varieties

Variability in ash gourd is limited except for size and shape of fruits. The improved varieties are:

KAU Local

Medium sized oval to oblong fruits with high flesh thickness, fruit length 45–55 cm, L:B ratio 2.05, fruit weight 6.1-8.1 kg, productivity 28.2 t/ha. developed by Kerala Agricultural University, it is high yielding, spreading and with vigorous growth habit. Medium sized oval to oblong fruits which are green at tender stage and wax coloured on maturity. Duration 105 -120 days.

Indu

Medium sized round fruits, tolerant to mosaic disease, average fruit length 24.3 cm, average fruit breadth 23.78, L:B ratio 1.02, average fruit weight 4.82 kg, productivity 24.5 t/ha. developed by Kerala Agricultural University. High yielding, spreading and vigorous growth habit, medium sized round fruits with flat stylar and pedicel end, fruits green at tender stage and wax coloured on maturity. Duration 105–120 days.

Co-1

Fruits round, average fruit weight 5-6 kg, duration 140 days. The mean yield of marketable fruits is 23.9 t/ha. Released from Tamil Nadu Agricultural University. They have more fruits/plant. They produce staminate and pistillate flowers at lower nodes, a narrow sex ratio of staminate to pistillate flowers (Kamalanathan *et al.*, 1970).

Co-2

Plant growth is moderately vigorous, less spreading and has light yellowish green hairy stem and round deep prominent lobed green leaves. The fruits are small sized with length 40–50 cm and breadth 20- 30 cm long spherical in shape with white ashy coating all over the fruit, weighing not more than 3 kg (2.5 to 3 kg), light green coloured flesh and less seeded (200 to 300 seeds weighing 25 to 40 g). It is an early maturing variety, takes on an average about 85 to 90 days for first harvest of marketable fruits from the date of its sowing. The period of availability of fruits extend from 90 days to 120 days of sowing. It is adaptable for high density planting with a spacing of 2.5 x 2.0 meters. Released from Tamil Nadu Agricultural University (Sundarajan and Muthukrishnan 1982.).

APAU Shakthi

Fruits long and cylindrical, Yield 30-35 t/ha in 140-150 days. Released from ANGRAU, Hyderabad.

Karikumbala

Local cultivar having dark green fruits without ash coat. Released from UAS, Bangalore.

Boodikumbala

Local cutivar where the fruits are covered with ashy coat. Released from UAS, Bangalore.

IVAG–502

Fruits oblong with average weight of 12–13 kg. Yield 30-35 t/ha. Released from IIVR, Varanasi.

In addition to above open pollinated varieties, a few F1 hybrids like MAH -1, MHAG - 2 are developed under private sector in the country.

'Okinawa No.1' and 'Kurokawa Early' are popular varieties in Japan. 'S1' is recommended for Punjab and 'Thudiyalur' is popular in Maharashtra.

Breeding for Resistance

Fusarium wilt (*Fusarium oxysporium*) is the most serious disease affecting ash gourd. In field tests on ground infected with *Fusarium oxysporium* of 4 self–rooted varieties of melon and two of those grafted to ash gourd. Rugeri (1968) observed that 90 per cent of plants belonging to each of the ungrafted varieties became infected while those grafted on ash gourd escaped infection. Ester and Gerlagh (1980) also found that out of different melons tested only ash gourd was resistant to race 1, 2 of *Fusarium oxysporium* f. sp. *melonis*. Polygenic resistance was observed as being present in the variety 'Koganenaski Golden Pear'. Kaur *et al.* (1985) isolated *B. hispida* varieties resistant to *Fusarium oxysporium, F. semitectum* and *F. equiseti* causing fruit rot. Providenti (1966) reported a few accessions of *B. hispida* resistant to viruses–'PI291544' and 'PI 391545' to virus WMV1 and tolerant to virus WMV2.

Crop Improvement

Ashgourd is an important warm-season cucurbit vegetable, grown for its succulent hairy fruits, used in confectionary and in ayurvedic medicinal preparations (Indira and Peter, 1987). Crop improvement in ashgourd has been much less compared to other cucurbits.

Genetic Cataloguing

Genetic cataloguing based on standard descriptors helps to easily describe the morphological features of a genotype and thus helps exchange of information about new accessions.

Morton (1971) described three cultivar types in wax gourd group on the basis of fruit characters namely, fruits nearly round and essentially hairless, fruits nearly round and hairy, fruits oblong and hairy. Walters and Walters (1989) proposed four major categories as cultivar groups in *Benincasa hispida viz.*, unridged winter melon group, ridged winter melon group, fuzzy winter melon group and wax gourd group.

Resmi (2004) genetically catalogued twenty five landraces of ash gourd collected from different agro climatic regions of Kerala, Tamil Nadu and Karnataka based on the descriptor list for cucurbits (IBPGR, 1983). The results revealed distinct variations among the landraces with respect to vegetative, flower, fruit and seed characters. Most of the landraces fall in moderate to high viny growth habit and short to long internodal length. Density of foliage hairs ranged from a few to dense. Leaf size varied from small to large with ovate, pedate or reniform shape. Most of the landraces had shallowly lobed leaves but exceptions of deeply lobed and lobeless cases were also found. Leaf dorsal surface pubescence was either soft hairy or bristle like.

Variability was more pronounced for flower and fruit characters. Flower and fruit size ranged from small to very large. Fruit form was either round, oval, globular or elongate bottle like. Skin colour was green in smooth textured fruits and white in waxy textured fruits. Fruit shape at stem end and blossom end ranged from deep round to taper point. Peduncle length ranged from short to long. Peduncle detachment from fruit was difficult in most of the landraces. Most of the landraces possess medium earliness of harvest. Fruit storage ability was more than six weeks in most of the landraces.

Seed quantity/fruit ranged from very a few to many with small to large seed size. Most of the landraces had yellow seed colour, but exceptions of brown and whitish yellow colour were also found. Glossy to dull seed surface lush was noticed. Seed separation from placenta was easy in most of the landraces.

Variability

Variability either naturally existing or created artificially forms the basis for any crop improvement programme. In India, a wide range of variability in vegetative and fruit characters is available in ashgourd (Sundararajan and Muthukrishnan, 1982; Peter *et al.*, 1991; Seshadri, 1993 and Mandal and Sirohi, 2003).

Plant Characters

George (1981) conducted biometrical studies to estimate genetic variability, correlations, path coefficients and selection indices in ash gourd. Thirty two ash gourd genotypes studied were significantly different for 28 polygenic characters.

Resmi (2004) reported high significant differences among the twenty five landraces of *B. hispida* for the characters namely, vine length, inter nodal length, number of primary branches, number of secondary branches, root shoot length ratio, days to first male flower, node to first male flower, days to first female flower, node to first female flower, sex ratio, days to first fruit harvest, fruit length, fruit girth, fruits/plant, average fruit weight, yield/plant, seeds/fruit, 1000-seed weight and mosaic incidence indicating sufficient diversity among the landraces. Such variation indicated the scope for improving the population for these characters as reported earlier by Randhawa *et al.* (1983), Madalgeri and Dharmatt (1989) and Parkash *et al.* (2000) in ashgourd.

In ash gourd, Resmi (2004) reported that vine length ranged from 283.00 to 875.00 cm, with a mean of 541.80 cm. Considerable variation among the landraces for inter nodal length was observed in ash gourd while number of primary branches showed a low range of variation ranging from 1.50 to 4.00 with a mean 2.62.

Fruit Characters

Wide genetic diversity among the cultivars for yield and yield related characters was reported by several workers in ashgourd (George, 1981; Randhawa *et al.*, 1983; Hamid *et al.*, 1989; Menon, 1998; Lovely, 2001; Mandal *et al.*, 2002 and Resmi 2004). Genetic diversity among the medicinal types of ash gourd (Neykumbalam) grown in Kerala was studied by Devi (2003).

Days to first male flower exhibited a range of 47.00 to 64.12. Node to first male flower ranged from 9.62 to 19.62. Among the landraces, days to first female flower ranged from 46.25 to 65.62 with an overall mean of 56.06. Range in node to first female flower among the landraces was from 18.00 to 28.37. Sex ratio had a range from 7.21 to 16.40 with an overall mean of 11.64. Days to first fruit harvest exhibited a range of 88.00 to 107.50 days with a mean of 95.78 days in ash gourd (Resmi, 2004). Wide variability in fruit weight was noticed in ashgourd (Lovely, 2001 and Resmi, 2004).

Changlin (1998) obtained a range of 12.15 to 33.21 cm for fruit length in ashgourd. Resmi (2004) reported a range of 13.65 to56.00 cm for fruit length in ash gourd. Girth of fruits varied significantly among the landraces from 22.20 to 78.05 cm. Fruit number varied considerably from 1.87 to 9.12 with an overall mean of 3.77. Range in average fruit weight among the landraces was from 0.27 to 9.50 kg. A wide range of variation was observed for yield/plant from 1.62 to 21.20 kg. Seeds/fruit observed a range from 73.50 to 1647.00 with an overall mean of 666.44. Among the landraces, 1000-seed weight ranged from 19.25 to 96.95 g.

The vulnerability index for mosaic incidence ranged from 27.50 to 75.00. The reaction of landraces towards mosaic incidence indicated that six landraces were resistant; sixteen landraces were moderately resistant and remaining three were susceptible to the disease (Resmi, 2004).

High genetic variance for fruits/plant was reported in ash gourd by Miniraj *et al.* (1993). High phenotypic and genotypic variances were reported for several characters including seeds/fruit and vine length. Wide variation was reported in phenotypic and genotypic variances among the characters.

A close association between phenotypic and genotypic variances was noticed for seeds/fruit, 1000-seed weight, yield/plant, average fruit weight and fruits/plant. For most of the characters, genotypic variance makes up the major portion of the phenotypic variance, with very little effect of environment.

Phenotypic and genotypic coefficients of variation observed were high for most of the characters. Seeds/fruit had the highest PCV and GCV followed by average fruit weight, fruits/plant and yield/plant. The lowest PCV and GCV were exhibited by days to first fruit harvest.

Heritability and Genetic Advance

Effectiveness of selection depends upon the heritability and genetic advance of the character studied.

In ashgourd, George (1981) found high heritability for fruits/plant. Average fruit weight had the highest value of genetic advance in the next cycle of selection, followed by weight of first mature fruit and yield/plant (George 1981). Menon (1998) noticed high values of heritability and genetic advance for primary branches in ashgourd. In ashgourd, Parkash *et al.* (2000) reported low heritability and genetic advance for vine length. High heritability and low genetic advance were noted for days to first male and female flower in ashgourd (Lovely, 2001).

High heritability and genetic advance for average fruit weight were also noticed in ashgourd (Menon, 1998). Fruit yield/plant was observed to have high heritability along with high genetic advance in ashgourd (Parkash *et al.*, 2000). High values of heritability and genetic advance were also reported for fruit yield/plant in ashgourd (Lovely, 2001). In the case of 1000-seed weight, high heritability coupled with high genetic advance was noted in ashgourd (Lovely, 2001).

Resmi (2004) reported high heritability coupled with high genetic advance for most of the characters, except days to first male and female flower, node to first female flower, sex ratio, days to first fruit harvest and mosaic incidence. Heritability estimates were high for most of the characters studied *viz.*, seeds/fruit, 1000-seed weight, average fruit weight and fruit girth. Sex ratio recorded the lowest but a moderate heritability.

Genetic advance was the highest for seeds/fruit, followed by average fruit weight and the lowest for days to first fruit harvest and days to first female flower. High heritability combined with high genetic advance was reported for fruit length, fruit girth, fruits/plant, average fruit weight, yield/plant and seeds/fruit in ash gourd by Resmi (2004).

Correlation Studies

Measurements of phenotypic, genotypic and environmental correlations between yield and other characters have been a matter of great importance.

Fruit yield/plant was positively correlated with main vine length, female flowers/plant, average fruit weight and weight of first mature fruit. Average fruit weight had maximum direct effect on fruit yield/plant, followed by main vine length. Length of main vine was positively correlated with node at which first female flower appeared, nodes on main vine and leaves/plant George (1981).

Fruit yield has positive correlation with internodal length in ashgourd (Menon, 1998). A positive association was observed between seeds/kg of flesh and fruit yield in ashgourd (Menon, 1998). Parkash *et al.* (2000) reported positive association between yield and fruit girth in ashgourd. Lovely (2001) reported a negative correlation for days to first female flower and fruits/plant in ashgourd.

Resmi and Sreelathakumary, (2006) reported high positive significant correlation between yield/plant and fruit length, fruit girth, average fruit weight, seeds/fruit and 1000-seed weight. Node to first male and female flower and fruits/plant was negatively correlated with yield.

Vine length had positive correlation with internodal length and number of primary and secondary branches in ash gourd. Internodal length had positive correlation with number of secondary branches.A positive correlation was observed between number of primary branches and number of secondary branches. Number of secondary branches showed negative correlation with sex ratio and seeds/fruit (Resmi, 2004).

Days to first male flower recorded high positive phenotypic correlation with days to first female flower, node to first female flower and days to first fruit harvest. Days to first female flower also recorded high positive correlation with days to first fruit harvest. Node to first male flower exhibited positive correlation with node to first female flower and fruits/plant and negative correlation with fruit girth, average fruit weight, seeds/fruit and 1000-seed weight. Node to first female flower had negative correlation with fruit length, average fruit weight and 1000-seed weight. Sex ratio was positively correlated with fruit girth, average fruit weight, seeds/fruit and 1000-seed weight, while negatively correlated with fruits/plant in ash gourd (Resmi and Sreelathakumary, 2006).

Fruit length observed high positive correlation with fruit girth, average fruit weight, seeds/fruit, 1000-seed weight and mosaic incidence, while it showed negative correlation with fruits/plant. Fruit girth had negative correlation with fruits/plant and positive correlation with average fruit weight, seeds/fruit and 1000-seed weight. Fruits/plant recorded negative correlation with most of the characters, the highest being with 1000-seed weight. Average fruit weight exhibited high positive correlation with seeds/fruit and 1000-seed weight. Seeds/fruit was positively correlated with 1000-seed weight (Resmi and Sreelathakumary, 2006).

Path Coefficient Analysis

Path analysis facilitates the partitioning of correlation coefficients into direct and indirect effects of various yield attributes. Average fruit weight and vine length exhibited maximum direct effect on fruit yield/vine in ashgourd (George, 1981). Menon (1998) observed that in ashgourd average fruit weight exhibited the highest positive direct effect on fruit yield followed by fruits/plant, female flowers/plant, vine length, internodal length and seeds/fruit. Parkash *et al.* (2000) in ashgourd reported that fruits/plant and average fruit weight had high direct effect on yield, while number of days for flowering had negative direct effect on yield. Lovely (2001) found that in ashgourd, days to first female flower, fruits/plant, fruit length and girth had positive direct effects, while vine length and seeds/fruit had negative direct effects.

Resmi (2004) reported that fruit length had the maximum positive direct effect on fruit yield followed by average fruit weight and fruits/plant. The direct effects of vine length, node to first female flower and sex ratio were negligible, but their indirect effect through fruit length and average fruit weight were consistently high. High negative correlation of fruits/plant with yield was due to high negative indirect effect through fruit length, while in node to first female flower, high negative correlation was through average fruit weight.

Selection Index

Selection of landraces based on a suitable index is highly efficient for any crop improvement programme. Selection index involves discriminant function analysis, which is meant for isolating superior landraces based on the phenotypic and genotypic correlations. Identification of superior genotypes of *B. hispida* based on discriminant function analysis was done by George (1981), Lovely (2001) and Resmi (2004) in ashgourd.

The discriminant function indicated maximum selection response for yield when main vine length, female flowers/plant, average fruit weight and weight of first mature fruit were used together in selection with equal weightage (George, 1981).

Fruit length and fruit girth are the important characters which should be taken for selection in improvement programme in ashgourd (Lovely, 2001). Resmi (2004) reported that yield, fruit quality, earliness in male and female flowering, narrow sex ratio and mosaic resistance are the important characters that should be considered for selection in improvement programme in ashgourd.

Genetic Divergence

Mahalanobis's D^2 statistic is one of the potent techniques for measuring genetic divergence at both intra and inter cluster levels and thus provides a basis for selection of genetically divergent parents in hybridization programme. Anand and Murthy (1968) have emphasized the merit of D^2 statistics for genetic grouping of germplasm. Genetic divergence was assessed by Lovely (2001) and Resmi and Sreelathakumary (2008) in ashgourd.

Genetic divergence studies by Lovely (2001) resulted in clustering of 25 genotypes of ashgourd into 8 group constellations. The maximum number of genotypes (8) were included in Cluster III, followed by cluster I (7), cluster II (4) and cluster IV (2). The clusters V, VI, VII and VIII had only one genotype each in them. The genetic distance was maximum between II and IV and minimum between VII and VIII. The character seeds/fruit contributed maximum to the total divergence.

Genetic divergence studies in ash gourd by subjecting 25 landraces of *B. hispida* to cluster analysis, based on eleven characters *viz.*, vine length, days to first female flower, node to first female flower, sex ratio, fruit length, fruit girth, fruits/plant, average fruit weight, seeds/fruit and mosaic incidence by Resmi and Sreelathakumary (2008) resulted in clustering the 25 landraces into seven gene constellations. The maximum number of landraces (8) were included in Cluster I, followed by cluster II with 4 landraces. Cluster III, IV, V and VI had three landraces each while cluster VII had one landrace.

Cluster VII (BH 8) comprised of landrace with the smallest fruits,the highest fruits/plant, shorter vine length, earliness in flowering and the lowest sex ratio, yield and seeds/fruit. Cluster V consisted of landraces with medium sized fruits with the highest mosaic resistance and seeds/fruit. Cluster III had the highest average fruit weight and yield/plant. Cluster VI comprised of large sized fruits with high mosaic incidence. Considering the cluster means for various characters studied, clusters III and V were superior for most of the biometric characters, whereas clusters I and VI were generally poor. Cluster II and VII were intermediate. For crop improvement programmes, intercrossing among landraces with outstanding mean performance for these characters would be effective. The intra cluster distance was the highest for cluster VI followed by clusters II and IV.

The highest inter cluster distance was observed for clusters V and VII followed by clusters V and VI and clusters I and V. The genetic distances (D) between clusters I, IV, VI and VII were the largest with cluster V. The minimum inter cluster distance was observed between clusters III and V indicating a close relationship among the landraces included.

Heterosis

Mandal and Sirohi (2003) reported heterosis in ash gourd. The observations were recorded on important characters like vine length, days to first male flower opening, days to first female flower opening, days to first fruit harvest, fruits/plant, fruit length, fruit diameter, fruit flesh thickness, fruit weight and total yield/plant. They reported that it is useful to utilize the parental lines having important characters like early maturity, longer vine length, higher fruits/plant, bigger fruit size and higher fruit weight associated with high yield in order to achieve higher gains in F1 hybrids.

Verma and Behra (2007) selected ten elite inbred lines of ash gourd (*Benincasa hispida*) based on their performance and were crossed in a diallel mating system (without reciprocal crosses) to generate 45 F1 hybrids. These hybrids along with the parents were grown to study heterosis in 10 yield related characters. The mid and better parent heterosis was observed to be as high as ~165 per cent for yield/vine in DAG-6 × DAG-11. The maximum negative heterosis over the best parent for days to fruit maturity was noticed in cross DAG-2 × DAG-9 (~–10 per cent) indicating that it can be successfully utilized in breeding for earliness in ash gourd. Two hybrids namely, DAG-1 × DAG-5 (34.33 kg) and DAG-4 × DAG-11 (31.67 kg) recording positive heterosis over the best parent to the extent of 23.5 per cent and 14.0 per cent respectively for yield/vine can be utilized for commercial cultivation.

Sureja *et al.* (2006) reported the molecular diversity and its relationship with hybrid performance and heterosis in ash gourd. Nine inbred lines of ash gourd were crossed to produce all (72) possible F_1 hybrids, including reciprocals, which were evaluated along with the nine parents for 13 growth- and yield-related traits and 14 quality traits, in a replicated field trial. A high level of heterosis was observed among the hybrids for most of the traits examined, including iron and zinc content. Significant reciprocal differences in heterosis were observed. The inbred lines were analysed using 130 RAPD primers, of which 26 were informative and amplified 163 DNA marker bands. A total of 47 polymorphic bands were obtained with a mean of 1.8/primer which, in combination, discriminated all the inbreds from each other. Pair-wise genetic distance measurements ranged from 0.056 to 0.179, suggesting a narrow genetic base for the inbreds. Although the results indicated significant positive correlations of genetic distance with hybrid performance and heterosis, the magnitudes of the correlations were not high enough to be of predictive value.

Molecular Characterization

Molecular markers are genotypic markers. Unlike morphological characters, molecular markers characterize diversity at the molecular level, and therefore are environmentally independent. The use of these markers provide a potential effective selection technique for crop improvement and has advantage over selection based on phenotype alone. Molecular markers are widely used in genetic analysis and diversity assessment in a number of plant species (Waugh and Powell, 1992; Bretting and Widerlechner, 1995; Staub *et al.*, 2004).

Molecular markers which reveal polymorphism at the DNA level are known as DNA markers. They provide an opportunity to characterize genotypes and to measure genetic relationships more precisely than other markers (Soller and Beckmann, 1983). Various types of molecular markers are utilized to evaluate DNA polymorphism and among them, the most important is polymerase chain reaction (PCR) based markers.

Polymerase Chain Reaction Based DNA Markers

PCR based DNA marker techniques are fingerprinting techniques which use an *in vitro* enzymatic reaction to specifically amplify a multiplicity of target sites in one or more nucleic acid molecules.

Among the PCR based marker techniques, the important ones are Random Amplified Polymorphic DNA, Amplified Fragment Length Polymorphism and Microsatellite.

Random Amplified Polymorphic DNA (RAPD)

RAPD is a multiplex marker system that conventionally uses single-primer PCR to amplify random DNA fragments.

RAPD technique is particularly well suited to high through-put systems required for germplasm assessment because of their simplicity, speed and relatively low cost (Hadrys *et al.,* 1992). RAPD markers are commonly used for molecular characterization studies despite disadvantages in reliability (Peteira *et al.,* 1999).

RAPD for Detection of Genetic Variability

Random amplified polymorphic DNA analysis provides a quick and efficient method for resolving genetic relationship.

Most of the genetic diversity studies in crop plants are carried out using morphological markers only. Now-a-days Polymerase chain reaction (PCR) based molecular markers have developed into powerful tools to analyse genetic relationships and genetic diversity. RAPD technique is one among them. The genetic variation as detected by RAPD analysis opens up the avenue for proper identification and selection of the genotypes which could be used for varietal identification and planning for future crop improvement programme.

Resmi and Sreelathakumary (2005) reported the extent of genetic diversity in 25 landraces of *B. hispida* based on RAPD markers, making use of arbitrary primers to amplify random DNA sequence in the genome. Isolation of genomic DNA of ashgourd was done using modified Murray and Thompson (1980) method. Tissues from young tender leaves were found to yield good quality DNA. The DNA yield for twenty five landraces of *B. hispida* ranged from 0.21 to 3.90. The purity of DNA (A_{260}/A_{280} ratio) ranged from 1.40 to 2.08 mg ml^{-1}.

To identify the promising primers for RAPD analysis, forty decamer primers of kit A and B were screened using the DNA of landrace BH 1. The procedure standardized by Staub *et al.* (2000) in *Cucumis melo* germplasm was tried for amplification. Twenty nine primers, out of the forty decamer primers yielded amplification products indicating presence of sequence complementary to these primer in the DNA of BH 1 landrace. A total of 83 RAPDs (average 2.08 bands/primer) was generated by the 29 primers, of which 92.77 per cent were polymorphic (77 bands) and six were monomorphic. Five primers showed high level of polymorphism. This could be explained by the capability of individual primers to amplify the less conserved and highly repeated regions of the genomic DNA. There is high possibility for the amplified fragments to contain repeated sequences.

For further amplification of DNA from twenty five ashgourd germplasm, three promising primers were identified for RAPD analysis based on performance in DNA amplification, production of the highest number of polymorphic bands as well as intense bands and reproducibility. They were OPA-01, OPA-07 and OPA-13. A total of 20 scorable bands (average of 6.66 bands/primer) were generated by the selected three primers of which 2 were monomorphic and rest, 18 were polymorphic (90.0 per cent). The number of bands ranged from 1 to 8 with an average of 3/primer. The primer OPA-07 was unique as it could distinguish maximum of the landraces tested. The highest number of scorable bands was given by OPA-13 of which one of the bands produced was monomorphic. The primer OPA-01 produced five scorable bands of which one band was monomorphic for all the landraces.

The estimation of Jaccard's similarity coefficients and construction of dendrogram by using UPGMA revealed the presence and extent of genetic similarities among the twenty five landraces of *B. hispida* examined. The overall similarity coefficients ranged from 0.14 to 1.00. Cluster analysis revealed that at about 0.35 similarity coefficient, the twenty five landraces of *B. hispida* grouped into two clusters. Landraces with morphologically distinct smooth and waxy textured fruits grouped into two major clusters with an exception of BH 19 falling in the first cluster. This substantiates the moderately broad distribution of genetic variability, which can be attributed to the broad genetic base in their origin.

The waxy textured group formed a more divergent cluster than smooth textured group. With in the group of waxy textured fruits, limited variation was detected among landraces with small sized fruits. Morphologically similar landraces BH 11, BH 14 and BH 24 are grouped together and showed 100 per cent similarity. Landraces with medium sized fruits also showed limited variability. They formed five sub clusters with in the waxy textured group with 100 per cent similarity for BH 2 with BH 23 and BH 3 with BH 9 respectively.

Further, landraces classified as belonging to the same morphotypic group did not always cluster together. This was evidenced from the results of RAPD analysis that morphologically similar landraces with large fruits form distinct clusters within the major clusters. BH 1, BH 5, BH 6, BH 7, BH 15, BH 18 and BH 25 with high average fruit weight formed distinct clusters under molecular study (Resmi and Sreelathakumary, 2007)

Ten inbred lines of ash gourd [*Benincasa hispida* (Thunb.) Cogn.] were crossed to produce 45 F_1 hybrids (without reciprocal) which were evaluated along with the parents for 20 growth- and yield-related traits, in a replicated field trial (Verma *et al.*, 2007). High level of heterosis was observed among the hybrids for most of the traits examined, including yield. These inbred lines were analysed by using 42 RAPD primers which produced 282 DNA marker bands. A total of 130 RAPD markers were obtained with a mean of 3.1/primer, which in combination discriminated all the inbreds from each other. Pair-wise genetic distance measurements ranged from 0.07 to 0.31, suggesting a wide genetic diversity for these inbreds. These inbreds were also analysed with five ISSR primers of which four were informative. Twenty-six ISSR marker bands were generated of which 11 were polymorphic with an average of 2.80/primers. The percentage of polymorphic bands produced were higher in ISSR markers (>80 per cent) than generated through RAPD markers (46 per cent). Although the results indicated significant positive correlations of genetic distance with hybrid performance and heterosis, the RAPD based genetic distance measures and use of limited ISSR markers in this present study could not effectively predict hybrid performance in this crop. The genetic variation among ash gourd inbred lines examined, herein, defined a marker array (combined ISSR and RAPD) for the development of a standard reference for further genetic analyses, and the selection of potential parents for predicting hybrid performance and heterosis.

References

Agarwal, V.S., 1997. *Drug Plants of India*. Kalyani Publishers, New Delhi, p. 217.

Chatterjee, A. and Prakashi, S.C., 1997. *The Treatise on Indian Medicinal Plants*, Vol. 5. National Institute of Science Communication, New Delhi, p. 287.

Changlin, W., 1998. *Waxgourd Varietal Trial*. AVRDC–ARC Research Report, Taiwan, 202 p.

Chopra, R.N., Nayar, S.L. and Chopra, J.C., 1996. *Glossary of Indian Medicinal Plants*. Council of Scientific and Industrial Research, New Delhi, p. 330.

CSIR, 1986. *The Useful Plants of India*. Publications and Information Directorate, Council of Scientific and Industrial Research, New Delhi, P. 98.

Devi, R., 2003. Evaluation of Neykumbalam [*Benincasa hispida* (Thunb.) Cogn.] in oil plantations. *M.Sc. (Hort.) Thesis,* Kerala Agricultural University, Thrissur, 121 p.

Dymock, W., Warden, C.J.H. and Hooper, D., 1976. *Pharmacographia India,* Vol. II. M/s Periodicals Experts, New Delhi, p. 643.

Ester, A. and Gerlagh, M., 1980. *Fusarium oxysporium* f. sp. *melonis*: A risk factor in melon culture. *Zaadbelangen* 34(5) : 135–136

George, T.E., 1981. Biometrical studies in ashgourd [*Benincasa hispida* (Thunb.) Cogn.]. *M.Sc. (Hort.) Thesis,* Kerala Agricultural University, Thrissur, 78 p.

Gopalakrishnan, R., 1957. Medicinal ash gourd: A new variety. *Madras Agric. J.*, 44: 345–346

Hadrys, H., Balick, M. and Schierwater, B., 1992. Applications of random amplified polymorphic DNA (RAPD) in molecular ecology. *Molecular Ecol.*, 1: 55–63.

Hamid, M.M., Sana, M.C., Begum, R.A. and Hussain, S.M.M., 1989. Physio-morphology and yield of different ashgourd [*Benincasa hispida* (Thunb.) Cogn.] lines. *Bangladesh J. Agric.*, 14: 51–55.

IBPGR, 1983. *Genetic Resources of Cucurbitaceae.* International Board for Plant Genetic Resources, Rome, 71 p.

Indira, P. and Peter, K.V., 1987. *Underexploited Tropical Vegetables*. Kerala Agricultural University, Thrissur, 68 p.

Kamalanathan, S., Sundarajan, S., Thamburaj, S. and Subbiah, R., 1970. A high yielding ashgourd strain, CO-1. *Madras Agric. J.*, 58, 25.

Kaur, S., Randhawa, K.S., Singh, H. and Arora, S.K., 1985. Comparative pathogenecity of three species of *Fusarium* causing fruit rot of ashgourd. *Indian Journal of Mycology and Plant Pathology*, 15: 308–310.

Kumari, V. and Bhadauria, S., 2002. Utilization of petha industry waste in making vermicompost by using earthworm (*Eisenia foetida*). *J. Agric. Sci. Res.*, 38: 63–65.

Lovely, B., 2001. Evaluation of genetic divergence in ashgourd. *M.Sc. (Ag.) Thesis*, Kerala Agricultural University, Thrissur, 76 p.

Madalgeri, S.B. and Dharmatt, D., 1989. Variability in fruit size of single plant progeny of BNF-3 ashgourd [*Benincasa hispida* (Thunb.) Cogn.]. *S. Indian Hort.*, 37: 249.

Mandal, J. and Sirohi, P.S., 2003. Heterosis in ashgourd [*Benincasa hispida* (Thunb.) Cogn.]. *Veg. Sci.*, 30: 169–179.

Mandal, J., Sirohi, P.S. and Behara, T.K., 2002. Genetical studies on flowering and fruit maturity in ashgourd. *Orissa J. Hort.*, 30: 40–42.

Mathewkutty, T., 1996. *Neykumbalam*. Mathrubhumi Daily Ltd., 30.1.1996.

Menon, M.P., 1998. Cataloguing and identification of promising ashgourd ecotypes in relation to season and maturity. *M.Sc. (Hort.) Thesis*, Kerala Agricultural University, Thrissur, 105 p.

Miniraj, N., Prasanna, K.P. and Peter, K.V., 1993. Ashgourd [*Benincasa hispida* (Thunb.) Cogn.]. In: *Genetic Improvement of Vegetable Crops*, (Eds.) B.O. Bergh and G. Kalloo. Pergamon Press, New York, pp. 235–238.

Morton, J.F., 1971. The Waxgourd: A year round Florida vegetable with unusual keeping quality. *Proc. Fla. St. Hort. Soc.*, 84: 104–109.

Murray, M. and Thompson, W., 1980. The isolation of high molecular weight plant DNA. *Nucl. Acids Res.*, 8: 4321–4325.

Nadkarni, K.M., 1927. *Indian Materia Medica*. Nadkarni and Company, Bombay, 76 p.

Parkash, C., Singh, K.P. and Kalloo, G., 2000. Variability analysis and cause and effect relationship in ashgourd [*Benincasa hispida* (Thunb.) Cogn.]. *Indian J. Pl. Genet. Resour.*, 13(3): 298–301.

Peter, K.V., 1998. *Genetics and Breeding of Vegetables*. Indian Council of Agricultural Research, New Delhi, 333 p.

Peter, K.V., Sadhankumar, P.G., Miniraj, N. and Prasanna, K.P., 1991. Genetic diversity in cucurbits– Ashgourd. *Indian Hort.*, 36: 2 and 35.

Peteira, B., Fernandez, E., Garcia, P., Miranda, E., Leon, O. and Miranda, I., 1999. Repeatability of RAPD markers in the genus *Lycopersicon*– Brief study. *Revista de Proteccion Vegetal*, 14(2): 75–79.

Providenti, R., 1986. Viral diseases of cucurbits and sources of resistance. *Technical Bulletin* 93. New York State Agricultural Experiment Station, Cornell, Geneva.

Randhawa, K.S., Singh, M., Arora, S.K. and Singh, P., 1983. Varietal variation in physical characters and chemical constituents of ashgourd fruits. *J. Res. Punjab Agric. Univ.*, 20: 251–254.

Resmi, J., 2004. Characterization of landraces of ashgourd [*Benincasa hispida* (Thunb.) Cogn.]. *M.Sc. (Hort.) Thesis*, Kerala Agricultural University, Thrissur, 100 p.

Resmi, J., Sreelathakumary, I. and Rajmohan, K., 2005. Detection of genetic variability in ashgourd [*Benincasa hispida* (Thunb.) Cogn.] using RAPD markers. In: *Proc. Natl. Symp. on Biotechnological Interventions for Improvement of Hortl. Crops: Issues and Strategies*, 10–2 January, KAU, Thrissur, Kerala.

Resmi, J. and Sreelathakumary, I., 2006. Variability, heritability and correlation studies in ashgourd [*Benincasa hispida* (Thunb.) Cogn.]. In: *Proc. 18th Kerala Science Congress*, 29–31 January, CESS, Thiruvananthapuram, p. 138–140.

Resmi, J., Sreelathakumary, I. and Rajmohan, K., 2007. Detection of genetic variability in ashgourd [*Benincasa hispida* (Thunb.) Cogn.] using RAPD markers. In: *Recent Trends in Horticultural Biotechnology*, (Eds.) R. Kesavachandran, P.A. Nazeem, D. Girija, P.S. John and K.V. Peter. New India Publishing Agency, New Delhi, pp. 571– 577.

Resmi, J. and Sreelathakumary, I., 2008. Divergence studies in ashgourd [*Benincasa hispida* (Thunb.) Cogn.]. In: *Proc. 20th Kerala Science Congress*, 29–31 January, Thiruvananthapuram, p. 174–177.

Robinson, R.W. and Decker, W.D.S., 1997. *Cucurbits*. Cab International, UK, 226 p.

Rugeri, D., 1968. Investigations on *Fusarium* wilt of melon (*C. melo*). *Phytopath. Mediterranean*, 7: 150–153.

Saimbhi, M.S., 1993. Agro-techniques for cucurbits. In: *Advances in Horticulture, Vol. 5: Vegetable Crops Part I*, (Eds.) G. Kalloo and K.L. Chadha. Malhotra Publishing House, New Delhi, pp. 401–432.

Santhakumar, R., 1991. *Chikithsavinjnanakosam*. Samrat Publishers, Thrissur, p. 735.

Seshadri, V.S., 1993. Cucurbits. In: *Vegetable Crops*, 2nd edn, (Eds.) T.K. Bose, M.G. Som and J. Kabir. Naya Prokash, Calcutta, pp. 40–124.

Simmonds, N.W., 1976. *Evolution of Crop Plants*. Longman, New York, p. 306.

Staub, J.E., Danin-Poleg, Y., Fazio, G., Horejsi, T., Reis, N. and Katzir, N., 2000. Comparative analysis of cultivated melon groups (*Cucumis melo* L.) using random amplified polymorphic DNA and simple sequence repeats markers. *Euphytica*, 115: 225–241.

Staub, J.E., Robbins, M.D., Chung, S. and Lopez-Sese, A.I., 2004. Molecular methodologies for improved genetic diversity assessment in cucumber and melon. *Acta Hort.*, 637: 41–47.

Soller, M. and Beckmann, J.S., 1983. Genetic polymorphism in varietal identification and genetic improvement. *Theor. Appl. Genet.*, 67: 25–33.

Sundararajan, S. and Muthukrishnan, C.R., 1982. A high yielding ashgourd. *Indian Hort.*, 27: 15.

Supreeja, A., 2002. The ash gourd. *Amruth.*, 4(20): 31–32.

Sureja, A.K., Sirohi, P.S., Behera, T.K. and Mohapatra, T., 2006. Molecular diversity and its relationship with hybrid performance and heterosis in ash gourd [*Benincasa hispida* (Thunb.) Cogn.]. *J. Hortl. Sci. and Bio.*, 81(1): 33–38(6).

Varghese, B.M., 1973. Studies on the cytology of south Indian cucurbitaceae. *Ph.D. Thesis*, Kerala University, Thiruvananthapuram, p. 206.

Verma, V. K., Behera, T.K., Munshi, A.D., Parida, S.K.and Mohapatra, T., 2007. Genetic diversity of ash gourd [*Benincasa hispida* (Thunb.) Cogn.] inbred lines based on RAPD and ISSR markers and their hybrid performance. *Scientia Horticulturae*. 113 (3), 231–237

Verma, V. K. and Behera, T.K., 2007. Heterosis studies in ash gourd [*Benincasa hispida* (Thunb.) Cogn.] for yield and related traits. *J. Trop. Agri.*, 45(1–2): 51–54.

Walters, T.W. and Walters, D.S.D., 1989. Systematic re-evaluation of *Benincasa hispida* (Cucurbitaceae). *Econ. Bot.*, 43: 274–278.

Waugh, R. and Powell, W., 1992.Using RAPD markers for crop improvement. *Trends in Biotech.*, 10: 186–191.

2011, **Biodiversity in Horticultural Crops Vol. 3** *Pages* **16–23**
Editor: **Professor K.V. Peter**
Published by: **DAYA PUBLISHING HOUSE, NEW DELHI**

Chapter 2

Satputia

B.R. Choudhary* and P.K. Singh
Indian Institute of Vegetable Research,
Sargatia, P.O. Seorahi, Kushinagar – 274 406, U.P.
**E-mail: choudharyiivr@gmail.com*

Satputia (*Luffa hermaphrodita*) is an important underutilized vegetable crop belonging to Cucurbitaceae family. It is a hermaphrodite form of ridge gourd and widely cultivated in Bihar and adjoining areas. Satputia is widely grown as mixed crop and as mono crop in both summer and rainy seasons. The fruits are available in varying sizes from round to long and borne in clusters. It is known for its nutritional and medicinal properties and is cooked as vegetable when fruits are quite tender. *L. hermaphrodita* is very nutritious and a good source of vitamin A, calcium, phosphorous, ascorbic acid and iron. It is often used against toothache is a disinfectant, antihelmintic, anti-diarrhea, anti-syphilitic, purgative, cardio tonic, laxative and also potential cure to diabetes and hypertension.

Luffa is an old world genus comprising 9 species. Out of them, 7 species including *Luffa acutangula*, *L. cylindrica*, *L. echinata*, *L. graveolens*, *L. hermaphrodita*, *L. tuberose* and *L. umbellate* are native to India (Pan, 1995). Among the available species; *Luffa acutangula*, *L. cylindrica* and *L. hermaphrodita* are economically important for their tender fruits consumed as vegetable (Chandra, 1995).

Botany

Satputia is an annual climber with branching at the base. Leaves are simple, smooth and five lobed. Tendrils are present on leaf axils. Flower is hermaphrodite, pale yellow to yellow in colour, borne in clusters (8.40-15.80) on every node and cross pollinated in nature. Calyx is 5 lobed and united at the base. Petals are 5, companulate and bright yellow in colour. The flower bears 3 anthers. Ovary is inferior, three carpellary and round to long in shape. Anthesis starts in the evening by 5.00 pm and continues up to 8.00 pm. Fruits are smooth to having longitudinal slight ridges. Seeds are many and become black in colour after attaining full maturity.

Cultivars

Despite presence of wide genetic diversity very little work is done for the improvement of satputia and still only one high yielding line is identified.

VRS-1

It was identified at IIVR Seed Production Centre, Kushinagar (UP) through inbreeding from a local material. It is sequential bearing and produces early fruiting (44.0-46.2 days for first picking) coupled with high fruit yield (5.2-6.3 kg/plant). This line bears hermaphrodite flowers which appear at 5.0-7.6 nodes. The fruits are tender and produced in clusters (4.8-5.6 fruits/cluster) on every node weighing 30.5-35.7g and 13.5-15.0cm long. It is characterized by 10 dark green, superficial and continuous longitudinal ridges (Choudhary *et al.*, 2009).

Breeding Objectives

For the improvement of satputia, knowledge on desired plant type and fruit characteristics are essential. The following point should be taken into account in any breeding programme of satputia.

1. Vigorous well branched plants having sequential bearing habit.
2. The lowest node number at which first hermaphrodite flower appears is an indicator of earliness.
3. High frequency of hermaphrodite flowers as it results in higher fruit yield.
4. Uniformly thick cylindrical, 10-15cm long green or dark green fruits.
5. Short intermodal length, bear more number of fruits/cluster, high number of fruits/plant and high fruit yield.
6. Tender and non-fibrous fruits at edible maturity.
7. Resistance to powdery mildew, downy mildew and insects.

Genetic Diversity

Genetic variability is the basis of all plant improvement programmes. Sufficient genetic variability if present can be exploited to develop superior cultivars. Vavilov (1951) was the first to realize that a wide range of variability in any crop provided a better chance of selection of desirable types. Due to cross pollination,wide range of variability exists in satputia for vegetative and fruit characters. Knowledge about the magnitude and nature of genetic variation in a specific population is of prime importance for the effective breeding programme (Debnath, 1988). Possibility of improvement in any crop is measured by available variability. Wider is the genetic variability, greater the chances of improvement for desirable traits. The efficiency of selection depends mainly on the nature and extent of variability in a population. Despite presence of sufficient variability in satputia, it has not received due attention for exploiting the existing variability. Ram *et al.* (2006) studied the genetic variability of satputia among 24 genotypes and reported wide variability for days to first hermaphrodite flower anthesis (40.67-55.67), days to first harvesting (50.00-69.33), vine length (4.13-7.20m), length of full grown fruit (5.47-10.67cm), diameter of full grown fruit (2.33-4.23cm), fruit weight at edible stage (17.37-41.30g), days to edible maturity from fruit set (7.33-17.33), number of fruits/plant (96.33-185.33), number of seeds/fruit (21.00-140.33), seed weight/fruit (1.80-12.06g), 100 seed weight (6.83-10.33g) and edible fruit yield (34.40-107.97q/ha). Genotype SP-59 expressed maximum fruit yield (107.97q/

ha) whereas DR/Y-1 showed earliness with respect to days taken to first hermaphrodite flower anthesis (40.67) and maximum number of fruits/plant (185.33). Choudhary *et al.* (2007) collected germplasm of satputia from different locations of Uttar Pradesh and Bihar and observed variability in respect of vine length, earliness, fruit shape, fruit length, fruit colour, number of fruits/cluster, harvest duration and yield/plant at farmer's field. The collected material was studied for quantitative variability and maximum variability was observed for fruit length, number of fruits/cluster, vine length, number of primary branches/plant, fruit weight, days to first picking and number of seeds/fruit. The fruit colour was also recorded from green to dark green. Choudhary *et al.* (2009) evaluated 35 genotypes and reported VRS-1, VRS-7 and VRS-11 as promising genotypes for horticultural traits. Wide genetic variability was observed for vine length (3.50-5.90m), internodel length (5.50-13.60cm), days to appearance of first hermaphrodite flower (41.13-52.67), node at which first flower appears (3.40-11.53), number of fruits/cluster (3.87-5.93), fruit length (3.60-16.40cm), days taken to first picking (48.07-57.67), number of fruits/plant (124.13-147.53), fruit weight (18.95-38.64g) and yield/plant The observed variability is a combined estimate of genetic and environmental factors, of which only the former one is heritable. Presence of rich variability in satputia offers excellent opportunities for breeding of high yielding varieties and F_1 hybrids of superior quality. Inspite of wide range of genetic variability in this crop, very little attention is given to exploit it.

Genetic Parameters

Information on heritability for different component traits are essential for any crop improvement programme, because the heritable component is the consequence of genotype and is inherited from generation to generation. However, the estimate of heritability alone does not provide an idea about the expected gain in next generation; therefore, it has to be considered in conjunction with genetic advance. Wright (1921) reported that heritability components comprised of additive and non additive portion and it was the former which responded to selection. Genotypic and phenotypic coefficients of variations are helpful in drawing meaningful comparison of the variation of several traits of plants. Estimation of expected genetic advance is important to have an idea of effectiveness of selection. Burton (1952) suggested that characters with high heritability coupled with high genetic advance would respond to selection better than those with high heritability and low genetic advance. Johnson *et al.* (1955) also reported usefulness of heritability values together with genetic advance in predicting the expected progress to be achieved through selection.

Ram *et al.* (2006) found the narrow difference between phenotypic and genotypic coefficient of variation indicating the possibility of lower environmental influence on expression of these characters. They recorded the genotypic and phenotypic coefficient of variation for number of seeds/fruit (44.10, 44.83), seed weight/fruit (42.53, 43.50), edible fruit yield/ha (30.41, 31.19), individual fruit weight at edible maturity (22.21, 22.92) and days to edible maturity from fruit set (20.05, 22.13) indicating maximum variability in these characters and have ample scope for effective improvement through selection. High heritability and high genetic advance for number of seeds/fruit and edible fruit yield/ha indicting presence of high amount of fixable variation among genotypes for these traits, which provide ample scope for varietal improvement through hybridization and selection in *Luffa hermaphrodita* (Ram *et al.*, 2006). High heritability suggests that characters might be governed by additive gene action and phenotypic value could provide fair and reliable measure of genotypic variance. Ram *et al.* (2006) also observed high heritability (96.6) along with high genetic advance (65.22) for number of seeds/fruit.

Figure 2.1: Hermaphrodite Flower and Fruits

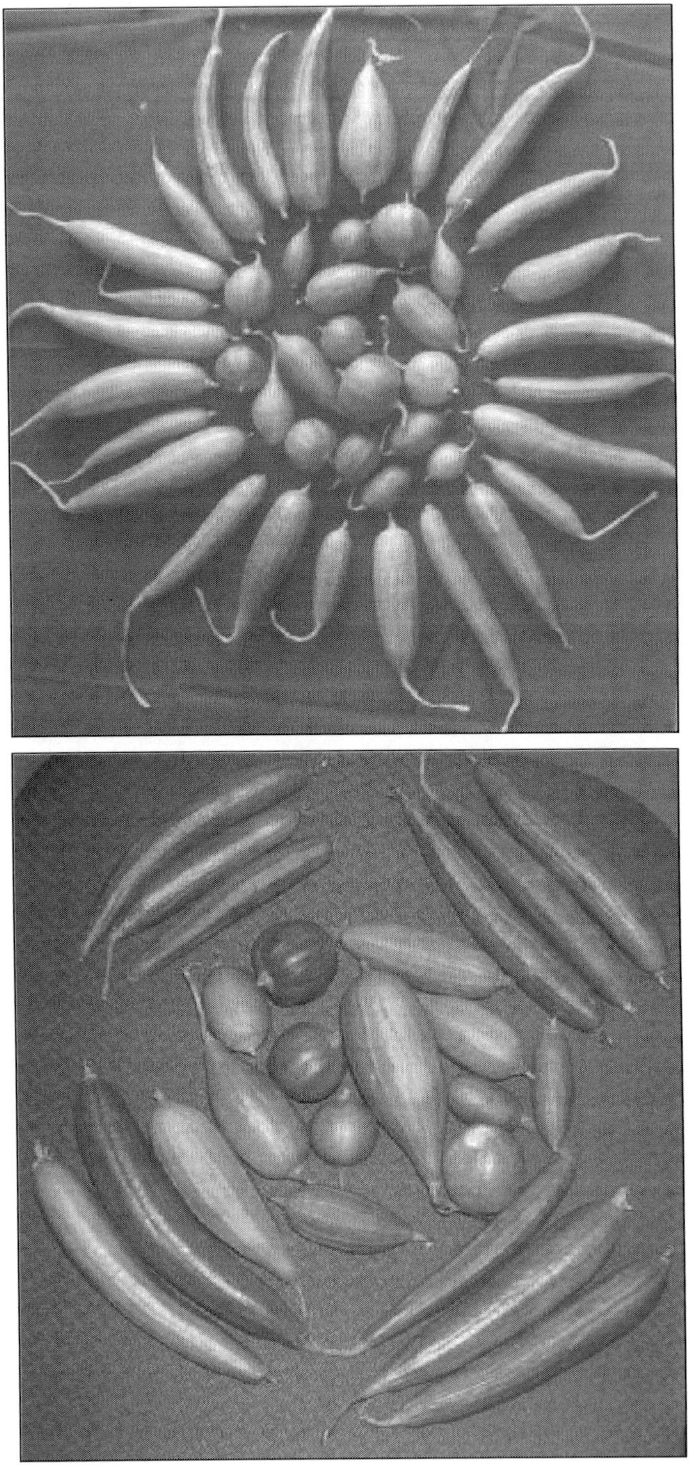

Figure 2.2: Variability in Fruits

Combining Ability Effects

Combining ability helps the breeders in selecting desirable parents for exploitation in breeding programme. General combining ability indicates the average performance of a line in a series of hybrid combinations, whereas, specific combining ability shows those effects in some definite combinations which significantly departed from what would be expected on the basis of the average performance of the lines involved (Sprague and Tatum, 1942). Combining ability studies are useful in assessing the nicking ability of parents and thus, help in selecting parents which, when crossed, would give rise to more desirable segregants during a hybridization programme. It also helps to characterize the nature and magnitude of a pragmatic breeding approach. Ram *et al.* (2004) identified VRST-2 as the best general combiner of satputia as it showed maximum and significant gca effect in the desired direction for most of the traits.

Correlation and Path Analysis

Correlation coefficient is a measure of degree of association between two traits worked out at the same time (Hayes *et al.*, 1955). Yield is a very complex entity; it is not only polygenic in nature but also affected by environment. Hence, the selection of superior plants based on performance of yield as such is usually not very effective. For selecting superior genotypes, the breeder has to choose from the material on the basis of its phenotypic expression. As most of the traits of economic importance are complex involving several related traits, the knowledge of degree of phenotypic and genotypic correlation of the traits is important (Robinson *et al.*, 1951). Correlation estimates between fruit yield and its components are helpful in developing suitable selection criteria for selecting desired plant types or developing high yielding cultivars. The nature and magnitude of correlation coefficient promote a rational choice of the character(s) in which selection can be exercised. Path coefficient analysis is simply a standardized partial regression, which may be useful in choosing the character(s) which have direct or indirect effects on yield. Such a study may be useful and effective in selection for simultaneous improvement of the component characters which contribute towards yield. Path analysis was initially suggested by Wright (1921) but was applied for the first time in plant breeding by Dewey and Lu (1959). Ram *et al.* (2006) studied the character association in *Luffa hermaphrodita* and found fruit weight and number of fruits/plant were positively and significantly associated with yield. Days to first harvest, vine length and fruit weight were found to be positively associated with first hermaphrodite flower anthesis. He also recorded positive and significant association of first fruit harvesting with vine length, individual fruit weight and days to edible maturity of fruit from fruit set.

Heterosis

The phenomenon of hybrid vigour in artificial plant hybrids was first studied by Koelreuter (1776). Shull (1914) was the first who coined the term heterosis and defined as the superiority of heterozygous genotypes with respect to one or more characters. Hull (1945) explained this phenomenon based upon overdominance. Hayes (1964) concluded that heterosis is the normal expression of a complex character when the genes are in highly heterozygous condition with dominance or partial dominance. In different crops, heterosis can be evaluated in comparison with the parental variety or the best variety. There is good scope for improvement of yield and other desirable characters through heterosis breeding. Ram *et al.* (2004) recorded the highest heterosis in satputia for fruit yield/plant in VRST-2 x VRST-16 (121.74 per cent) followed by VRST-62 x VRST-16 (108.20 per cent) and VRST-57 x VRST-53 (86.30 per cent) in a set of 15 F_1 hybrids. He also observed significant heterosis for days to germination (VRST-2 x VRST-66), days to flowering (VRST-57 x VRST-53) and days to first fruit harvesting (VRST-57 x VRST-53) over better parent. As heterosis was noted in most of the traits, exploitation of heterosis in the form of hybrids is advocated. Ram *et al.* (2004) found the ratio of genetic variance less than unity, which indicated the predominance of non-additive gene action in 11 traits of economic importance and suggested the possibilities of heterosis in satputia. He suggested that an ideotype which has high fruit weight and more number of fruits/plant is expected to be better suited to improve fruit yield of *Luffa hermaphrodita* through selection.

References

Burton, G.W., 1952. Quantitative inheritance in grasses. In: *Proc. 6th Intl. Grassland Cong. J.*, 1: 277–283.

Chandra, U., 1995. Distribution, domestication and genetic diversity of *Luffa* gourd in Indian sub-continent. *Indian J. Plant Genet. Resour.*, 8: 189–196.

Choudhary, B.R., Pandey, S., Singh, P.K., Singh, R. and Rai, Mathura, 2009. VRS–1: A promising line of satputia (*Luffa hermaphrodita*). In: *International Conference on Nurturing Arid Zones for People and the Environment: Issues and Agenda for the 21st Century*, CAZRI, Jodhpur from November 24–28, p. 88.

Choudhary, B.R., Pandey, Sudhakar, Yadav, D.S. and Rai, Mathura, 2007. Satputia (*Luffa acutangula* var. *hermaphrodita*): A high potential cucurbit for eastern India.In: *IInd Indian Hort. Cong.*, ICAR Complex for NE Region, Barapani, Meghalaya from 18–21 April, p. 183.

Debnath, S.C., 1988. Nature and estimation of genetic components of variation. *Acta Agronomica*, 37: 337–351.

Dewey, D.R. and Lu, K.H., 1951. A correlation and path coefficient analysis of components of created wheat grass seed. *Agron. J.*, 51: 515–518.

Hayes, H.K., 1964. Development of heterosis concept. In: *Heterosis*, (Ed.) J.W. Gowen. Hafner Publishing, New York, 1964. Chap. 3.

Hayes, H.K., Immer, F.H. and Smith, D.C., 1955. *Methods of Plant Breeding*. Mc Graw Hill Book Co., Inc., New York, p. 439.

Hull, F.H., 1945. Recurrent selection and specific combining ability in corn. *J. Amer. Soc. Agron.*, 37: 135–145.

Johnson, H.W., Robinson, H.F. and Comstock, R.E., 1955. Estimation of genetic and environmental variability in soybeans. *Agron. J.*, 47: 314–318.

Koelreuter, J.G., 1776. *Vorlanfigen Nachricht, Von einigen das Geschlacht der Pflanzen hetreflenden Versuchen and Boebachtungen*. p. 266 Liepzig (cited by Shull, G.H. in Heterosis, p. 14, Iowa State College Press).

Pan, X., 1995. Discovery of a loofah plant with both male and female flower in same inflorescence. *Cucurbit Genet. Resour.*, 1: 51.

Ram, D., Rai, Mathura, Verma, A.K. and Pandey, Sudhakar, 2004. Heterosis and combining ability in satputia (*Luffa hermaphrodita*). *Veg. Sci.*, 31(2): 129–134.

Ram, D., Rai, M., Verma, A. and Singh, Y., 2006. Genetic variability and association in *Luffa* sp. *Indian J. Hort.*, 63(3): 294–297.

Robinson, H.F., Constock, R.E. and Harvey, P.H. 1951. Genotypic and phenotypic correlations in corn and their implications in selection. *Agron. J.*, 43: 262–267.

Shull, G.H., 1914. The composition of yield of maize. *Rpt. Amer. Breeder's Assoc.*, 4: 296–301.

Sprague, G.F. and Tatum, L.A., 1942. General v/s specific combining ability in single crosses of corn. *J. Amer. Soc. Agron.*, 34: 923–932.

Vavilov, N.I., 1951. The origin, variation, immunity and breeding of cultivated plants (Trans. from Russian Chester, K.S.). *Chronica Botanica*, 13: 361–364.

Wright, S., 1921. Correlation and causation. *J. Agric. Res.*, 20: 557–585.

2011, Biodiversity in Horticultural Crops Vol. 3 *Pages* **24–42**
Editor: **Professor K.V. Peter**
Published by: **DAYA PUBLISHING HOUSE, NEW DELHI**

Chapter 3

Watermelon

B.R. Choudhary* and P.K. Singh
Indian Institute of Vegetable Research,
Sargatia, Kushinagar – 274 406, U.P.
**E-mail: choudharyiivr@gmail.com*

Watermelon (*Citrullus lanatus* (Thunb.) Matsam. and Nakai) is an important crop belonging to Cucurbitaceae family with (2n=2x=22). It is one of the most widely cultivated crops in the world (Huh *et al.*, 2008). Its global consumption is more than that of any other cucurbits. It accounts for 6.8 per cent of the world area devoted to vegetable production (Goreta *et al.*, 2005). China is leading in production of watermelon followed by Turkey, United States, Iran and Republic of Korea (Huh *et al.*, 2008). Watermelon is grown for its fleshy, juicy and sweet fruit. Mostly eaten fresh, they provide a delicious and refreshing dessert in hot weather. It is no longer just a summer fruit and is becoming an everyday fruit like apple, bananas and oranges. The major nutritional components of the fruit are carbohydrates (6.4g/100g), vitamin A (590 IU) and lycopene (4100μg/100g), an anti-carcinogenic compound found in red flesh of watermelon. Lycopene is classified as useful in the human diet for prevention of heart attacks and certain types of cancer. Watermelon seeds are rich in fat (about 45 per cent edible oil) and protein (30-40 per cent). The seeds are powdered and baked like bread in India. The seed kernels are also used in various sweets and other delicacies. The unripe fruits are also cooked as a vegetable in some parts of India.

Species, Origin and Distribution

Watermelon is native to Tropical Africa chiefly the Kalahari Desert (the current nations of Namibia and Botswana) where wild forms are still found (De Candolle, 1882). Now it spread throughout the tropics and the Mediterranean (Tindall, 1983). Watermelon is domesticated in Africa at least 4000 years ago and now grown worldwide, particularly in regions with long, hot summers (Robertson, 2004). A good amount of variability is also found in South East and North Africa, Southern and South Central Asia, Iran, Pakistan and India. China is the largest watermelon producer in the world. The other major watermelon producing countries are Turkey, Iran, Egypt, the United States, Mexico and Korea. Whitaker (1933) considered wild watermelon (*Citrullus colocynthis*) as probable ancestor of watermelon which is a native of arid soils in Africa. Morphologically the *colocynth*, has similar characters with *vulgaris*, but fruits are bitter and seeds small. However, the bitter forms of *C. lanatus* were considered the probable ancestor of watermelon by others. *C. lanatus* is one of four known diploid (n=11) species belonging to the xerophytic genus *Citrullus* Schard. ex Eckl. and Zeyh., found in temperate regions of Africa, Central Asia and the Mediteranean (Whitaker and Davis, 1962; Jaffery, 1975 and Whitaker and Bemis, 1976). The genus *Citrullus* Schard. ex Eckl. and Zeyh. has the following four diploid species:

Table 3.1

Sl.No.	Species	Distribution	Growth Habit	Reference
1.	*Citrullus lanatus* (Thunb.) Matsam. and Nakai includes the cultivated watermelon (*Citrullus lanatus* var. *lanatus*) also called '*egusi*' melon, and the preserving melon (*C. lanatus* var. *citroides*) also called '*tsamma*' melon.	Tropical and subtropical climates worldwide including West and South Africa	Annual	Whitaker and Bemis (1976)
2.	*C. colocynthis* (L.) Schard- 'bitter apple'	Sandy areas throughout northern Africa, south-western Asia and the Mediterranean	Perennial	Zamir *et al.* (1984); Burkill, (1985) and Jarret *et al.*, (1997)
3.	*C. ecirrhosus* Cong.	Desert regions of Namibia	Perennial	Meeuse (1962)
4.	*C. rehmii* De Winter	Desert regions of Namibia	Annual	Meeuse, (1962) and De Winter, (1990)

Botany and Floral Biology

Watermelon differs from other economically important cucurbits by having pinnatifid leaves. The hairy stem is thin, angular and grooved bearing branched tendrils. Watermelon is an annual vine or creeper. It has an extensive root system so that it can be well adapted for river bed cultivation to utilize subterranean moisture. Stems are branched and prostrate. It is monoecious (both male and female flowers occur on the same vine) in sex expression with very high male to female (7:1) sex ratio, but the ratio ranges from 4:1 to 15:1. However, andromonoeciuos (staminate and perfect flowers on the same plant) sex form is also found in watermelon (Choudhary, 2009a). The plant is classified as a naturally cross pollinated crop. The pistillate flowers normally occur in every 7th leaf axil while the intervening axils are occupied by the staminate flowers. Flowers are unisexual, cross pollinated and light yellow in colour. Corolla is showy yellow in colour, united and 5 in number. Stamens are three and attached to the calyx tube. Ovary is inferior. The edible part of the fruit is the endocarp (planceta). Fruit is pepo (berry with hollow cavity, thick rind and seeds with parietal placentation). Botanically it varies in shape, size, colour and taste. Fruits are large, round to oblong or cylindrical fruits measure as long as 60 cm. Usually watermelon produced single fruit/node but Choudhary (2009b) observed two fruits/node. Flesh is red or pink in colour on full ripening. In some cultivars, yellow flesh is also seen. The rind may be dark green to light green in colour and it may be with stripes or without stripes. Seeds are many, dark brown to black in colour. The seed coat is warty.

The whole period of bud developmental stage is completed in 12-16 days by the male and in 11-13 days by female bud. The rate of increase in size is associated with bud development and this increase in size is more rapid in earlier stages *i.e.* up to 8-12 days. Thereafter it decreases and during the post developmental period remains more or less constant. The anthesis starts in the early hours of morning at 5.30 am and continues up to 7.30 am with peak between 6.30-7.00 am. The dehiscence of anthers starts $1^{1/4}$ hours before anthesis *viz.*, 4.45 am and continues up to 6.30 am. Pollen is visually evident in sticky masses adhering to anther. The stigma becomes receptive 2 hours before anthesis and continues up to 3 hours after anthesis. From the date of pollination and fertilization, it takes 30-40 days to full maturity and ripening of fruits. Watermelon is a diclinous plant with large flowers and hence pollination control is easy. At the time of flowering, the staminate flowers are pinched from the vine and pistillate flowers isolated/bagged. The pollen collected from the desired pollen parent can be used to pollinate isolated pistillate flowers. Yield of hybrid can be appreciably enhanced by suppressing the appearance of staminate flowers. Application of chemicals like ethephon, NAA, GA_3 and 2, 3-dichloropropionic acid can control the staminate flowers when applied before the initiation of flowering.

Figure 3.1: Male Flower

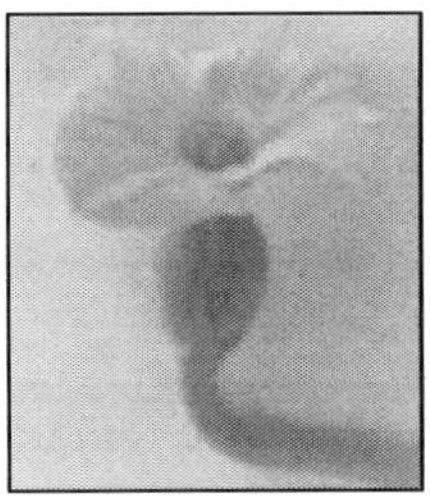

Figure 3.2: Female Flower

Figure 3.3: Andromonoecious Flower

Figure 3.4: Cluster Bearing Watermelon

Varieties/Hybrids

Depending on the cultivar, watermelon fruits are produced in different sizes (ice box, small, medium, large or giant), different shapes (round, oval, blocky or elongate), different rind patterns (gray, narrow stripe, medium strip, wide strip, light solid or dark solid), different flesh colours (white, yellow, orange or red) and different types (seeded or seedless). Commercially, the most popular seeded cultivars are red flesh, blocky shape and large sized (8-10kg). High yield is a major goal for watermelon breeders. The successful cultivars of watermelon must have traits other than high yield. High fruit quality is of major importance and includes bright flesh colour, firm texture, high sugar content and proper fruit shape. In several countries demand is also there for mini watermelons (up to 3.5 kg/fruit). Several cultivars and F_1 hybrids are developed in watermelon in India. Under changing agro climatic conditions, its cultivation has become very difficult since farmers are encountering several fungal and viral diseases such as *Fusarium* wilt, anthracnose, powdery and downy mildews and bud necrosis virus. Developing suitable varieties/hybrids resistant against *Fusarium* wilt and watermelon bud necrosis diseases is necessary. Following varieties are developed by adopting several breeding methods which are described as follows:

Sugar Baby

It is originally introduced by M. Hurdin, Geary, Oklahoma, USA in 1956. From USA, it was introduced to India by IARI, New Delhi. The fruits are slightly small in size weighing 3-5 kg, round having bluish-black rind and deep pink flesh, small seed with 11-13 per cent TSS. The fruits ripe in 85 days. Yield potential is 150q/ha.

Asahi Yamato

It is a mid-season Japanese introduction by IARI, New Delhi. Fruits are medium sized weighing 6-8 kg, rind light green, flesh deep pink with 11-13 per cent TSS. Fruits ripe in about 95 days.

Improved Shipper

An introduction from USA and released by PAU, Ludhiana. Fruits are big sized weighing 8-9 kg, dark green and oval to oblong with moderate sweetness (8-9 per cent TSS).

Arka Jyoti

A mid-season F_1 hybrid developed by crossing IHR-20 x Crimson Sweet at IIHR, Bangalore. It produces round fruits weighing 6-8 kg with light green rind and dark green stripes. Flesh colour is crimson with excellent texture, excellent flavour, high TSS (11-13 per cent) and high yield (500-600 q/ha).

Arka Manik

Developed at IIHR, Bangalore by crossing a local selection IHR-21 from Rajasthan and Crimson Sweet from USA followed by modified backcross method. Fruits are round to oval weighing 6 kg with green rind and dull green stripes. Flesh is deep-red, very sweet (12-13 per cent TSS) with a few small seeds. It stands well in transport and storage. It is resistant to powdery mildew, downy mildew and tolerant to anthracnose. It produces 400-500q fruits/ha in 100-110 days.

Pusa Bedana

It is a seedless triploid variety developed at IARI, New Delhi by crossing Tetra-2 (tetraploid) x Pusa Rasal (diploid). Fruits are dark green weighing 5-6 kg, somewhat triangular in shape with tough rind, red fleshed, seedless with 12-13 per cent TSS. It takes 115-120 days for first fruit harvest.

Durgapura Meetha

It is a late maturing variety (125 days) developed from indigenous germplasm at Agricultural Research Station, Durgapura, Rajasthan. Fruits are round with light green thick rind and good keeping quality. Flesh is red with 10-11 per cent TSS. First picking starts in 110-120 days. Average fruit weight is 6-8 kg having 400-500q/ha yield potential. It possesses good keeping and transport quality.

Durgapura Kesar

It is a late variety developed at Agricultural Research Station, Durgapura, Rajasthan by crossing Durgapura Basanti x Durgapura Meetha. Fruits are slightly ovicular shaped with light green rind. Flesh is light saffron in colour, juicy and sweet (10-11 per cent TSS) with large seeds. Average fruit weight is 8-10 kg and fruit yield is 350 q/ha.

Durgapura Lal

It was developed at Agricultural Research Station, Durgapura, Rajasthan from the cross Sugar Baby x KS-3566. Fruits are round, dark green with dark lining with thin and hard rind weighing 4-5 kg. Flesh is crisp, dark red with 10-11 per cent TSS. First picking starts at 105-110 days and produces 350-450 q fruit yield/ha. It is moderately resistant to blight, powdery mildew and bud necrosis under field conditions (Luthra and Yadav, 2004).

Besides, there are a number of extensively cultivated hybrids being marketed by private companies. They are Madhu, Milan, Mohini, MHW-4, MHW-5, MHW-6, MHW-11, MHW-12, MHW-15, NS-246, NS-295, Suruchi, Samtrupti, Amruth, Century-2 and Nath-102. Appealing appearance, good quality flesh and transportability and resistance/tolerance to wilt are important characters available in most of these hybrids. There are several other cultivars locally grown which are named after the region in which they are grown, such as 'Farrukhabadi', 'Moradabadi', 'Faizabadi' of Uttar Pradesh. 'Charleston Gray' an introduction of watermelon from USA, characterized by its oblong to cylindrical shape, is a leading cultivar suited to long distance transportation.

Heterosis Breeding

Although attempts for utilizing heterosis in watermelon date back to 1949 in Japan, significant progress was recorded only during the 1960s. Watermelon can inbreed indefinitely without inbreeding depression (Porter, 1933). Both diploid and triploid hybrids are now fast replacing open pollinated cultivars. The impact is significant in Japan, China and the USA, as hybrid occupies significant acreage. Popularity of hybrids is due to their higher yield, better quality, fruit uniformity, good storage and disease resistance.

Watermelon, having predominant monoecious sex expression, is a highly cross pollinated crop and hence highly suitable for hybrid seed production. Hybrid seed production is done through hand pollination all over world. Isolation planting of male and female lines helps in their maintenance while crop pollination is done by covering mature buds in the evening in both lines and collecting pollen from such covered flower of male line to pollinate the flowers of female line in the morning 6. 30 to 9.30 am and covering the pollinated flower of female line again. At the time of pollination, touching of ovary should be avoided. Depending upon gca and sca estimation, the cultivars/lines namely Charleston Gray, Crimson Sweet, Asahi Yamato, Durgapura Meetha and Arka Manik are good combiners for most of the important traits of consideration. Controlled plant growth through pruning should be allowed to bear one fruit/vine in case of large fruited varieties and two fruits/vine in case of small to medium fruited varieties to get good amount, size and vigour of seed.

Nath and Dutta (1970) recorded more than 50 per cent heterosis for yield in the cross-combinations IHR-6 x Charleston Grey, IHR-20 x Crimson Sweet and Sugar Baby x Crimson Sweet. High heterosis to an extent of 87 per cent for yield was attributed to more number of fruits/vine and fruit size, fruit number contributed significantly. Work carried out by Nandpuri *et al.* (1974) led to the identification of heterotic cross-combination Shipper x Striped Klondik for yield and number of fruits and Sugar Baby x Verona for early bearing of female flowers and TSS content. Combining ability studies conducted by Sidhu and Brar (1977) revealed that F_1s between low x high and medium x high general combiners produced productive hybrids. Brar and Sukhija (1978) pinpointed importance of greater diversity in origin of parental lines for obtaining enhanced magnitude of heterosis. Uniformity of the hybrids depends upon the level of genetic purity of the combining parents. For the expression of potential heterosis, prevailing environmental factors also play a significant role.

Hybrids also exhibit increased sugar content and low fiber content (Zhang, 1987). Sidhu and Brar (1988) observed non-consistency between parental divergence and magnitude of heterosis in the F_1 hybrids. Gill and Kumar (1988) identified superior hybrid combination *viz.*, Sugar Baby x Durgapura Meetha for yield and number of fruits/vine, and also good performance for earliness. Heterosis is exhibited for early seed germination; fast seedling growth; profuse root system; thick, dark green and strong leaves; strong stem growth, good branching and better adaptability (Li, 1992 and Pai, 1992). Hybrids exhibit better tolerance to diseases like anthracnose. Zhang and Rhodes (1992) recorded 33.3

to 86.9 per cent heterosis for number of fruits/vine and average fruit weight. Pichaimuthu and Swamy (2009) evaluated 10 F_1 hybrids at IIHR, Bangalore during 2006-07, of which WM-35 gave the highest fruit yield of 66.21t/ha having average fruit weight of 11.2kg. The flesh is red, juicy and sweet (11.33 per cent TSS). Singh *et al.* (2009) noted Arka Manik x Durgapura Selection, Durgapura Selection x Asahi Yamato and Arka Manik x Asahi Yamato to be the best performing hybrids for number of fruits/plant and fruit yield/plant which showed significant heterosis over standard parent (Arka Jyoti). Asahi Yamato x Sugar Baby showed desirable heterosis for early appearance of first female flower (-5.10 per cent) and GP-7 x AHW-19 for anthracnose resistance (-42.30 per cent).

Development of F_1s

In general, there is no inbreeding depression in watermelon and also data to show yield superiority of F_1 are lacking. Still, development of hybrid cultivars of water melon is receiving increasing attention due to the following reasons:

1. Availability of male sterile mutant to facilitate F_1 seed production.
2. Availability of marker gene, non-lobed leaf type distinguishable in young plant stage where heterozygotes can be recognized by their intermediate lobing.
3. Large seed production by single fruit through manual means leading to not too prohibitive cost of hybrid seed.
4. Easy development of dominant genes conditioning resistance to diseases (anthracnose) provided the F_1 is good in other respects.
5. Easy manipulation of intermediate fruit shape by crossing long x round fruited types.

For development of diploid hybrids, monoecious parental lines are developed through 3-4 generations of inbreeding followed by selection of specific combinations based on performance of good F_1s. It should be ensured that one of the parental lines should have dominant resistance gene against *Fusarium* and preferably the female parent should have marker character (non-lobed leaf).

All the staminate flowers from female parent should be removed daily and fruit should be allowed to set on female parent through cross pollination from the male parent. The seed so harvested is supposed to be F_1 seed between the two parental lines having indeterminate fruit shape between typical long and round type and will produce young plants with intermediate leaf shape between non-lobed and normal segmented type leaves. The accidental self could be identified by non-lobed leaf marker phenotypes.

Disease Resistance Breeding

Orton (1911) in a classical investigation found that resistance to *Fusarium* wilt is determined genetically. His work indicated that the semi-wild, non-edible stock melon or citron was resistant to *Fusarium* wilt. The resistant stock melon or citron was crossed with the cultivar Eden, susceptible to diseases. The F_1 from this cross was exceptionally vigorous and resistant to *Fusarium*. F_2 progenies were grown and tested for resistance over a period of several years. Finally a selection was chosen that combined a high level of resistance with satisfactory horticultural characters. The result was the cultivar 'Conqueror'. Unfortunately, 'Conqueror' never justified the high hopes of its sponsor, chiefly because of poor quality. Orton's work, however, did trigger a vast amount of research aimed at developing wilt resistant cultivars.

Most watermelons are resistant to races of powdery mildew (*Sphaerotheca fuliginea*), but a single recessive gene (*pm*) for susceptibility is found in the accession PI 269677. A high level of resistance to zucchini yellow mosaic virus available in four landraces of *C. lanatus*, is governed by a single recessive gene. The cultivar Arka Manik carries high resistance to powdery mildew and moderate resistance to anthracnose. Krishna Prasad *et al.* (2002) identified *Fusarium* wilt resistant inbred lines, *viz.*, IIHR-40-2-3, IIHR-40-2-1-1-3, IIHR-51-1-1-2, IIHR-51-1, IIHR-51-1-1-2-1, IIHR-114-1, DHR-118 and IIHR-119-5-2 which fell in well separated clusters, *i.e.*, II, IV, V and VI respectively. This indicated existence of divergence among *Fusarium* wilt resistant inbreds and such type of diverse inbreds as parents will be helpful in producing transgressive segregants through effective breeding programme.

PI 296341 is reported to contain resistance to 0, 1 and 2 races of *Fusarium* wilt (Netzer and Martyn, 1989; Martyn and Netzer, 1991 and Levi *et al.*, 2001a). The six PIs (PI 203551, PI 270550, PI271778, PI 189225, PI 299379 and PI 271779) are resistant to race 2 of anthracnose. PI 189225 and PI 271778 were reported resistant to gummy stem blight disease (Sowell and Pointer, 1962; Sowell, 1975 and Levi *et al.*, 2001a). Levi *et al.* (2001a) tested several watermelon accessions for downy mildew and watermelon mosaic virus (WMV) resistance of which only two (PI 169289 and PI 169290) found resistant. They also reported PI 162667, PI 165451, PI 185635, PI 185636, PI 203551 and PI 270306 resistant to races 1 and 3 of anthracnose. Most watermelon PIs resistant to race 1 were also resistant to race 3, but not to race 2 of anthracnose.

Based on the gene list given by Guner and Wehner (2004), resistances to race 1and 3 of anthracnose (*Colletotrichum lagenarium)* are governed by a single dominant gene. Resistance to race 2 of anthracnose is also controlled by a single dominant gene. Resistance to race 1 of *Fusarium oxysporum* f. sp. *niveum* is controlled by a single dominant gene. Rolania *et al.* (2004) reported the lowest severity of watermelon mosaic virus in Asahi Yamato (16.90 per cent), severity of powdery mildew in RW-187-2 (2.1 per cent) and severity of anthracnose in GP-7 (8.8 per cent) which could be used in breeding programme to improve watermelon. RW-177-3 a promising non-lobed leaf marker genotype identified by Yadav and Luthra (2005) possesses moderate resistance against blight, powdery mildew and Ubias virus. Gusmini *et al.* (2005) selected PI 279461, PI 482379, PI 245744, PI 526233, PI 482276, PI 271771, PI 164248, PI 240019, PI 296332 and PI 490383 as the most lines cultigens against gummy stem blight which can be used in further breeding efforts. Davis *et al.* (2006) identified PI 525088-PMR line of watermelon having resistance to melon race 1 of powdery mildew. It does not have desirable fruit quality however; it could be a useful parent for introducing race one of powdery mildew (*Sphaerotheca xanthii*) resistance into adapted watermelon cultivars. Davis *et al.* (2007) found that wild type accessions tended to be more resistant against *Sphaerotheca xanthii* race 1W often than the cultivated species of *C. lanatus* var. *lanatus*. Among 1573 genotypes, none were immune, eight exhibited high level of resistance and 86 demonstrated immediate resistance.

Genetics

Information on the genetic make-up of various qualitative and quantitative traits are required for determining the most appropriate breeding procedure to be adopted for watermelon. Information on the genetic make-up of various qualitative and quantitative traits are required for determining the most appropriate breeding procedures to be adopted for watermelon improvement. Variety Arka Manik released by IIHR, Bangalore carries high degree of resistance to powdery mildew, downy mildew and anthracnose (Nath *et al.*, 1973). Screening work done at Punjab Agricultural University, Ludhiana led to the identification of varieties "Asahi Yamato" and "Sugar Baby" to be moderately resistant to powdery mildew (Waraitch *et al.*, 1975). Genotype x environment interaction plays a great role in the expression of characters *viz.* days to first female flower, days to maturity, yield/vine, fruit weight and number of fruits/vine (Gill and Kumar, 1989). Sugar Baby, Arka Jyoti and Special No. 1 are identified to exhibit average stability. It was also observed that late maturing varieties, though high yielding but had poor stability.

Guner and Wehner (2004) published the latest version of the gene list for watermelon. Three major genes which control seed coat colour are 'r' for red, 't' for tan and 'w' for white seed coat. These genes interact to produce six phenotypes. Two genes *viz.* 's' for short and "l" for long seed length control seed size. Several genes govern leaf or foliage traits of watermelon. Non-lobed leaf has entire leaf rather than the lobed leaf of the typical watermelon. Four dwarf genes are identified that affect stem length and plant habit. Golden yellow mutant is governed by a single recessive gene (*go*). One benefit of this gene is that the fruit becomes golden yellow as it matures, possibly useful as a maturity indicator. The flesh colour changes from pink to red.

Four genes for male sterility are reported. Glabrous male sterile is unique, with sterility associated with glabrous foliage. All male sterile genes reduce female fertility as well. These mutants are used in hybrid production but are not as successful as hoped, since they often have low seed yield. A new, spontaneous male sterile mutant (*ms*-2) with high normal seed set is identified, and will be more useful for hybrid production. A single recessive gene (*su*) eliminates bitterness in fruit of *C. lanatus* and appears same as the dominant gene (*Su*) for bitter flavour in the fruit of *C. colocynthis.* Fruit shape (elongate, oval or spherical) is governed by a single incompletely dominant gene. Thirteen genes are identified that affect fruit colour either in the skin or flesh. Watermelon flesh colour is controlled by several genes to produce red, orange, salmon yellow, canary yellow or white.

The watermelon genes were originally organized by Poole (1944). Guner and Wehner (2004) provided an update of the known genes of watermelon, with 163 total mutants grouped into seed and seedling mutants, vine mutants, flower mutants, resistance mutants, proteins (isozyme) mutants, DNA (RFLP and RAPD) markers and cloned genes. Robinson *et al.* (1976) listed the quantitative genes of watermelon which are given in Table 3.2.

Male Sterility and Marker

Male sterility can be induced in watermelon by applying maleic hydrazide (MH) before anthesis (Rehm, 1952). However, chemicals are not suitable for inducing transient male sterility as they exert adverse effect on female fertility and vegetative growth. Watts (1962) reported the occurrence of male sterile line in watermelon variety Sugar Baby by irradiating it with 64 Krad x-rays. The male sterile plant so induced has glabrous leaves. Male sterility and glabrous leaf trait were governed by a single recessive gene (Watts, 1967).

A watermelon mutant without indentation and lobing on the leaves was discovered by Mohr (1953). Glabrous seedling marker facilitates early identification and roguing of male fertile sibs from the seed parent line. However, this genetic male sterility could not be commercially exploited due to poor female fertility associated with the system (Love *et al.,* 1986). Spontaneously occurring male sterile plants are controlled by a single recessive gene (Xia *et al.,* 1988). Female sterility of this male sterile line is good and hence has promising commercial potential. Zhang *et al.* (1996) developed a procedure with seedling markers to expeditiously incorporate *ms* into breeding lines to produce hybrid seed. Yadav and Luthra (2005) also reported a simple and non-lobed leaved character in RW-177-3 genotype which is dominant in nature over lobed leaves and it can be used as marker gene in hybridization programme. Choudhary and Singh (2009a) also observed a non-lobed mutant which can be identified right from the seedling stage and could be used as female parent in the development of F_1 hybrids.

Table 3.2: Preferred Genes Symbols in Water Melon

Preferred Symbol	Synonym	Character
A		Andromonoecious, recessive to monoecious.
Ar	(B) Ge	Anthracnose resistance, resistance to *Giometella cingulata* var. *orbiculare.*
C		Canary yellow flesh, dominant to pink.
D		dotted seed coat, black dotted seed when dominant for r, t and w.
dw-1		dwarf-1, short internodes, due to fewer shorter cells than normal.
dw-2		dwarf-2, short internodes, due to fewer cells.
E	(t)	explosive rind, thin, tender rind, bursting when cut.
F		furrowed fruit surface, recessive to smooth.
G	(D)	light green skin, light green fruit, recessive to dark green.
g^s	(d^s)	striped green skin, recessive to dark green but dominant to light green skin.
Go	(c)	golden, yellow colour of older leaves and mature fruit.
Gms	(ms_g)	glabrous male sterile, foliage lacking trichomes, male sterile.
L		long seed, long recessive to medium length of seed; interacts with s.
M		mottled skin, greenish white mottling of fruit skin.
N		non-lobed leaves, leaves lack lobbing, dominance incomplete.
O		Oval fruit, incompletely dominant to spherical.
P		pencilled lines on skin, incinspicous stripes, recessive to netted fruit.
Pm		powdery mildew susceptibility, susceptibility to *Sphaerotheca fuliginea.*
R		red seed coat, interacts with w and t.
S		short seeds, epistatic to l.
Su	(su^{Bi})	suppressor to bitterness, non-bitter fruit.
T	(b^t)	tan seed coat, interacts with r and w.
W		white seed coat, interacts with r and t.
Wf	(W)	White flesh, dominant to red.
Y	(r)	yellow flesh, recessive to red.

Genetic Diversity

Cucurbits are very similar in above ground development, but they have high genetic diversity for fruit shape and other fruit characteristics, resulting in a variety of uses (Bisognin, 2002). It is vital for plant breeding programmes to have sufficient diversity available to allow for the production of new varieties with high crop productivity and ability to withstand damage from biotic and abiotic factors (Querol, 1987). Watermelon contains a fruit part and a plant part. Each part contains different traits desired by consumers and/or growers, including such traits as flavor, texture, disease resistance and appearance traits like shape and colour (Wehner *et al.*, 2001). Under changing agro climatic conditions, cultivation of watermelon has become very difficult due to attack of *Fusarium* wilt, anthracnose, powdery mildew, downy mildew, mosaic and bud necrosis virus. Although many watermelon cultivars are

Figure 3.5: Non Lobed Watermelon

developed but there is a need for improvement, especially for disease and pest resistance coupled with high TSS, good shipping quality and high yield. Genetic variability is the basis of all breeding programmes. There is a need for breeders to restructure the genetic variability to combat with emerging challenges. The existing variability can be utilized to develop suitable cultivars/hybrids of watermelon for ever changing agro climatic scenario. The building up of a wide and comprehensive germplasm is well recognized as an essential pre-requisite to plant breeding. The next step is to study the diversity in respect of quantitative characters which are of economic importance to break up the observed variation in desirable characters into heritable and non-heritable components. Selection of desirable traits in unimproved accessions especially in the landrace has priority for the future breeding programmes. The low genetic diversity among commercial cultivars emphasized the need to broaden the genetic base of available accessions to develop superior genotypes.

Mohr (1953) discovered a watermelon mutant without indentation and lobing on the leaves. Yadav and Luthra (2005) identified a promising non-lobed leaf marker genotype (RW-177-3) having moderate resistance to blight, powdery mildew, Ubias virus and fruit fly. It bears dark green round fruits with thick and hard skin, which helps in good keeping quality and easy transportation. The flesh is crisp and dark red in colour with 10-11 per cent TSS. It starts flowering in 48 days and picking in 105-110 days after sowing. The average fruit weight is 4-5kg and average yield is 350-450q/ha. The simple and non-lobed leaves unlike other varieties are dominant in nature over lobed leaves which help the breeders to distinguish the genotype easily and this distinguishes leaf character that can be used as marker gene in hybridization programme. Thakur and Nandpuri (1974) assessed the variability among 25 varieties with respect to economic quantitative character and observed wide range of variability for yield/plant (2.43-6.60kg), weight/fruit (2.29-5.95kg), number of fruits/plant (0.64-1.85), fruits free from blossom end rot (53.3-100.0 per cent), number of days taken to first picking (81.5-99.2), length of plant (2.64-4.84m), number of branches/plant (5.34-7.65), sex ratio (15.7:1-26.9:1), TSS (6.17-8.74 per cent), number of seeds/kg of fruit weight (53.2-260.3) and 100 seed weight (2.55-13.85g). Variety 'Shipper' was significantly superior and gave the highest yield. Special No. 1 was the earliest in maturity and had the maximum number of fruits. The biggest fruit size was recorded in Texas Giant. They concluded that number of fruits/plant and fruit weight seemed to be the most reliable criteria for selection and further improvement.

Prasad *et al.* (1988) recorded highly significant varietal differences with wide range of variation for the node at which first female flower appeared (5.75-13.50), number of fruits/plant (1.04-3.26), vine length (1.41-3.26m), days required for 50 per cent appearance of female flowers (49.33-60.00), days taken to first picking (76.33-89.67), number of branches/plant (3.17-3.50), average weight of fruit (0.54-3.09kg), length of fruit (12.18-18.99cm), girth of fruit (10.30-17.83cm), rind thickness (0.77-1.43cm), TSS (5.77-12.27 per cent), number of seeds/fruit (179.43-588.37), 100 seed weight (4.12-11.11g) and yield/plant (1.46-3.59kg). Arka Manik was the highest yielder with a yield of 3.59kg/plant under the poor experimental conditions. The characters showing high range of variation indicate good scope for improvement. The magnitudes of heritability were quite high for days taken to first picking and number of branches/plant. High values of heritability suggest the role of genetic constitution in the expression of the character and such traits are considered dependable from breeding point of view. Krishna Prasad *et al.* (2002) studied genetic variation and genetic divergence among 48 watermelon inbreds of diverse geographical sources and observed wide variation in node at which female flower appeared (3.50-15.00), days to appearance of female flowers (43.00-61.00), node at which male flowers appeared (2.00-4.50), days to appearance of male flowers (41.50-53.00), number of nodes (17.50-45.50), number of branches (3.00-13.00), vine length (1.45-4.10m), fruit length (13.15-22.90cm), fruit girth (13.20-21.65cm), rind thickness (0.65-5.05cm), TSS (4.25-11.00 per cent), number of fruits/plot (3.00-16.50) and yield/plot (4.00-18.60kg). They observed high heritability coupled with high genetic advance for yield/plot, number of nodes, days to female flowers appeared and number of fruits/plot and advocated selection based on yield and its attributes may be effective in breeding programme. The 48 inbred lines have also been grouped into 10 clusters of different sizes based on Tochers method and results revealed that maximum range for number of fruits/plant (5.00-10.66) from cluster X to VI. The inbreds in cluster I took the least number of days taken for flowering (46.12 days) and selection can be bestowed on the inbred lines, *i.e.*, IIHR-86-3, IIHR-178-3-1, IIHR-189 and IIHR-190 for earliness in breeding programme.

Rolania *et al.* (2004) determined genetic variation for severity of watermelon mosaic virus (16.90-38.47 per cent), powdery mildew (2.1-19.1 per cent) and anthracnose (8.8-16.67 per cent). The lowest severity of watermelon mosaic virus, powdery mildew and anthracnose was recorded in Asahi Yamato (16.90 per cent), RW-187-2 (2.1 per cent) and GP-7 (8.8 per cent), respectively which could be used in breeding programme to improve watermelon. The severity of watermelon mosaic virus (97.4 per cent) and powdery mildew (97.9 per cent) had high heritability coupled with high genetic advance (9.30,

9.32), which indicated the role of additive gene effects and thus, could be improved through selection. High heritability (82.6 per cent) along with lower genetic advance (3.77) of anthracnose severity might be attributed to the non-additive gene effects and this could be improved through hybridization. Yadav and Asati (2005) evaluated 22 accessions for quantitative and qualitative characters and found Wm 99-13, Wm 99-21, Sugar Baby and Arka Manik as promising genotypes on the basis of fruit shape, sweetness and TSS. They observed variability in qualitative traits *viz.*, fruit shape (round, oval, oblong, flat round, round oblong and cylindrical), rind colour (green, light green, dark green, whitish green, pale green, whitish, deep blue and bluish black), stripes (light green, deep green, black, light green, bluish green, dull green and absence of stripes), flesh colour (pink, deep pink, light pink, red, light red, crimson, bright crimson, light crimson, dark green, pale pink and yellowish), juiciness (poor juicy, mild juicy, juicy and highly juicy), flesh firmness (loose, compact, mild compact and highly compact) and sweetness (poor, mild, sweet and very sweet). Similarly they also recorded wide variability for

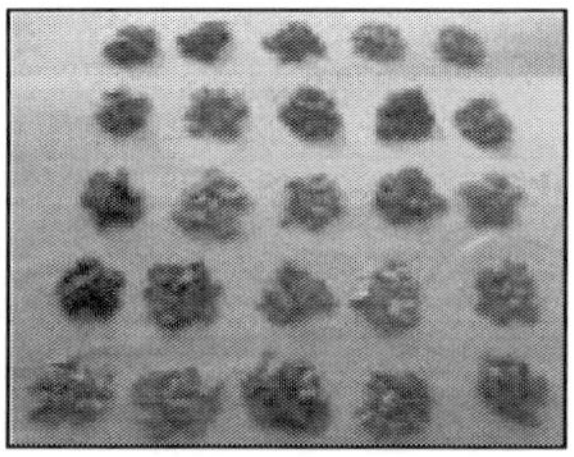

Normal Varieties

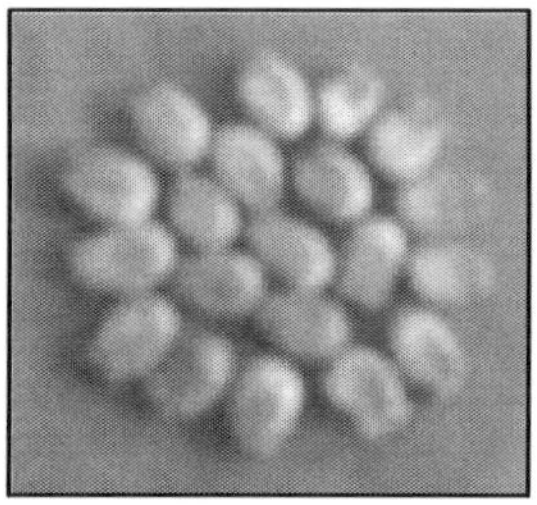

Tetraploid line VRWT-1-5

Figure 3.6: Variability in Seed

Figure 3.7: Variability in Fruit Colour, Shape and Size

Figure 3.8: Variability in Leaf

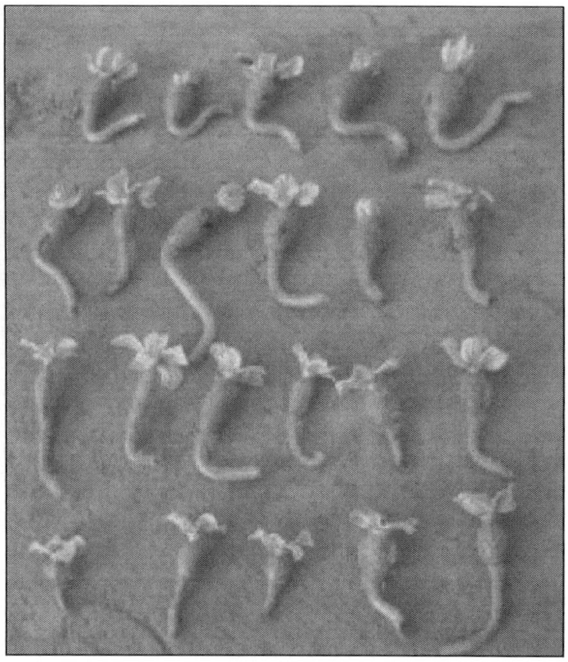

Figure 3.9: Variability in Ovary

Figure 3.10: Variability in Flesh Colour

Figure 3.11: VRW-3–A Yellow Rinded Watermelon

fruit weight (3.0-9.8kg), polar diameter of fruit (12.0-35.0cm), equatorial diameter of fruit (10.0- 31.5cm), flesh thickness (0.2-0.6cm) and TSS (5.0-13.0 per cent).

Choudhary *et al.* (2009a) characterized 20 genotypes of watermelon and recorded considerable variability for fruit shape (round, oblong, oval, elliptical), rind colour (green, light green, dark green, gray, yellow), rind pattern (stripped, non stripped, solid, mixed), flesh colour (red, yellow, pink, white, cream white, scarlet red), seed colour (brown, dark brown, black, tan, white), vine length (2.47-6.10m), number of primary branches/plant (3.13-10.27), node at which first female flower appeared (7.33-28.73), days to appearance of first female flowers (46.67-58.47), days to first fruit harvest (74.47-95.33), number of fruits/plant (1.73-5.93), fruit weight (2.60-10.80kg), rind thickness (6.13-18.47mm) and TSS (6.00-11.2 per cent). Among the genotypes, VRW-3 was a surprise which produced yellow rinded fruits. Choudhary *et al.* (2009b) also identified VRW-3 (IC No. 583400); a yellow rinded line of watermelon with bright red flesh at IIVR Seed Production Centre, Sargatia, Kushinagar (UP). It was isolated from the local material collected from Shahjahanpur (UP). It is characterized by yellow coloured rind, early maturity (75-80 days), short vine length (3.2-3.6m), high TSS (11 per cent) and profuse bearing (5-7 fruits/plant). The yellow colour of fruits appears right from the ovary stage. Fruits are small in size weighing 3.0-3.5 kg which suits the icebox segment which is the need of the hour for nuclear families. The fruits of VRW-3 are very attractive and delicious in taste with a unique aroma. Fruits contain 6.7 mg/100g ascorbic acid, 4.95 mg/100g carotenoids, 0.192 per cent citric acid and 10.0 per cent oxalic acid. Choudhary and Singh (2009b) identified a tetraploid line of watermelon (VRWT-1-5) having grooved seed with colchicine treatment to seed as seed soaking.

Gichimu *et al.* (2009) assessed genetic diversity and relatedness among three most common commercial watermelon cultivars in Kenya, one newly introduced commercial cultivar from the US, one Kenyan landrace and one wild (*Citrullus colocynthis*) accession. The cluster analysis results demonstrated high morphological diversity (8-27 per cent) between unimproved accessions (wild accession and landrace) and quantitative characters of cultivated accessions demonstrated highly significant variation between accessions. Results of the principle component (PC) analyses for the six quantitative traits indicated that the first two PCs explained 68 per cent and 29 per cent (a total of 97 per cent) of the total variation. They emphasized the need to expand the genetic base of the cultivated watermelon which has low morphological diversity among commercial cultivars. They observed extensive variation in qualitative characteristics *viz.*, leaf shape (pentalobate and trilobite), petal colour (light yellow and deep yellow), fruit shape (round and elliptical), rind colour (green, gray, light green and dark green), rind colour pattern (solid, stripped and mixed), rind strip colour (no stripes, light green, medium green and dark green), flesh colour (red, pink, yellow, white and cream-white), seed shape (elliptical and roundish), seed colour (white, tan, brown and dark brown) and quantitative characteristics *viz.*, main vine length (200.55-447.66cm), branches/plant (5.11-11.33), fruits/plant (0.89-5.66), fruit weight (1.43-3.01kg), rind thickness (7.58-13.22mm) and number of seeds/fruit (126.38-372.33).

Genetic Diversity Using Molecular Markers

Identification of watermelon cultivars and determination of their genetic purity and relatedness relay mainly on fruit characteristics. Molecular markers can be an effective means to determine genetic relatedness among cultivars and among selections used in watermelon breeding programmes. Previous studies designed to examine genetic diversity and polygenetic relationships among watermelon cultivars using isozymes, the most isozymes tested produced monomorphic patterns (Zamir *et al.*, 1984 and Biles *et al.*, 1989). In contrast, RAPD procedure provided a sufficient number of informative markers that could distinguish among watermelon cultivars (Hashizume *et al.*, 1993 and Zhang *et al.*, 1994). Navot and Zamir (1987) found little isozyme variation among watermelon accessions and suggested that this is due to the domestication of watermelon outside of its center of origin. Katzir *et al.* (1996) could not detect any polymorphism among the watermelon cultivars Sugar Baby and Malali (diploids) and Tri-X-313 (triploid) using seven simple sequence repeat (SSR) markers. However, Jarret *et al.* (1997) could detect genetic diversity among *Citrullus* PIs using the same SSR markers and found PIs of *C. lanatus* var. *citroides* slightly more divergent than PIs of *C. lanatus* var. *lanatus*. RAPD markers can also be used in the construction of an initial genetic linkage map for watermelon (Hashizume *et al.*, 1996), and to determine genetic relatedness among Asian watermelon cultivars and breeding lines (Lee *et al.*, 1996).

Levi *et al.* (2000) examined genetic diversity and relatedness among US Plant Introduction accessions (PIs) of *C. lanatus* var. *lanatus*, *C. lanatus* var. *citroides* and *C. colocynthis* using RAPD analysis. Levi *et al.* (2000) also studied genetic diversity and relatedness among 17 *C. lanatus* var. *lanatus* PIs, 12 *C. lanatus* var. *citroides*, 13 *C. colocynthis* and 5 cultivars (Allsweet, Charleston Gray, Ironsides, Mickylee and New Hempshire Midget) using RAPD analysis. In that study, the five watermelon cultivars grouped in a distinct cluster, indicating that they were derived from common parents. However, the low genetic diversity has been observed among *C. lanatus* var. *lanatus* PIs. Levi *et al.* (2001a) estimated genetic diversity among 42 US Plant Introduction (PI) accessions of the genus *Citrullus* (of these, 34 PIs have been reported disease resistant) and 5 watermelon cultivars using 30 RAPD primers. These primers produced 662 RAPD markers which could be rated with high confidence. The analysis delineated three major clusters. The first cluster consisted of a group of five watermelon cultivars, a group of *C. lanatus* var. *citroides* accessions which contained some *C. lanatus* var. *citroides* genes. The second cluster consisted of *C. lanatus* var. *citroides* accessions while the third cluster consisted of *C. colocynthis* cluster at the level of 58.8 per cent and 38.9 per cent genetic similarity, respectively. Assessment of genetic diversity among accessions which are reported to have disease resistance to either anthracnose, downy mildew, powdery mildew, or watermelon mosaic virus is found among all major groups of *Citrullus* PIs. Additionally, resistance to gummy stem blight or *Fusarium* wilt may exist among *C. lanatus* var. *citroides* PIs. This study demonstrated that molecular markers can be useful in assessing genetic diversity and in sorting *Citrullus* PIs into phylogenetic groups prior to their evaluation for disease or pest resistance.

Levi *et al.* (2001b) assessed genetic diversity and relatedness among 46 American cultivars of watermelon (*Citrullus lanatus* var. *lanatus*) and 12 US Plant Introduction accessions (PIs) of *Citrullus* sp. using 25 randomly amplified polymorphic DNA (RAPD) primers. Theses primers produced 288 distinct reproducible bands which could be scored with high confidence among cultivars and PIs. Based on the RAPD data, genetic similarity coefficients were calculated and found to have high genetic similarity values (92 per cent to 99.6 per cent) among watermelon cultivars and among PIs of *C. lanatus* var. *lanatus* (88 per cent to 95 per cent) considered the progenitor of the cultivated watermelon. Lower genetic similarity values were found among the *C. lanatus* var. *citroides* and among the wild species *C. colocynthis* PIs (65 per cent to 82.5 per cent, and 70.5 per cent, respectively). They emphasized the need to expand the genetic base of cultivated watermelon having low genetic diversity.

References

Bisognin, D.A., 2002. *Origin and Evolution of Cultivated Cucurbits*. Ciencia Rural, Santa Maria, 32(5): 715–723.

Biles, C.L., Martyn, R.D. and Wilson, H.D., 1989. Isozymes and general proteins from various watermelon cultivars and tissue types. *Hort. Sci.*, 24: 810–812.

Brar, J.S. and Sukhija, B.S., 1978. Hybrid vigour in intervarietal crosses in watermelon (*Citrullus lanatus* (Thunb.) Mansf.). *Indian J. Hort.*, 34: 277.

Burkill, H.M., 1985. *The Useful Plants of West Tropical Africa*, 2nd edn. Royal Botanic Gardens, Kew, p. 1.

Choudhary, B.R., 2009a. Occurrence of andromonoecious sex form in watermelon. Unpublished Data.

Choudhary, B.R., 2009b. Cluster bearing watermelon. Unpublished Data.

Choudhary, B.R. and Singh, P.K., 2009a. A non-lobed leaf mutant in watermelon. Unpublished Data.

Choudhary, B.R. and Singh, P.K., 2009b. Tetraploidy in watermelon. Unpublished Data.

Choudhary, B.R., Pandey, S., Singh, P.K., Singh, R. and Rai, M., 2009a. Characterization of watermelon genotypes for horticultural traits. In: *Intl. Conf. on Horticulture for Livelihood Security and Economic Growth* held at Bangalore from November 9–12, p. 57.

Choudhary, B.R., Pandey, S., Singh, P.K., Singh, Ramesh and Rai, Mathura, 2009b. VRW 3, yellow-rinded watermelon. *ICAR News* (July–September), p. 13–14.

Davis, A.R., Levi, A., Wehner, T. and Pitrat, M., 2006. PI 525088-PMR, a melon race 1 powdery mildew-resistant watermelon line. *Hort. Sci.*, 41(7): 1527–1528.

Davis, A.R., Levi, A., Tettech, A., Wehner, T., Russo, V. and Pitrat, M., 2007. Evaluation of watermelon and related species for resistance to race 1W powdery mildew. *J. Amer. Soc. Hort. Sci.*, 132(6): 790–795.

De Candolle, A., 1882. *Origin of Cultivated Plants.* New York.

De Winter, B., 1990. A new species of *Citrullus* (Benincaseae) from the Namib. *Bothalia*, 20: 209–211.

Gichimu, B.M., Owuor, B.O., Mwai, G.N. and Dida, M.M., 2009. Morphological characterization of some wild and cultivated watermelon (*Citrullus* sp.) accessions in Kenya. *ARPN J. Agril. and Biol. Sci.*, 4(2): 10–18.

Gill, B.S. and Kumar, J.C., 1988. Combining ability analysis in watermelon [*Citrullus lanatus* (Thunb.) Mansf.]. *Indian J. Hort.*, 45: 104–109.

Gill, B.S. and Kumar, J.C., 1989. Phenotypic stability in watermelon (*Citrullus lanatus*). *Indian J. Agric. Sci.*, 59: 145–148.

Goreta, S., Perica, G., Dumicic, G., Bucan, L. and Zanic, K., 2005. Growth and yield of watermelon on polyethylene mulch with different spacings and nitrogen rates. *Amer. J. Hort. Sci.*, 40(2): 366–369.

Guner, N. and Wehner, T.C., 2004. The genes of watermelon. *Hort. Sci.*, 39(6): 1175–1182.

Gusmini, G., Song, R. and Wehner, T.C., 2005. New sources of resistance to gummy stem blight in watermelon. *Crop Sci.*, 45: 582–588.

Hashizume, T., Sato, T. and Hirai, M., 1993. Determination of genetic purity of hybrid seed in watermelon (*Citrullus lanatus*) and tomato (*Lycopersion esculentum*) using random amplified polymorphic DNA (RAPD). *Jpn. J. Breed.*, 43: 367–375.

Hashizume, T., Skimamoto, I., Harushima, Y., Yui, M., Sato, T., Imai, T. and Hirai, M., 1996. Construction of a linkage map for watermelon [*Citrullus lanatus* (Thunb.) Matsum. and Nakai] using random amplified polymorphic DNA (RAPD). *Euphytica*, 90: 265–273.

Huh, Y.C., Solmaz, I. and Sari, N., 2008. Morphological characterization of Korean and Turkish watermelon germplasm. 1 Cucurbitaceae 2008, *Proceedings of the IXth EUCARPIA Meeting on Genetics and Breeding of Cucurbitaceae*, (Ed.) M. Pitrat. INRA, Avignon (France), May 21st–24th.

Jarret, R.L., Merrick, L.C., Holms, T., Evans, J. and Aradhya, M.K., 1997. Simple sequence repeats in watermelon [*Citrullus lanatus* (Thunb.) Matsum. and Nakai]. *Genome*, 40: 433–441.

Jeffery, C., 1975. Further notes on Cucurbitaceae. III. Some African taxas. *Kew Bull.*, 30: 475–493.

Katzir, N., Danin-Poleg, Y., Tzuri, G., Karchi, Z., Lavi, U. and Cregan, P.B., 1996. Length polymorphism and homologies of microsatellites in several Cucurbitaceous species. *Theor. Appl. Genet.*, 93: 1282–1290.

Krishna Prasad, V.S.R., Pichaimuthu, M. and Dutta, O.P., 2002. Adaptive responses and diversity pattern in watermelon. *Indian J. Hort.*, 59(3): 298–306.

Lee, S.J., Shin, J.S., Park, K.W. and Hong, Y.P., 1996. Detection of genetic diversity using RAPD-PCR and sugar analysis in watermelon [*Citrullus lanatus* (Thunb.) Mansf.] germplasm. *Jpn. J. Breed.*, 43: 367–375. *Theor. Appl. Genet.*, 92: 719–725.

Levi, A., Thomas, C.E., Keinath, A.P. and Wehner, T.C., 2000. Estimation of genetic diversity among *Citrullus* accessions using RAPD markers. *Acta Hort.*, 510: 385–390.

Levi, A., Thomas, C.E., Keinath, A.P. and Wehner, T.C., 2001a. Genetic diversity among watermelon (*Citrullus lanatus* and *Citrullus colocynthis*) accessions. *Genet. Resour. and Crop Evol.*, 48: 1566–2001.

Levi, A., Thomas, C.E., Wehner, T.C. and Zhang, X., 2001b. Low genetic diversity indicates the need to broaden the genetic base of cultivated watermelon. *J. Amer. Soc. Hort. Sci.*, 36(6): 1096–1101.

Li, Y., 1992. The progress of utilization of heterosis in diploid watermelon in China. *China Watermelons*, 1: 6–7.

Love, S.L., Rhodes, R.B. and Jugent, P.I., 1986. Controlled pollination transfer of a nuclear male sterility gene from a diploid to a tetraploid watermelon line. *Euphytica*, 35: 633–638.

Luthra, J.P. and Yadav, V.S., 2004. Durgapura Lal : A new watermelon. *Indian Hort.*, 49(1): 32.

Martyn, R.D. and Netzer, D., 1991. Resistance to races 0, 1 and 2 of *Fusarium* wilt of watermelon in *Citrullus* sp. PI 296341-FR. *Hort. Sci.*, 26: 429–432.

Meeuse, A.D., 1962. The Cucurbitaceae of Southern Africa. *Bothalia*, 8: 1–111.

Mohr, H.C., 1953. A mutant leaf from watermelon. In: *Proc. Assoc. Agric. Work*, 50: 129–130.

Nandipur, K.S., Kumar, J.C. and Dhillon, G.S., 1974. Heterosis in watermelon. *Punjab Hort. J.*, 14: 75–83.

Nath, P. and Dutta, O.P., 1970. Heterosis in watermelon. *Indian J. Hort.*, 27: 176.

Nath, P., Subramanyam, S. and Dutta, O.P., 1973. Improvement of cucurbitaceous crops in India: A review. *SABRAO Newsletter*, 5: 71–79.

Navot, N. and Zamir, D., 1987. Isozymes and seed protein phylogeny of the genus *Citrullus* (Cucurbitaceae). *Plant Syst. Evol.*, 156: 61–67.

Netzer, D. and Martyn, R.D., 1989. PI 296341, a source of resistance in watermelon to race 2 of *Fusarium oxysporum* f. sp. *niveum*. *Plant Disease*, 73: 518.

Orton, W.A., 1911. The development of disease resistant-varieties of plants. In: 4th *Int. Conf. Genet.*, p. 247.

Pai, L.D., 1992. Watermelon gene index. *China Watermelon and Melons*, 2: 1–4.

Pichaimuthu, M. and Swamy, K.R.M., 2009. Development of F_1 hybrids for high yield and good qualities in watermelon [*Citrullus lanatus* (Thunb.) Matsum and Nakai]. In: *Intl. Conf. on Hort. 'Horticulture for Livelihood Security and Economic Growth'* held at Bangalore from November 09–12, p. 56.

Poole, C.F., 1944. Genetics of cultivated cucurbits. *J. Hered.*, 35: 122–128.

Porter, D.R., 1933. Watermelon breeding. *Hilgardia*, 7: 585–624.

Prasad, L., Gautam, N.C. and Singh, S.P., 1988. Studies on genetic variability and character association in watermelon [*Citrullus lanatus* (Thunb.) Mansf.]. *Veg. Sci.*, 15(1): 86–94.

Querol, D. 1987. *Genetic Resources: A Practical Guide to their Conservation*. Zed Books Ltd., London and New Jersey, 52: 55–57.

Rehm, S., 1952. Male sterile plants by chemical treatments. *Nature*, 170: 38–39.

Robertson, H., 2004. *Citrullus lanatus* (Watermelon, Tsamma). Museums Online South Africa. Iziko Museums of Cape Town Online Publication: http:/museums.org.za/bio/index.htm.

Robinson, R.W., Munger, H.M., Whitaker, T.W. and Bohn, G.W. 1976. Genes of the cucurbitaceae. *Hort. Sci.*, 11: 554–568.

Rolania, S., Sarites, Arpana and Choudhary, M., 2004. Genetic variation, correlation and path coefficient analysis for disease resistance in watermelon [*Citrullus lanatus* (Thunb.) Mansf]. *Udyanika*, 10(1): 22–25.

Sidhu, A.S. and Brar, J.S., 1977. Heterosis and combining ability of yield and its components in watermelon (*Citrullus lanatus* (Thunb.) Mansf.]. *J. Res. Punjab. Agric. Univ.*, 14: 52–58.

Sidhu, A.S. and Brar, J.S., 1985. Genetic divergence and hybrid performance in watermelon. *Indian J. Agric. Sci.*, 55: 459–461.

Singh, S.P., Dadwadia, G. and Annapurna, 2009. Analysis of heterosis and combining ability status among diallel set of hybrids for yield and quality traits in watermelon (*Citrullus lanatus* Thunb). *Veg. Sci.*, 36(3 Suppl.): 323–326.

Sowell, G. Jr., 1975. An additional source of resistance to gummy stem blight in watermelon. *Plant Dis. Reptr.*, 59: 413–415.

Sowell, G. Jr. and Pointer, J.R., 1962. Gummy stem blight resistance of introduced watermelons. *Plant Dis. Reptr.*, 46: 883–885.

Thakur, J.C. and Nandpuri, K.S., 1974. Studies on variability and heritability of some important quantitative characters in watermelon (*Citrullus lanatus* Thunb Mansf.). *Veg. Sci.*, 1: 1–8.

Tindall, H.D., 1983. *Vegetables in the Tropics*. The McMillan Press Limited, London, pp. 150–152.

Waraitch, K.S., Munshi, G.D. and Brar, J.S., 1975. Varietal evaluation of watermelon against *Alternaria* blight caused by *Alternaria cucumerina* (Ell. and Ev.) Elliot and its chemical control. *Pb. Veg. Grower*, 10: 12–15.

Watts, V.M., 1962. A marked male sterile mutant in watermelon. *Proc. Am. Soc. Hort. Sci.*, 81: 498–505.

Watts, V.M. 1967. Development of disease resistance and seed production in watermelon stocks carrying msg gene. *J. Am. Soc. Hort. Sci.* 91: 579–590.

Wehner, T.C., Shetty, N.V. and Elmstrom, G.W., 2001. Breeding and seed production. In: *Watermelons: Characterization, Production and Marketing*, (Ed.) D.N. Maynard. ASHS Press, Alexandria, Va. pp. 27–73.

Whitaker, T.W., 1933. *Bot. Gaz.*, 94: 780–790.

Whitaker, T.W. and Davis, G.N., 1962. *Cucurbits: Botany, Cultivation and Utilization*. Interscience Publishers, Inc., New York.

Whitaker, T.W. and Bemis, W.B., 1976. Cucurbits. In: *Evolution of Crop Plants*, (Ed.) N.W. Simmonds. Longman, London, pp. 64–69.

Xia, X., Lin, Y., Liu, W. and Chen, A., 1988. Selection of watermelon (*Citrullus vulgaris*) male sterile line G_{17}AB. *J. Shenyang Agric. Univ.*, 19: 9–13.

Yadav, R.K. and Asati, B.S., 2005. Correlation among fruit characters in indigenous germplasm lines of watermelon. *Haryana J. Hort. Sci.*, 34(1–2): 135–136.

Yadav, V.S. and Luthra, J.P., 2005. A promising simple non-lobed leaf marker watermelon genotype: RW-177-3. *Haryana J. Hort. Sci.*, 34(1–2): 137.

Zamir, D., Navot, N. and Rudich, J., 1984. Enzyme polymorphism in *Citrullus lanatus* and *C. colocynthis* in Israel and Sinai. *Plant Syst. Evol.*, 146: 133–137.

Zhang, X.P., 1987. Heterosis and its utilization in watermelon. *Cucurbits Comm.*, 4: 7–9.

Zhang, X.P. and Rhodex, B.B., 1992. Watermelon variety improvement in China. *Cucurbit Genet. Rep.*, 15: 76–79.

Zhang, X.P., Rhodes, B.B. and Skorupska, H.S., 1994. RAPD molecular markers in watermelon. *Cucurbit Genetics Coop. Rpt.*, 17: 116–119.

Zhang, X.P., Rhodes, B.B., Baird, W.V., Bridges, W.C. and Skorupska, H.T., 1996. Development of genic male-sterile lines with juvenile albino seedling marker. *Hort. Sci.*, 3: 426–429.

2011, Biodiversity in Horticultural Crops Vol. 3 *Pages* **43–73**
Editor: **Professor K.V. Peter**
Published by: **DAYA PUBLISHING HOUSE, NEW DELHI**

Chapter 4

Cranberry (*Vaccinium macrocarpon* Ait)

Hilary A. Sandler*[1]*, Carolyn J. DeMoranville*[1]*,
***Yu Zhou*[2]*, Robert R. Boulanger, Jr*[2] *and Bal Ram Singh**[2]

[1]*UMass Cranberry Station,*
College of Natural Sciences, University of Massachusetts Amherst,
1 State Bog Road, P.O. Box 569, E.Wareham, MA 02538 USA
[2]*Department of Chemistry and Biochemistry,*
University of Massachusetts Dartmouth, 285 Old Westport Road,
N. Dartmouth, MA 02747 USA
**E-mail:bsingh@umassd.edu*

The cranberry (*Vaccinium macrocarpon* Ait, Ericaceae) is a native North American fruit. Its native range extends in temperate climate zones from the East Coast to the Central United States and Canada and from Southern Canada in the north to the Appalachian Mountains of the United States in the South. That range is approximately between 39° and 48° N. Latitude and 93° and 56° W. Longitude. The fruit is grown commercially in other areas of North and South America with similar climate conditions. Most of the world's cranberries are produced in the United States on approximately 16,000 ha. Cranberries are also commercially cultivated in other countries including Canada (~3500 ha) and Chile.

Cranberries are predominantly used as a food crop, either as fresh produce or processed into sauces, juices, or sweetened dried fruit. They are also used in nutraceutical products, primarily in powdered form and in cosmetic products. Historically, the cranberry was also used as a dye and as a food preservative.

The plant is a low-growing, trailing, woody vine with a perennial habit. Cranberries produce stolons or runners from 30 cm to 2 m long. Short vertical upright branches, known as uprights, form from the buds along the runners. The uprights have a vertical (non-trailing) growth habit and form the terminal buds which contain the flower buds. Most of the fruit is formed from the flowers on the uprights, with some berries arising from flowers on the runner ends. During the growing season, the leaves are dark green and glossy, turning reddish-brown during the dormant season. The vines form a thick mat over the surface of a cultivated bed.

Cranberries are a perennial plant, producing one crop of fruit each year. Figure 4.1 shows the 15-month production cycle of cranberry from the initiation of the flower bud (bud set) to the harvest of the fruit (mature fruit). Figure 4.2 shows the cranberry flower (A) and fruit (B).

The soil in a commercial cranberry bog is unique in that it consists of alternating layers of sand and organic matter. Dead leaves accumulate over the course of time and sand is added to the bed surface every 2-5 years to encourage upright production and maintain productivity. In contrast to normal agricultural soils, cranberry soil requires no tilling, remains undisturbed over time, and little mixing of sand and organic matter occurs.

As a wetland-adapted plant, cranberries are tolerant of flooded soils. However, during the active growing season (generally from March through October in the Northern hemisphere), good drainage in the soil is essential for proper root growth and function. During that part of the season, commercial cranberry bogs are managed with drained soil and are generally not flooded for extended periods. Flooding is mainly confined to winter protection, harvest and several specialized pest-control floods.

Uses

Food Uses

About 95 per cent of cranberries produced in North America are processed into products such as juice drinks, sauce, powders, and sweetened dried cranberries. The remaining 5 per cent is sold fresh to consumers. Cranberry fruits are often served as a compote or jelly, known generically as cranberry sauce. Such preparations are traditionally served with roast turkey meat and are considered by some to be a staple of English Christmas dinners, and the Canadian and United States holiday Thanksgiving. The berry is also used in baking (muffins, scones and cakes) but, unlike many other berries, is normally considered too tart to be eaten as is. Fresh cranberries can be frozen, and will keep up to nine months. Cranberry juice is the major use of cranberries; either sweetened to reduce its natural severe tartness and make "cranberry juice cocktail" or blended with other fruit juices. Many cocktails, including the Cosmopolitan, are made with cranberry juice. Cranberry wines and cordials are made from whole cranberries, cranberry juice or cranberry juice concentrate.

The USDA Nutrient Database is a searchable index of food nutrient composition (http://www.nal.usda.gov/fnic/foodcomp/search/ — search keyword "cranberries").

Medicinal Uses

Cranberry (*Vaccinium macrocarpon*) is used as both a food and a medicine for centuries. It is native to North America and was used by Native Americans to treat bladder and kidney diseases. Early settlers from England learned to use the berry both raw and cooked for many ailments, including appetite loss, digestive problems, blood disorders, and the vitamin C deficiency, scurvy (Upton, 2002).

Cranberry is the best known as a preventive treatment for urinary tract infections (UTIs), commonly caused by bacteria known as *Escherichia coli (E. coli)*. Originally, scientists thought cranberry worked by making urine acidic enough to kill the bacteria. Now, studies show that cranberry actually prevents bacteria from attaching to the walls of the urinary tract (Avorn *et al.*, 1994; Ahuja *et al.*, 1998). However, evidence suggests that cranberry is not as effective against bacteria once they have attached to cells in the urinary tract. Cranberry is more effective at preventing UTIs than treating them.

A preliminary study suggests that cranberry may also prevent the bacteria *Helicobacter pylori* from attaching to stomach walls (Burger *et al.*, 2002). *H. pylori* can cause stomach ulcers so it is possible that cranberries may play a role in the prevention of this type of ulcer.

In some test tube studies, cranberry appears to inhibit the growth of cancer cells (Yan *et al.*, 2002). It is too early to say whether the fruit will have the same effect in humans (Prasain, 2008).

Cranberries are rich in antioxidants such as Vitamin C, flavonoids, and phenols. These antioxidants in cranberry may protect from heart disease by lowering LDL ("bad") cholesterol, relaxing blood vessels, and preventing plaque from building up in arteries (McKay and Blumberg, 2007)

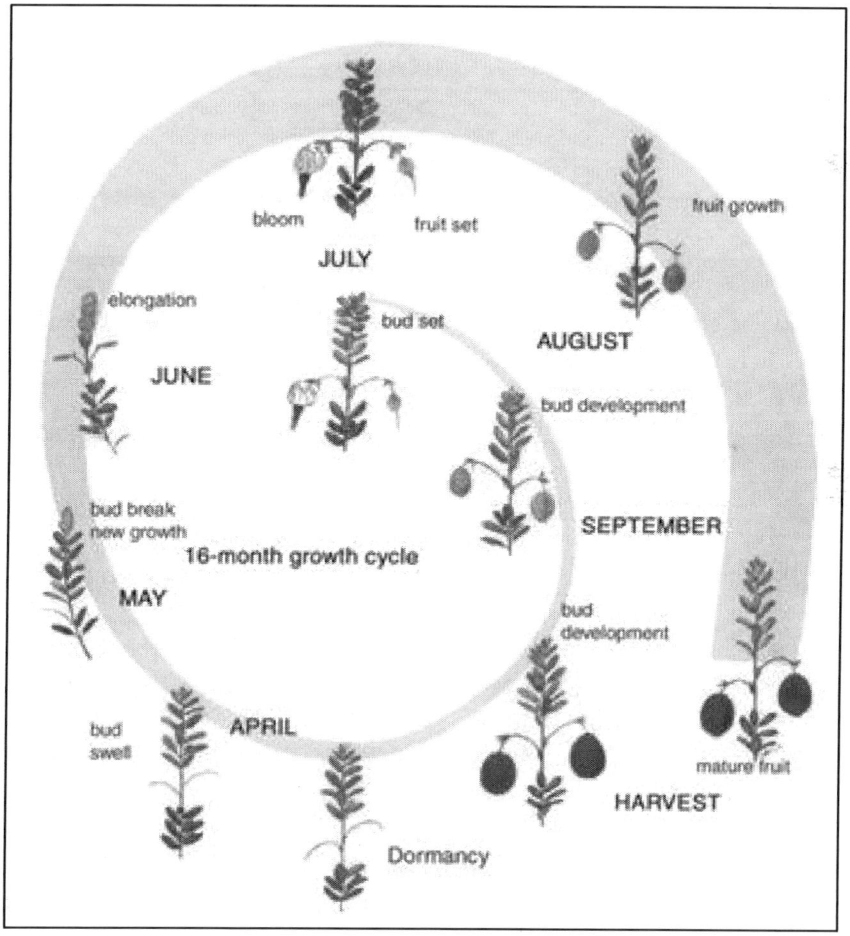

Figure 4.1: The Production Cycle of Cranberry

Figure 4.2: The Cranberry Flower (A) and Fruit (B)

Other Uses of Cranberry

Cranberries have been used to dye fabric and add into soaps and other cosmetic products. Cranberry honey is produced by the bees used in the pollination of cranberry flowers. Cranberry seed oil is marketed as a source of omega fatty acids. Cranberry powder is used as a food preservative.

Climate and Soil

Cranberry is a temperate plant with a requirement for moderate temperatures during the growing season and chilling temperatures in the winter. Commercial cranberry growing areas of North America tend to fall within the range of locations where July daily average maximum temperature is 29.5°C (Pilcher 1985). Cranberry fruit development is optimum when temperatures in July and August are moderate; minimum daily temperature above 15.5°C and maximum daily temperature below 30°C (DeMoranville *et al.,* 1996a).

Cranberry flower buds are initiated in the year before the crop is harvested. The signal to initiate flower buds is a combination of day length (photoperiod) and temperature (Pilcher 1985). By late summer, the initial stages of the flower bud can be observed. During the fall, the flower buds become dormant until the following spring. The signal to enter dormancy is most likely a combination of low temperatures and short days. In common with other perennial fruit crops, the cranberry plants must accumulate a critical number of chilling units in order to break dormancy in the spring and initiate flowering for the new season. The exact length of chilling at temperatures below 7.2°C required to satisfy the plant requirement depends on local climate conditions but is at least 600 hours and may be as long as 1800 hours.

Cranberry plants may be injured by extreme temperatures. The plants are hardy, tolerating temperatures as low as -40°C during the dormant period. In the spring, the plants lose hardiness and temperatures below -1.4°C will injure the flower buds. As the fruit matures in the fall, hardiness increases and fully mature berries may tolerate temperatures as low as -6.7°C. Injury during high temperatures generally is associated with inadequate soil moisture but wilting of new growth has been observed when temperatures exceed 32°C early in the growing season.

As a wetland-adapted plant, cranberries are tolerant of flooded soils. However, during the active growing season (generally from March through October in the Northern Hemisphere), good drainage in the soil is essential to proper root growth and function. During that part of the season, commercial cranberry bogs are managed with drained soil and are not flooded for extended periods as a rule. Flooding is mainly confined to winter protection, harvest, and several specialized pest-control floods. During the growing season, cranberries do require adequate moisture, on average the plant uses 2.5 cm/week, with actual demand varying between 1.2 and 5 cm. This requirement is met by rainfall and/or irrigation.

Native cranberry plants thrive on the special combination of soils and hydrology found in wetlands. In commercial production, cranberries are planted in a range of soils, from sand to peat. In wetland soils, the underlying material of the bed is peat, and often an iron-containing hardpan. In mineral soil beds, the underlying layer is formed of compacted clay, compacted silt, or compacted topsoil. In all cases, the underlying layer serves to hold water in the bed for flooding practices. The underlying layer also serves to separate the bed from ground water. Regardless of soil type, a pH between 4 and 5 is ideal.

Botany

The cultivated cranberry is a low-growing, trailing, woody, broadleaf and non-deciduous vine. Vines form a thick, continuous mat over the entire surface of a cultivated bed. Stolons, often referred to as runners, will range from 0.3-1.8 m long. The leaves are leathery, subsessile, and elliptical-oblong (5-15 mm), and rounded at the tip (Gleason and Cronquist 1991). They are reddish-brown during the dormant season (October through April) and dark green during the growing season. The cranberry root system is made up of very fine, fibrous roots which develop within the upper 7.5-15 cm of soil. Cranberry roots do not have root hairs.

Short vertical branches 5-20 cm high, called uprights, originate from the axillary buds on the runners and grow for several years (Eck 1990; Shawa *et al.*, 1984). The uprights are distinguished from runners by the whorled arrangement of their leaves and their vertical growth habit. Cranberry uprights can produce two types of buds: flowering (fruit) and vegetative. Flowering buds, also known as mixed buds, are easily recognized by their large size and plump appearance; in contrast, the vegetative bud is more pointed. Flowering buds are formed terminally on the uprights and become easily visible during summer or early fall. Each fruit bud may contain two to seven flowers as well as leaves and a growing point.

The plants come out of dormancy in April (depending on weather conditions and nutritional status) and begin to develop new leaves in late May. The flowering period begins during the middle of June and lasts from three to six weeks. The first berries are visible in late June or early July. The cranberry flower is self-fertile. Cross-pollination (pollen from one flower fertilizes the egg in a different flower) produces bigger fruit with increased seed count (Marucci and Filmer 1964).

Berries are almost always produced on the uprights, although in some varieties under certain conditions, fruit may be produced on runners. Berries (and flowers) mature from the 'bottom up', so the largest fruits will be found towards the bottom of the upright and the smallest fruits will be towards the top. Percent fruit set varies among cranberry varieties but typically falls in the range of 25-45 per cent. Berries reach maturity about 80 days after full bloom. Harvesting typically begins around mid-September and continues through early November.

Cropping System and Planting

Peat-based cranberry bogs differ from natural bogs in that the upper layers of natural vegetation have been removed, the soil has been modified by the addition of a sand layer, and cranberry plants have been introduced. Not only are the cranberry plants separated from the natural groundwater table by an impervious layer, they are often also separated from the stagnant water in the underlying peat. Thus, cranberry bog soils are subjected to desiccation as would any other agricultural soil.

When bogs are constructed on mineral soils, the site is engineered to provide suitable hydrology and soil characteristics to mimic those in traditional wetland settings. Adapting the existing site hydrology to one that supports cranberry production may require manipulation of the water table, soil permeability, soil texture and soil organic carbon content. The objective is to create a cranberry bog that can be managed using many of the same techniques used on peat-based bogs. To accomplish this, a slowly permeable subsoil layer (water confining layer) is placed so that a 'perched' water table is created at some distance above the true water table. An organic layer is placed above the confining layer with the sand planting medium on top (DeMoranville, 2006).

Cranberry bog soil, in commercial production, is a man-made substrate consisting primarily of sand. In the root zone of well-established Massachusetts cranberry bogs, organic matter makes up less than 3.5 per cent of the soil; silt and clay combined may account for as much as another 3 per cent; the remaining particles are sand. In these mature bogs, the soil is stratified or layered, with layers of almost pure sand alternating with organic layers composed of partially decomposed leaf litter and non-functional roots (Deubert and Caruso 1989). This layering arises due to the common cultural practice of applying sand to the bogs periodically to improve vigor and control pests. Newly established or renovated cranberry bogs have very little organic matter in the soil - they are virtually all sand in the root zone (DeMoranville 2008b).

Cranberry beds in North America range from less than one acre to more than one hundred acres in size. The bog area is typically the lowest part of the landscape. It is comprised of perimeter and interior drainage ditches and dikes that can readily contain water. Due to the periodic need for flooding, beds are always associated with nearby water bodies likes ponds, rivers, or man-made reservoirs (Sandler 2008).

Planting Cranberry Vines

The establishment of a new planting (or renovation) and its associated activities are among the most expensive operations performed by cranberry growers. Typical activities include removal of existing vines by bulldozer, laser leveling of the bog surface, addition of a deep sand layer approximately 10-20 cm, fumigation, repairing or replacing irrigation systems, purchasing and planting of new vines, and the application of fertilizers and herbicides (DeMoranville *et al.*, 1996b; DeMoranville *et al.*, 2001). Vines are typically planted at densities between 2200-4500 kg/ha, depending on cost and availability (Sandler *et al.*, 2004).

Vines can be planted as long (30-45 cm) unrooted cuttings or as small rooted plugs. Unrooted cuttings are uniformly scattered on the ground and then disked into the ground. This has been the conventional approach and is quite effective for propagating established varieties. Use of unrooted cuttings allows growers to mow or prune established plantings, collect the vines, transport them to the new area and scatter the cuttings. Vines are available for purchase but use of one's own vines saves money (Sandler and DeMoranville, 2009).

In comparison, the new hybrid varieties from Rutgers University breeding program are being propagated and sold as rooted plugs. Plugs are more expensive but since the vines already have roots when they are placed in the ground, their survival and colonization rates are high. The use of plugs reduces the initial introduction of weeds with the vine material. If off-site weed seed banks are minimal, labor and material inputs needed to manage weeds in the first a few years could be lower than with the use of cut vine.

Site Preparation and Establishment

Many decisions are involved in the preparation of the cranberry farm for renovation and planting and no single list fits every situation. Growers must decide how to deal with the existing plant material, both vines and weeds. If weed pressure is the main reason for renovation, growers may use fumigants to minimize re-infestation. Most use laser leveling to ensure the evenness of the grade of the planting surface. Sand must be obtained and transported to the site and distributed onto the surface. Conventionally, growers apply a thick layer of sand (ca. 15 cm), but recently growers are exploring the use of thinner sand layers or no sand. Labor is needed to scatter the vines or plant the plugs. Most growers take the opportunity during renovation to reconfigure and/or upgrade the irrigation system and correct drainage problems.

Fertilizers are used to stimulate initial growth of the vines. At planting, phosphorus fertilizer is added to encourage good rooting. In the first year, vines are fertilized frequently with nitrogen to promote runner growth. Runner growth is needed to enable vines to colonize the surface. Preemergence herbicides can be used once root growth has started (usually 3 weeks). Irrigation in the first month is critical; it is not unusual for plants to need irrigation at least twice per day until the roots are established. Two years are typically needed to achieve adequate vine colonization; fruit production is emphasized in the third year (DeMoranville *et al.*, 2001).

Flowering Behaviour and Fruiting

Cranberries are a perennial plant and produce a single crop each year. Figure 4.1 shows the 15-month cycle from flower bud initiation through harvest. Flower buds, within the terminal buds, are initiated in the year before the crop, beginning in mid-summer (July in the Northern Hemisphere). The buds continue to develop throughout rest of the summer and fall. By August, the initial stages of the flower bud can be found and visible changes in the buds continue until at least some time in October, later in mild growing areas. Eventually the flower buds become dormant until the following spring.

The dormant state lasts until the plant has been exposed to sufficient chilling hours to complete the dormant cycle, at which time events leading to bud break and flowering may begin. Once dormancy has broken in response to increased temperature and day length, completion of the developmental cycle (vegetative growth, flowering, and fruiting) occurs. The terminal bud begins to lose dormant color and swell during April. By late in May, buds have begun to grow and new leaves and flower buds are visible, growing from the terminal end of the uprights.

During June, the uprights continue to increase in length (new leaves formed) and the first flowers open. From late June to early July, bees pollinate the cranberry flowers and tiny fruit form (fruit set). Berry size within a variety is proportional to average number of seeds per berry, although the statistical relationship is weak. In hand pollinated flowers, the deposition of at least eight pollen tetrads was required for maximum fruit set to occur. If deposition was greater, fruit size increased only slightly although seed number continued to increase proportionally (Cane and Schiffhauer 2003).

When insect pollination was supplemented with hand pollination, fruit set could be increased from 30 to 38 per cent but yield did not increase significantly (Birrenkott and Stang 1989). The limitation to fruit set is likely due to limitation of carbon resources. The lower flowers along an upright stem open and are pollinated before the upper flowers. Once these lower fruits set, the remaining flowers often do not set fruit. However, if the lower flowers were removed, fruit set in the upper flowers increased to 45 per cent compared to 25 per cent when the lower flowers were left in place (Birrenkott and Stang 1990). Carbon resources are low during this period (Botelho and Vanden Heuvel 2006) and developing fruit must compete with previously set fruit as well as with vegetative growth (Vanden Heuvel and DeMoranville 2009).

From fruit set until harvest, the growth cycle overlaps the beginning of a new cycle for the following year. During the first three weeks following fruit set, the fruit acquires most of their mineral components. From that point through harvest, fruits grow by the accumulation of carbohydrates and water and carbohydrate resource allocation shifts from leaf dry weight accumulation to fruit growth (Vanden Heuvel and DeMoranville 2009). By September, the fruit begin to develop their characteristic red color through the production of anthocyanin pigments. Full fruits maturity occurs approximately 80 days after full bloom but varies by cultivar. Cranberry fruit continues to increase in weight, albeit at a slower rate in September compared to that in July and August, until late September for early varieties and mid-October or longer for late varieties. Cranberry harvest extends from mid-September through October.

Varieties

The American cranberry is native in bogs from Newfoundland south to North Carolina and west to Minnesota (Dana 1990). Cultivation of the American cranberry began in the early 1800's with the selection of vines from the wild that possessed favorable qualities. These vines were usually transplanted to a swampy area where they were cultivated and the berries were eventually harvested. The cranberries discovered in North America were initially divided into three groups based on general berry shape: (1) bell, (2) bugle, and (3) cherry (Eastwood 1856).

Initially, there was great diversity of cultivars used throughout North America (Table 4.1). As the performance of certain cultivars proved consistent over an extended time period (whether considering production, pest resistance, color, quality, or other factors), more growers planted these cultivars and the acreage increased. Current estimates of cultivars with significant acreage in different growing areas are listed in Table 4.2 (Caruso 2008a).

Table 4.1: Varieties Popular during the Early-mid 20th Century

Massachusetts:	Bugle, Centennial, Early Black, Holliston, Howes, Mathews, McFarlin, Round Howes, Shaw's Success, Vose's Pride.
New Jersey:	Centennial, Champion, Early Black, Howard Bell, Howes, Natives (Jerseys).
Wisconsin:	Bennett Jumbo, Berlin, Howes, McFarlin, Metallic Bell, Natives (Bell, Cherry), Prolific, Searles.
British Columbia, Oregon, Washington:	McFarlin.

Table 4.2: Percentages of Acreage of Varieties Grown in Various Cranberry Regions in the U.S., Canada and Chile

Location[z]	*Variety (Per cent of Acreage)*									
	Early Black	*Howes*	*Stevens*	*Ben Lear*	*McFarlin*	*Pilgrim*	*Searles*	*Crowley*	*Bergman*	*Others*
MA	40	30	15	5	–	–	–	–	–	10
WI	–	–	60	10	5	5	12	–	–	8
NJ	50	3	28	12	–	–	–	–	–	7
WA	–	–	45	–	40	8	–	–	–	7
OR	–	–	77	–	7	3	–	4	–	9
Br.Columbia	–	–	65	5	6	4	–	–	17	3
Quebec	–	–	76	13	–	7	–	–	–	4
N. Brunswick	–	–	90	5	–	4	–	–	–	1
Nova Scotia	6	–	58	15	–	–	–	–	12	9
Chile	–	–	X	X	–	X	–	–	X	–

z: Small acreages exist in Maine, Michigan, Ontario, Prince Edward Island and Newfoundland, planted with the cultivars listed above.

X: Varieties grown but exact percentages unknown.

–: Either not grown or negligible.

In most cases, cranberries are not available from nurseries in true genetic stock as are blueberries, grapes, raspberries, strawberries or other small fruits. The methods by which the vines are moved around the industry (intrastate and interstate) and by which they are propagated, means that it is highly likely that there is much genetic diversity for different strains are purported to be the same cultivar. Until recently, cranberry cultivars could only be identified by their qualitative traits, such as fruit shape, fruit size, fruit color, and vine and leaf texture. DNA analysis has proven to be a useful means of differentiating between cultivars of a wide variety of different plants. Vorsa and Novy (Vorsa and Novy 1995) used Random Amplified Polymorphic DNA (RAPD) technology to target cranberry DNA from individuals under analysis. This technology will likely be used in the future to formally characterize varieties from the field and greenhouse.

Early Black

A native Massachusetts selection, the fruits are blackish-red with a 90-130 cup count (an industry standard for comparing fruit size in which one determines the number of fruit that fits into a standardized cup; the number of fruit decreases with increasing fruit size). Ripening early in the season, it has good production, coloring in storage and keeping quality. The fruits are pyriform and the stem end is pointed. The uprights are short, and fine in texture with small light green leaves.

Howes

A native Massachusetts selection, this late-maturing variety has medium-red fruit with an 80-115 cup count. It has good production and coloring in storage and excellent keeping quality. The fruits have no outstanding attributes and are oblong-oval. The uprights are long, coarse in texture with large dark leaves.

Ben Lear

A native Wisconsin selection, the fruits are deep red with a 70-90 cup count. Ripening early, these fruits can have a light to medium bloom (waxy exterior coating) on them. It has very good production, good coloring in storage and poor keeping quality. The fruits are pyriform with a pointed stem end. The uprights are medium-tall with medium texture and medium dark leaves.

McFarlin

A native Massachusetts selection, this late-maturing variety has deep red fruits with a 60-80 cup count. The bloom can be quite heavy on these fruits. They have good production and keeping quality, but poor coloring in storage. The fruits have a prominent calyx region and are round-oblong. The uprights are short, coarse in texture, and have light green, medium leaves.

Stevens

A cross of McFarlin and Potter, Stevens ripens mid-season and has a cup count of 65-85. The deep red fruits have an occasional light bloom but usually not. They have excellent production, fair coloring in storage and good keeping quality. The fruits are round-oval with the stem end broadly rounded. The uprights are tall, coarse in texture with medium green and large leaves. This is the most commonly planted hybrid cultivar in North America and produces well in all growing areas.

Pilgrim

A cross of McFarlin and Prolific, the deep red fruit ripens late and has a cup count of 43-66. The fruit may have moderate to heavy bloom. They have very good production, and good coloring in

Figure 4.3: Example of cv. Ben Lear Fruit Photo courtesy C. DeMoranville.

Figure 4.4: Example of cv. Early Black Fruit Photo courtesy C. DeMoranville.

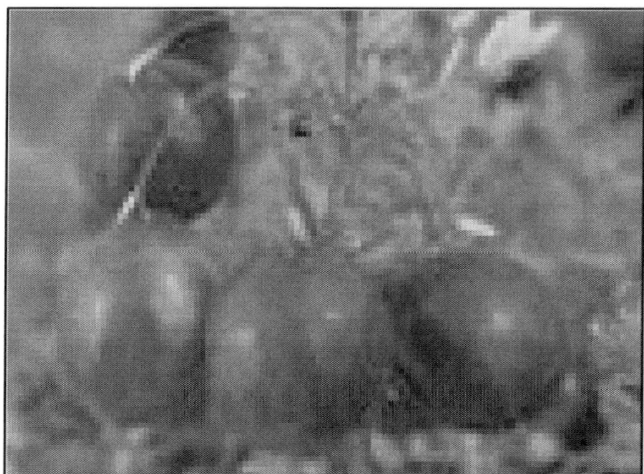

Figure 4.5: Example of cv. Crimson Queen Fruit Photo courtesy N. Vorsa.

Figure 4.6: Example of cv. Mullica Queen Fruit Photo courtesy N. Vorsa.

Figure 4.7: Example of cv. Howes fruit.
Photo courtesy C. DeMoranville.

Figure 4.8: Example of cv. Stevens fruit.
Photo courtesy C. DeMoranville.

storage and keeping quality. The fruits are oval with the stem end broadly rounded. The uprights are medium to tall, coarse in texture, with large dark leaves.

Rutgers Hybrids

Three new cultivars bred at Rutgers were released in 2006. Crimson Queen, a cross between Stevens and Ben Lear, was first established in the field in 1993. It was specifically selected for its high yield, good anthocyanin production and high stolon vigor. Mullica Queen resulted from a cross between LeMunyon and No. 35 and represents a genetic composition totally unrelated to Crimson Queen. This cultivar offers excellent yield potential with good anthocyanin production, excellent runner vigor and fruit rot resistance equal or better than Stevens. Demoranville, a cross between Franklin and Ben Lear, offers excellent fruit production, superior anthocyanin levels, but may have the least fruit rot resistance (N. Vorsa, pers. comm.).

Canopy Management

The cranberry is a perennial plant and established commercial plantations may sustain productivity for many years, if the plant canopy is properly managed. Cranberry plants produce additional length on upright stems as each year's terminal bud expands and grows. Over time, these elongated uprights fall over and a dense mat of stems forms along the soil surface. Additionally, lateral buds on upright stems may grow to produce runners (stolons) that trail within or above the canopy. Canopy management techniques are implemented to overcome the negative impact of these growth traits on productivity and to facilitate the mechanical harvest of the crop.

The primary canopy management practices are pruning and sanding (the application of a 1-3 cm layer of fresh sand to the soil surface). Pruning is used extensively to remove runner growth from within the canopy and to thin choked stands of uprights. Sanding is used to cover the mat of stems formed by toppled uprights, thus encouraging the production of new roots and new uprights from the stem base, but may also serve to anchor runners to the soil surface. The latter is important during bed establishment when runners are encouraged as a way to establish the stand. The layer of sand encourages the runners to root at the nodes and produces new upright stems from lateral buds.

The canopy environment may be improved with either sanding or pruning. Although both practices will give physical improvements such as increased light penetration, better aeration, and improved vine vigor, each practice offers unique benefits and consequences.

Pruning

Cranberries are pruned to remove excess runners and old, long upright shoots and to facilitate use of dry harvesting equipment. In fact, some modern dry harvesters, notably the Furford Harvester, combines pruning action with harvesting. When runners are present and upright stands become dense, light penetration to the individual plants is limited. This light limitation leads to declines in yield either due to decreased flower bud initiation or limitations on pollinators reaching the flowers to set fruit or both. A dense canopy also provides a moist micro-climate for the growth and spread of fungal fruit rot disease (Caruso and Ramsdell 1995).

Cranberry vines are pruned during or just after harvest or early in the spring. Cranberry pruning machines are usually a series of vertical knife blades set at an angle to the direction of movement and spaced at 30-cm intervals on a rotating frame. This device is mounted on a buggy or small tractor. Such pruners move through the bog, removing runners and some uprights. The severity of pruning relates to knife spacing and speed of operation. Alternatives to these pruning machines are knife rakes (hand pruning) or mechanical harvesters such as the Western Picker and Furford Harvester which prune during dry harvest. These harvesters can be used for pruning after water harvest or for additional pruning of dry harvest bogs by making a second pass across the bog.

Regardless of pruner used, severe pruning or mowing greatly reduces yield. In some instances, even light pruning can cause some reduction in yield. However, in Oregon, studies (Strik and Poole 1992) showed that light pruning (up to 2000 kg/ha) in alternate years resulted in larger *cumulative* yield compared to no pruning or severe pruning, even if crop was reduced in the year of the light pruning. In a Massachusetts study, pruning at ~500 kg/ha increased yield in the year of pruning in replicated plots (Suhayda *et al.*, 2009), and at grower sites, light pruning was the most often associated with increased or unaffected cumulative yield in the 2 years following pruning.

Sanding

Historically, sanding was probably the first practice used in cultivating wild cranberry plants. Henry Hall of Dennis, MA is credited with being the first to observe the beneficial effects of sand on cranberry vines in 1816. He noted that sand blown from nearby dunes which partially covered the vines was associated with increased growth, vigor and yield of cranberries.

Sanding covers the leaf litter on the surface of the cranberry bog. This has several benefits, including stimulation of organic matter decomposition (nitrogen release and relief of root congestion), suppression of fruit rot inoculum, and limitation of the habitat of cranberry girdler larvae which feed at the base of exposed stems (Cross and Demoranville 1978; Tomlinson 1937). Uniform applications of sand can also suppress dodder germination (Sandler *et al.*, 1997). Sanding improves soil drainage and may physically strengthen peat soils so that mechanical operations on the farm are easier.

Many of the pest management benefits of sanding are dependent upon the deposition of uniform layers of sand. When sanding depth was measured in the spring after the sand had settled into the vine canopy, uniform layering was rarely achieved (Hunsberger *et al.*, 2006). The nonuniform application of sand to the production surface could explain the variable results seen when growers apply sand for pest management.

Sand can be applied directly onto dry vines by ground rigs which ride on the vines, applied during winter on top of frozen flood waters (ice sanding), sprayed onto the vines via sand-water slurry (hydrosanding) or delivered via a floating barge in shallow flood waters (barge sanding) during the spring or fall. Yield impacts are variable depending on method of application, cranberry variety and thickness of the sand layer (DeMoranville *et al.*, 1996c; Strik and Poole, 1995).

Sand with a few fine particles (fine sand, silt and clay) gives the best result when sanding a bog. An ideal bog sand has >70 per cent of particles in the coarse range (0.5-2 mm) and less than 3 per cent silt or clay. The amount of sand to be applied depends on the sanding method chosen. Generally, a layer 1.3 to 1.9 cm thick is recommended if ice sanding or dry sanding, but at least 2.5 cm if barge sanding. Applying layers greater than 2.5 thick can cause significant yield loss (Suhayda *et al.*, 2009). The interval between applications of sand varies but every 4 years is common.

Sanding is of particular importance in the management of new plantings. Cranberries are traditionally planted as unrooted cuttings and then heavily fertilized to promote the production of runners (trailing stems) from axillary buds on the stems. As these runners cover the soil surface, thin layers (1.3 cm) of sand are applied to anchor the runners, promoting rooting at the nodes and leading to the production of upright stems that will then bear the crop. At minimum, new plantings should be sanded after the second season and may need to be sanded after the first season as well, depending on how much growth has occurred.

Nutrient Management

Cranberries are adapted through evolution for growth on acid, sandy soils. These soils have little nutrient content and the plants such as cranberries and blueberries which evolved on them have correspondingly low nutrient needs. Each season nutrients are removed from the bog during harvest and detrashing (removal of fallen leaves from the bog surface). When the fruit is harvested, the elements removed in the largest quantities are nitrogen (N), potassium (K), and calcium (Ca). The amount of nutrient removal increases with increasing crop load and is less when crops are small. In addition, hybrid cultivars tend to have larger leaves and thicker stems so that more nutrients are used to produce plant parts and more are removed with the fallen leaves. In a 22,4000 kg/ha crop of Early Black cranberries (including the fallen leaves and stems removed during harvest) 26.4 kg N, 4.8 kg phosphorus (P), 23.3 kg K, and 17.7 kg Ca/ha were removed from the bed (DeMoranville 1992).

It is to compensate for nutrient removal that fertilizer is used. Most fertilizer added to producing cranberry beds contains N, P, and K. While P removal is low, some P is included in the mixture to maintain nutrient balance and because much of the P in cranberry bog soil is not available to the plants at crucial growth stages. Table 4.3 compares the N, P, and K recommendations for cranberry production to those for other fruit and agronomic crops. In general, cranberry production requires less fertilizer than that of other crops.

Table 4.3: Standard Recommendations (kg/ha per year) for Fertilizer Nitrogen, Phosphorus, and Potassium Rates for Several Crops***

Crops	*Nitrogen*	*Phosphorus*	*Potassium*
Cranberry	22–67	no more than 22	45–135
Blueberry	50–75	39–50	45–90
Apple	56–67	95–195*	140–280*
Peach	90	95–195*	140–280*
Corn	180–225	28–73**	97–135**
Wheat	85–125	50–90**	78–123**

*: When required based on soil and tissue tests, every 3–5 years.

**: Based on potential yield and soil test

***: Data for cranberry from UMass Extension recommendations (DeMoranville 2008a), agronomic crop data from Ohio State University Extension recommendations (Vitosh *et al.*, 1995), tree fruit (Hanson 1996) and blueberry (Hanson and Hancock 1996) recommendations from Michigan State University Extension.

However, during establishment of a new planting or renovation the recommended rates for N and P are higher than those for a producing cranberry bog. At planting, 22 kg/ha P and 22-34 kg/ha slow-release N are applied to the fresh sand to encourage plant rooting. During the first season of a new planting, N is applied at the rate of 5-11 kg/ha every two to three weeks until late in the summer, alternating N-only products with N-P-K products with a 1:1:1 ratio. This regimen stimulates robust growth and the production of runners which spread quickly to cover the soil surface. Rapid filling-in of cranberry plants discourages weed infestations. In the second season, fertilizer use is reduced at around the time of bloom to encourage the formation of flower buds for the third season. Generally, an established-bog fertilizer schedule is used in the third and subsequent years.

Fertilizers Used on Cranberries

The predominant fertilizers applied to cranberries are complete N-P-K materials with varying ratios of the three elements, applied based on seasonal N requirements. During the season, 22-67 kg/ha N are applied depending on cultivar and weather conditions. N-P-K materials are chosen so that no more than 22 kg/ha P is applied. K application rates in N-P-K products are generally 1-2 times the rate of N. Additional K may be applied as a supplement. N, P, and K are discussed further below.

While cranberries require many other mineral elements, often these are in sufficient supply in the soil to satisfy the plant needs. When testing shows that these other elements are lacking in the plants or in the soil, they are then applied on an as-needed basis. Ca is applied as gypsum (calcium sulfate) since the addition of lime can adversely affect the soil pH. Magnesium is added as magnesium sulfate or in combination with K. Sulfur is not used as a fertilizer in cranberry production but may be used to lower soil pH if necessary.

Minor elements such as copper, zinc, iron, and manganese are very available in acid soils. For this reason, cranberries seldom suffer minor element deficiencies, nor do they require minor element fertilizers in general. One exception is the use of minor element supplements in fertilizer blends during the first season of a newly planted mineral soil bog. Another exception is the use of calcium-boron supplements at bloom. For bogs with poor yield histories, such calcium-boron supplements may increase fruit set.

Nitrogen

The single most important nutrient element in cranberry production is N. N is required by cranberry plants for the production of vegetation (new leaves and stems), roots, and fruit (crop). As a critical constituent of protein, N is a controlling element in the plant's nutrition. The production of chlorophyll is regulated in part by the availability of N.

Adequate N is important in the production of cranberry fruit both directly and indirectly. N makes up a portion of the dry weight of each fruit and adequate supply during fruit formation is required. However, much more of the fruit dry weight is made up of various carbohydrates. Early in the season, the presence of N promotes the growth of the new uprights. It is photosynthetic activity in the leafy tissue above the developing fruit on a fruiting upright that is the source for the carbohydrates in the fruit. If early season N is lacking and the upright growth is stunted, the plants will lack the ability to support a large crop. Further, lack of N in the early season can lead to uprights that are pale greenish-yellow, indicating a lack of chlorophyll. This too will limit photosynthesis.

Each season N is removed from the bog in the crop and in fallen leaves. To compensate for this loss cranberry plants get N from the soil, from water, or from fertilizers added to the bog. The amount of N that must be added as fertilizer depends on how much N is supplied by the soil (soil organic matter and weather dependent), cultivar requirements, and general status (length, color) of the cranberry plants.

The average recommended seasonal rate of N for producing cranberry beds in Massachusetts varies from 11-67 kg/ha depending on plant vigor and variety. Best yield results were obtained in research plots when seasonal N application rate was divided into four applications corresponding to the periods of peak demand: 20-25 per cent at roughneck stage (1.3-2.5 cm new growth from the terminal bud), 30-35 per cent at bloom, 30-35 per cent at fruit set (about 2-3 weeks after bloom), and ~20 per cent at bud development/fruit sizing (early August). Split timing allows for in-season rate adjustment as conditions warrant. Additional N fertilizer is added if the cranberry plants show signs of N deficiency - poor growth, loss of leaf greenness, and/or low nitrogen content in the leaf tissue.

In a recent study (Davenport and DeMoranville 2004), the predominant factors that determined N release due to mineralization were organic matter content of the soil and soil temperature. Sandy soils (newer plantings) released less N compared to soils from older beds with layers of sand and organic matter. In the early spring as the soil dried after the winter flood, bacterial activity increased and there was a 'flush' of N release in the soil despite cool temperatures (10-13°C). Mineralization rates were similar at temperatures from 13-21°C. The rate increased dramatically when soil temperature rose to 24°C. The ammonium-N released in mineralization is readily used by the cranberry plants. However, once mineralization occurs, the resulting ammonium-N may be converted to nitrate-N in a second bacteria-mediated reaction, nitrification. The bacteria which mediate nitrification are sensitive to soil pH. In common cranberry soil with pH 3-4.5, activity of these bacteria was minimal. However, the same soils adjusted to pH 6.5 released large amounts of nitrate-N and had high populations of nitrifying bacteria (Davenport and DeMoranville 2004). Nitrate-N is not a good source of N for cranberries.

Unlike most horticultural crops, cranberries appear to preferentially use ammonium-N (Greidanus *et al.,* 1972; Smith 1993). In solution culture, cranberries can take up some nitrate-N, but uptake was only substantial in the presence of ammonium (Rosen *et al.,* 1990; Smith 1993). Recently, cranberry plants in culture inoculated with ericoid mycorrhizae did show the ability to take up and transport nitrate-N (Kosola *et al.,* 2007). These specialized mycorrhizal fungi colonize cranberry roots in the field and when present, may allow uptake and utilization of nitrate.

However, based on the predominant evidence that cranberries can use ammonium-N quite effectively and since nitrate-N can leach and pose environmental hazards, ammonium is the recommended form of N for cranberry production. Ammonium is readily taken up by the plants (10 times greater uptake compared to nitrate in solution culture), is only slowly converted to nitrate in acid soils, and is lower in leachability compared to nitrate. Urea, organic fertilizers, and many slow release fertilizers deliver N as ammonium during their breakdown. An additional benefit of ammonium-N is that as it is taken up by the cranberry plant, the plant releases acid equivalents to the soil, helping to maintain the cranberry-preferred low pH.

In the soil solution of cranberry bogs, N predominantly occurs in the dissolved organic form. Ammonium-N is the next most common form, followed by nitrate-N (Kosola *et al.,* 2007; Stackpoole 2008). Other research (Stribley and Read 1974) showed that cranberries could absorb and utilize organic N forms from the soil. Therefore, the dissolved organic pool of N in cranberry soil may also be a source for cranberry N nutrition even in the absence of mineralization to ammonium-N.

Soil pH and organic matter should be tested at least once every three-five years. The biological conversion of cranberry-useable ammonium-N to less-desirable nitrate-N with increasing pH is the most pronounced in beds with high organic matter soil. Soil pH on cranberry bogs with soil organic matter content of 0-5 per cent should be between pH 4.0 and 5.0, while soils with organic matter content greater than 5 per cent should have a pH of 4.5 or less.

Phosphorus

P plays many roles in plant metabolism. P is involved in energy transfer as part of the ATP molecule and is a primary constituent of DNA. P plays a regulatory role in photosynthesis and starch synthesis, active transport of materials across membranes, root growth and function and hormonal balance. This last is critical to floral induction. While only modest amounts of P are removed from cranberry bogs in fallen leaves and fruit, it is essential that soil P be available to the cranberry plants to support seasonal growth and flowering. Common characteristics of cranberry bog soil affect P availability.

Cranberry soils are high in iron and have low pH. This chemistry leads to conditions where P is tightly bound in the soil and is to a large extent unavailable to the cranberry plants (Davenport *et al.*, 1997). Cranberry plants with tissue P at or below the critical level (0.1 per cent dry matter) are often found growing on soils with high P test values. When fertilizer is applied to these soils, the P in the fertilizer dissolves into the soil water and quickly becomes bound to iron - only a small percentage of the P remains dissolved and available for plant use. Of the bound P, some portion may later be released and available for uptake by the plants depending on the aerobic status of the soil.

Phosphorus uptake and release in cranberry soils of varying organic matter content were investigated under flooded (anaerobic), dry (aerobic), and transitional conditions (Davenport *et al.*, 1997). Sand soils readily released P that had been previously applied and bound to the soil. However, the results were quite different for peat and layered (sanded cranberry) soils. In the layered soil, P was released from the bound state at the highest rate as the soil moved from the flooded to the seasonal dry state (field capacity). Once the soil reached seasonal dryness (late spring), P was only released if a certain threshold amount was present in the soil, indicating the need for fertilizer applications under those conditions. This pattern was even more pronounced in highly organic (peat) soil.

Despite the presence of bound P in the soil, research has shown that cranberry yield increases in response to the addition of P fertilizer (DeMoranville and Davenport 1997; Greidanus and Dana 1972). However, the response is not linear, yield was greater with the addition of 22 kg/ha P/season (compared to no P), higher rates did not significantly improve the response. Further research with a broader P rate range in Wisconsin and Massachusetts have confirmed that there is no experimental evidence for a cranberry yield response to P rates above 22 kg/ha and in many cases, good yield response was found with even lower rates (DeMoranville *et al.*, 2009; Roper 2009). The addition of more than 22 kg/ha P in a season is only justified if tissue P is <0.1 per cent or during the establishment of new or renovated plantings.

Most cranberry growers add inorganic phosphate to the soil in N-P-K fertilizer or as triple superphosphate. The P in the N-P-K fertilizers is generally derived from ammonium polyphosphate (used in ammoniated fertilizers), monoammonium phosphate (MAP) or diammonium phosphate (DAP). Fertilizers containing MAP are an excellent choice for cranberry production as the fertilizer particles form an acid zone in the soil, helping to maintain the low pH preferred for cranberry production. Slow release P forms are available. They perform similarly to soluble granular forms — at the same rate of applied P, yield and tissue P were similar with either fertilizer type (DeMoranville and Davenport 1997; Roper 2009). Since in producing cranberry beds, P is added in N-P-K materials and the material rate is selected based on N requirement, the N:P ratio in the fertilizer is critical if no more than 22 kg/ha P is to be applied. P is generally applied with N and therefore is split-applied in up to four applications as described above. This corresponds to the period of seasonal soil dryness when existing soil P is tightly bound to soil iron and poorly available for plant uptake. If slow release P materials are used, they may be applied in a single application early in the season (roughneck stage). Prior to roughneck stage there is no need for P applications as bound P is released from the soil as it transitions from saturation to seasonal dryness in the early spring.

Potassium

K is the only major element with no structural role in the plant. However, K is involved in the movement of sugars and starch in the plant and may play a role in resistance to disease, drought and cold temperatures. K also has a major role in preserving plant turgor (water relations) and in osmoregulation (regulating water movement across plant membranes). Cranberries have a much higher per cent of K in the fruit and seeds than in the leaf tissue. As a result, seasonal removal of K in crop and fallen leaves is about equal to removal of N.

Due to its role in plant turgor, K fertilizer is added to cranberry bogs when vines are brittle and dry, most often in the spring. Otherwise, K is generally added in the N-P-K fertilizer applied to satisfy N needs. Common cranberry fertilizers supply K in a 1:1 ratio with N. Seasonal rates of K applied to cranberry bogs are in the range of 45-135 kg/ha. Field plot research did not show any measurable benefit to the addition of higher K rates (Roper 2009).

Fertilizer K is the most commonly available as sulfate or chloride. There exists a body of evidence that cranberries are sensitive to chloride. Salt injury to cranberry vines is observed following coastal hurricanes and in areas which receive highway treatment overspray in the winter. In both instances, the salt in question is sodium chloride (NaCl). In a greenhouse study in sand culture, cranberries exposed to 250 mg/L chloride in irrigation water showed leaf reddening with Cl provided as NaCl *or* KCl. At lower concentrations, runner production was stimulated and at 250 mg/L as KCl, many plants died.

Fertilizer is applied to cranberry bogs using ground rigs (spreaders and seeders), helicopters (aerial application), and the irrigation system (fertigation). Fertilizer is applied in split doses if water-soluble materials are used. As previously mentioned, the dose for the total season is split over 3-4 applications. This lessens the potential for leaching of the material below the root zone. Cranberries are shallow-rooted. This, combined with the limited ability of cranberry soils to hold nutrients, makes split applications essential. Overloading of soluble materials would be unsound economically as well as ecologically. As an alternative to split applications of soluble materials, some growers use organic or inorganic slow-release materials.

Liquid or foliar fertilizers are also used when a quick response (generally to correct problems) is desired. These are low-analysis materials designed to be taken up quickly by the plants. When used at the recommended rates, they have little potential for movement into water supplies. Due to nonuniform application with irrigation systems, only low rates of fertilizer are applied by this method. Otherwise, the plant stand grows unevenly, leading to difficulties in harvesting and other management tasks.

Water Use and Irrigation

Water management practices on cranberry beds differ from those used for other forms of agriculture because of the variety of ways that water is used in cranberry culture. Water is used for disease and insect control, frost and heat protection, sanding, harvesting, and protection from winter desiccation and cold injury. Because of the periodic need for sizable amounts of water, impoundment of water adjacent to the beds is a normal farming practice in cranberry production. In addition to storage ponds and sumps, components of a typical water management system for a cranberry bog include irrigation systems, wells, flood gates and flumes, lift pumps, and drainage ditches and pipes.

Water and irrigation management are crucial to cranberry cultivation (Eck 1990). Cranberries can use up to 50-65 mm of water per day during the hottest, driest, windiest weather. This would amount to 2.5 cm in four to five days under the most severe conditions. A common practice is for vines to receive 2.5 cm per week from either rain, capillary action from groundwater, irrigation, or some combination of these. If provided by irrigation, it should be in at least 1.25 cm increments to ensure good infiltration into the soil. Bogs that are low in organic matter, or that have a thick sand layer like those that have been renovated, may need more frequent irrigation. Excessively wet soils increase the likelihood of Phytophthora root rot, while excessively dry soils can promote fairy ring disease. Further, soil moisture can affect the ability of the plants to acquire nutrients from the soil. When cranberry beds are either too wet or too dry, it can lead to inadequate rooting and leaf chlorosis, a condition known as yellow-vine syndrome, which occurs in part due to nutrient imbalances.

In a typical cranberry bog, water can wick up through the soil to the roots of the plants from a water table depth of up to 45 cm (Lampinen *et al.*, 2000). By measuring the depth of the water table, the water level float allows the grower to quickly adjust that depth to maintain it between the 45-cm limit and 15 cm (the recommended depth to avoid waterlogging in the roots). Depth can be adjusted by using solid-set sprinklers to provide surface irrigation or by moving water into the drainage system to provide water from beneath (subirrigation). Generally, a combination of the two is the best.

Plant Protection

The principle challenge for managing pests in cranberries is simply the vast number of organisms that can cause damage to the vine or the fruit or both. Over 20 insects cause injury to the cranberry and three are direct fruit pests (Averill and Sylvia 1998). The most serious insect pests include cranberry fruitworm (CFW) (*Acrobasis vaccinii*), *Sparganothis* fruitworm (*Sparganothis sulfureana*), cranberry girdler (*Chrysoteuchia topiaria*), cranberry weevil (*Anthonomus musculus*) and black-headed fireworm (BHF) (*Rhopobota naevana*). Fruit rot is the most serious yield-limiting disease problem for Massachusetts and New Jersey and is associated with more than 10 causal agents (Caruso 2008b; Oudemans *et al.*, 1998). More than 80 species of weeds are described as problematic (Demoranville 1984; Sandler 2004).

Chemical control is a critical component of pest management for cranberries and chemigation (chemical application via the sprinkler system) is the primary delivery mechanism in Massachusetts. According to a recent summary report, 32 different pesticides were used in Massachusetts in 2003. These included seven fungicides, nine herbicides, and 16 insecticides (J. DeVerna, pers. comm.). Chlorothalonil (*e.g.*, Bravo) was the most widely used fungicide, followed by the ethylene bisdithiocarbamate (EBDC) fungicides and the copper fungicides. For post emergence herbicides, glyphosate (*e.g.*, Roundup) was used most often. The top two pre emergence herbicides used were pronamide (Kerb; 46 per cent) and dichlobenil (Casoron or Norosac; 23 per cent). Diazinon was the most widely applied insecticide (84 per cent), followed by carbaryl (*e.g.*, Sevin; 72 per cent) and thiamethoxam (Actara; 54 per cent). Biological products such as *Bacillus thuringiensis* (Bt) and beneficial nematodes are also used for insect management. Fungal pathogens were tried but not successfully used for dodder control.

Several cultural practices (flooding, sanding and pruning) are utilized to help with pest management. Holding a late water flood (*i.e.*, reflooding the bog from mid-April to mid-May) can decrease the inoculum potential of the fruit rot fungi, cause a general reduction of annual weeds, suppress the spread of *Rubus* sp. as well as suppress populations of certain insects and mites (Averill *et al.*, 1994; Averill *et al.*, 1997). Short spring floods can control BHF (Cockfield and Mahr 1992) and dodder (Sandler 2003; Sandler and Mason 2004). Short (3 to 7 days) late summer floods can also be used for management of cranberry girdler (Beckwith 1925; Fitzpatrick 2007), and longer floods (held for 3-4 weeks after harvest of the fruit) can reduce CFW emergence from hibernacula and suppress growth of dewberries (DeMoranville *et al.*, 2005).

Sanding, *i.e.*, the application of a thin (1.3-5 cm) layer of sand on the production surface at 2 to 5 year intervals, can suppress fruit rot inoculum by burying infected leaves (Tomlinson 1937). Uniform applications of sand on a regular interval may reduce infestations of cranberry girdler and green spanworm (Franklin 1913; Tomlinson 1937). Uniform sand applications can also inhibit emergence of dodder seedlings (Sandler *et al.*, 1997). Pruning has indirect effects on pest populations but provides overall benefits to vine vigor and is an important cultural practice. Periodic pruning of vines improves aeration in the vine canopy and makes the environment unfavorable for fruit rot infection (Caruso and Ramsdell 1995).

Harvest, Post-harvest and Yield

In North America, cranberries are harvested from early September until early November. The exact harvest dates vary by region, weather conditions, and cultivar being harvested. In addition, some consideration must be given to whether the fruit will be sold in the fresh market, used in white juice products, or used for other processing. With the exception of white harvest, the fruits are harvested at full maturity with good color (anthocyanin content) but prior to the fruit becoming overripe. Timing of harvest is important for fresh-market fruit so that the berries are sufficiently red but retain good storage quality, while fruit for the process market ideally has maximum anthocyanin.

Fruit Development

During fruit development, acids in the cranberry reach a maximum level prior to the appearance of red color. The predominant acids in cranberry fruit are (in descending order) citric, quinic, malic, and benzoic. These acids are the source of the tart and astringent flavors of cranberry. As the fruit color develops, the sugar content of the fruit increases. At full maturity, the cranberry fruit is 88 per cent water, 4.2 per cent sugar, and 2.4 per cent acid. The remaining constituents of the mature fruit are pectins, other structural carbohydrates and minerals (Fellers and Esselen 1955).

Cranberry marketing companies produce 'no color added' products, so the color at harvest is of great importance in processed berries. Since, the primary pigments, anthocyanins, are antioxidants, color also has implications in the health benefits of cranberry products (Neto 2007; Reed 2002). However, the most studied health benefit associated with cranberry is related to proanthocyanidins in the berries.

Cranberries contain unique proanthocyanidins (PACs) which can prevent the adhesion of certain of bacteria, in particular *E. coli*, associated with urinary tract infections to the urinary tract wall (Howell *et al.*, 2001). The anti-adhesion properties of cranberry may also inhibit the bacteria associated with peridontal disease (Weiss *et al.*, 2002) and stomach ulcers (Burger *et al.*, 2002) from forming biofilms associated with these disorders. Cranberry PACs are present early in fruit ripening prior to full color development.

Color development in cranberry fruit varies with climactic conditions and so differs from year to year and from region to region (Sapers *et al.*, 1986). In addition, various cultivars develop color at different rates. All of these are considerations when a grower plans a harvest schedule. If the fruit is being harvested for the fresh market, ability to develop further color in storage is also of importance and may vary with cultivars and developmental stages at harvest. Of the four major cultivars grown in Massachusetts, Ben Lear and Early Black develop color the earliest (by early to mid-September),

Stevens develop full color from mid- to late-season (early to mid-October) and Howes are the latest (mid-October). Factors which may slow color development are warm temperatures, particularly at night, and poor penetration of light to the berries (thick canopy).

Cranberry Yields

In North America, the standard measure for cranberries is the 100 pound barrel (45.4 kg fresh weight). In 2008, the total United States production was 7.885 million barrels on 38,100 acres for a production average of 19,264 kg/ha (172 barrels per acre). Hybrid cultivars routinely yield >30,000 kg/ha.

Harvesting of Cranberries

There are two basic methods of harvesting cranberries. The first dry harvesting dates back to the origins of cranberry cultivation. The second system, flood or water harvesting dates to the 1920's and was first mechanized in the mid-1950's (Dana 1990; Eck 1990).

Dry-harvesting

The earliest harvesting was done on dry bogs by workers who handpicked the fruit into wooden boxes and barrels. By the turn of the 20th century, wooden rakes were being manufactured for hand-scooping cranberries. These hand scoops, consisting of wooden or metal tines or teeth set 1.3 cm apart in a wooden catch frame, became the industry standard by the 1930's. In modern cranberry production, hand scoops are only occasionally used to harvest bed edges or experiment plots.

Mechanized dry harvesters were introduced in 1920's. The machines were based on mechanizing a cranberry scoop and relied on tines that stripped the fruit from the vines. In the late 1940's the Western Picker was introduced in Oregon. This machine differed from previous harvesters in that a pruning blade was part of the design so that harvesting and pruning were accomplished simultaneously. In the 1960's, the Furford Harvester advanced the speed of harvest while retaining the pruning function.

Due to climactic conditions on the East Coast, fruit from this region that is bound for the fresh market must be dry harvested. However, dry harvesting has fallen from favor for process-market fruit due to the general inefficiency of the machines. It is common for dry harvesters to leave up to 20 per cent of the fruit on the bog. The fruit that is harvested often sustains bruising during the dry harvest operation. Finally, dry harvesting is quite labor-intensive. Each Furford harvester covers only 0.5-1 hectare per day, other models cover even less ground. The fruit is collected in boxes or bags which must be removed from the bed and transferred into bulk containers. Fruit is then delivered to packing houses where it is stored in common storage until cleaning and separating operations are carried out prior to pack-out. Fruit containing chaff (leaf litter and stems) delivered in bulk bins, is rough screened into smaller wooden boxes prior to storage. Cranberries may be held for two to three months in common storage if fall temperatures are not too warm. Under refrigeration, they may store well for several months. Bruising, physiological breakdown and storage rots (due to fungal pathogens introduced in the field) can all limit storage longevity.

Water Harvesting

Water harvesting overcomes two of the problems associated with dry harvesting: uncollected fruit and high labor costs associated with the slower dry harvest process. Water harvesting takes advantage of the buoyancy of cranberries — the fruit float in the flood water and are accessible to the harvester. A single water harvester covers many more acres than a dry harvester in a day, so the crop is brought in more quickly and labor costs are reduced.

Early water harvesting in Wisconsin was done using hand rakes with long handles. A shallow (~ 1.3 cm) flood was put on the bog and the raked fruit was easily removed. A few berries were left behind as they were floating above the surface of the soil. Vine injury was also minimized. In the 1940's and 1950's the process was mechanized by mounting a rake with either fixed or retracting tines on the front of a self-propelled machine that also had a conveyor to move the fruit into waiting float-boats. Fruit harvested by these methods was dried and sold on the fresh market.

In the early 1950's, the current industry standard water harvester was introduced. This machine is known as the water reel or beater and consists of horizontal bars mounted on a shaft that rotates as the machine moves forward. The horizontal bars are held 2.5-5 cm above the surface of the bog so that they hit against the upright shoots, causing the fruit pedicels to break and releasing the buoyant fruit to float to the surface. This type of rough handling precludes the possibility of drying these fruit and selling them fresh.

Fruit removed by the water reel is left floating in a shallow layer (~1.3 cm) of water. In order to remove this fruit from the bed, the water level is increased until the fruit float free of the tips of the vines. The fruit is then gathered to one edge of the bed using corralling booms and taking advantage of the prevailing wind. Once the fruit is corralled at the bed edge, it is lifted into trucks using conveyors or hydraulic pumps. As the fruit is sent to the truck in the conveyor system, trash (leaves and stems) is removed by passing the harvested mass over a coarse grating. The berries and small debris pass through the grating onto an inclined belt, down which the berries roll, while the debris clings to the belt and is carried away, often to a second, smaller truck.

Alternately, a low-pressure berry pump lifts the fruit mass through a large-bore hose onto an inclined grate. A second pump feeds water to cleaning nozzles. The washing spray pushes debris through the grate into a trash truck, where the trash is de-watered (the water is diverted back onto the bed). The fruits roll down the surface of the grate into the waiting delivery trailer.

Water-harvested berries are delivered to receiving stations where they are washed and placed into bulk containers for freezing or immediately used in processing (juice concentrate, sweetened dried cranberries, sauce). The frozen berries are used for subsequent processing.

Cranberry Chemistry

A number of flavonoids and anthocyanins are reported to be present in cranberries. The red color of cranberry fruit is due to presence of anthocyanins, the largest subclass of flavonoids (Harborne and Grayer 1988), which have important therapeutic values, including antitumor (Kamei *et al.*, 1995; Koide *et al.*, 1996), antiulcer (Cristoni and Magistretti, 1987), antioxidant and anti-inflammatory (Wang *et al.*, 1999) activities. In cranberry fruits, the anthocyanins are cyanidin and peonidin-based chromophores, whereas the flavonols are myricetin and quercetin-based chromophores (Figure 4.9). Six major anthocyanins in cranberry fruit are cyanidin-3-galactoside, cyanidin-3-glucoside, cyanidin-3-arabinoside, peonidin-3-galactoside, peonidin-3-glucoside, and peonidin-3-arabinoside (Fuleki and Francis, 1968; Sapers and Hargrave, 1983). The five different flavonols present in cranberries are myricetin-3-arbinoside, myricetin-3-diglactoside, quercetin-3-galactoside, quercetin-3-arabinoside and quercetin-3-rhamnoside (Fuleki and Francis, 1971; Lees and Francis, 1971).

Biosynthesis of flavonoids is well characterized in a number of plant systems (Holton and Cornish, 1995; Schwinn *et al.*, 1993), and their production in cranberry plants are assumed to be quite similar. The initial steps in the biosynthetic pathway of anthocyanins are identical to that of flavonols. Both are synthesized from phenylalanine or tyrosine, acetyl-Co A, and carbon dioxide. The enzymes which

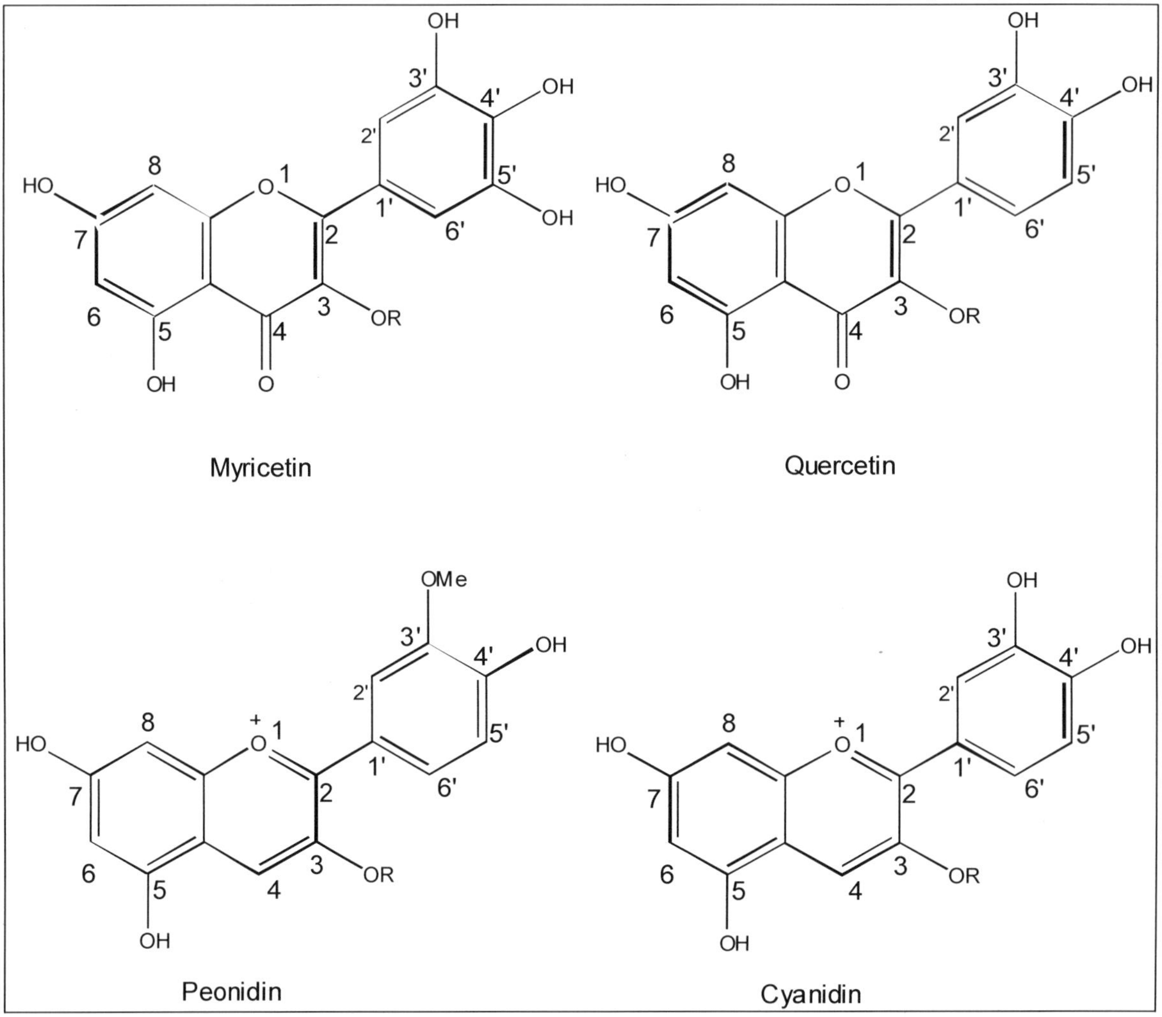

Figure 4.9: The Structural Differences between the Different Types of Anthocyanins and Flavonols Present in Cranberry Plants. Myricetin and quercetin are flavonols and cyanidin and peonidin are anthocyanins. R represents where the sugar groups are attached to the chromophores.

are used for both flavonol and anthocyanin biosynthesis are acetyl-Co A carboxylase, phenylalanine lyase, chalcone isomerase, flavanone-3-hydroxylase, and flavonoid-3-hydroxylase. These enzymes lead to the production of dihydroquercetin. After the formation of dihydroquercetin the pathway branches to either the formation of myricetin, quercetin or two other anthocyanidins (Figure 4.10). It is, therefore, possible that the biosynthetic pathways anthocyanins and flavonols compete for the common precursors. Thus selective biosynthesis of one of these flavonoids (*e.g.*, anthocyanins) may retard the production of the other (*e.g.*, flavonols).

The effects of light on individual as well as total anthocyanin levels in cranberry fruit have been documented, suggesting there may be differences in the composition of anthocyanins in cranberry crops grown under different light conditions (Zhou and Singh, 2002; 2004).

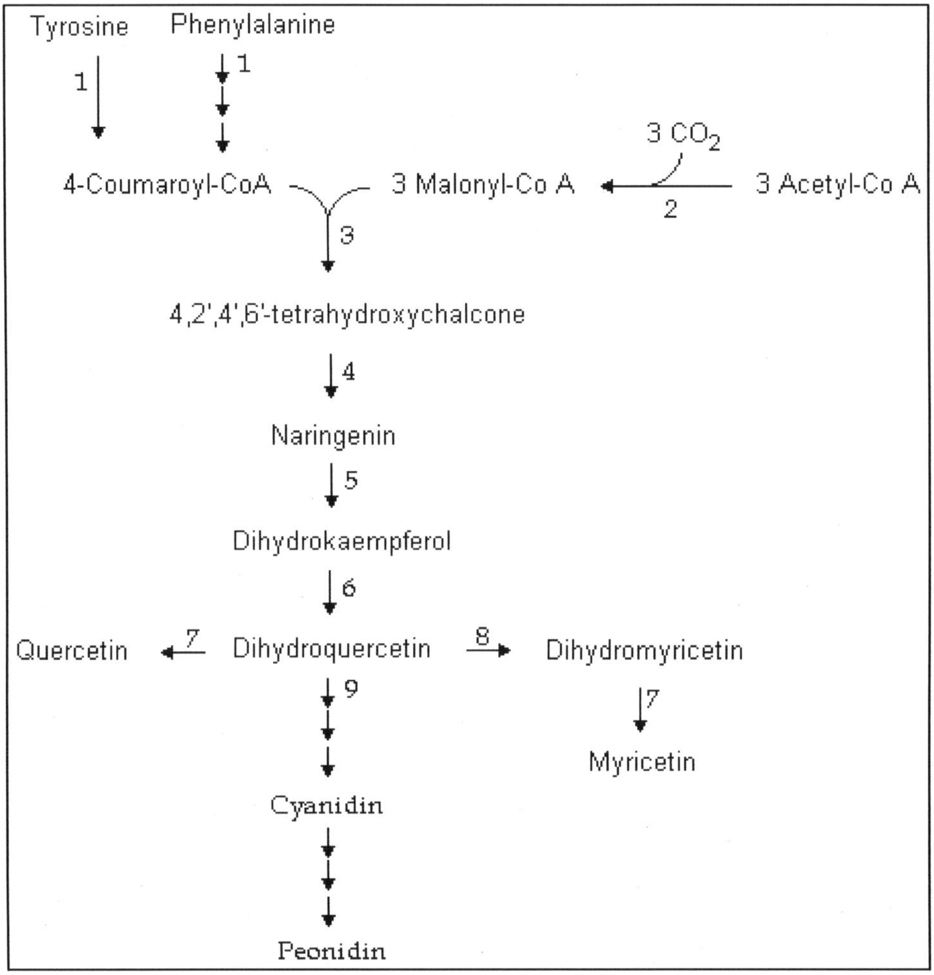

Figure 4.10: Biosynthetic Pathway of Peonidin, Cyanidin, Myricetin and Quercetin

(1) phenylalanine lyase; (2) acetyl-Co A carboxylase; (3) chalcone synthase; (4) chalcone isomerase; (5) flavanone 3-hydroxylase; (6) flavonoid 3'-hydroxylase; (7) flavonol synthase; (8) flavonoid 3',5'-hydroxylase; (9) dihydroflavonol 4-reductase.

Cranberry Cell Culture Studies

Cranberry callus is demonstrated to produce anthocyanins (Madhavi *et al.*, 1995). The effects of light, temperature, and pH on anthocyanin production in cranberry callus have been studied (Zhou and Singh, 2007). Interestingly it was found that anthocyanin-producing cranberry callus was obtained only under conditions of continuous light exposure. Red light and UV light exposure of the callus enhanced anthocyanin content by 41.3 per cent and 29.3 per cent, respectively. Light-dependent anthocyanin production in the callus was regulated by temperature; anthocyanin content in the callus decreased 81.1 per cent at 42°C, 58.9 per cent at 37°C, 47.0 per cent at 30°C, and increased 10.4 per cent at 4°C, compared to the callus maintained at 25°C after 48 hours of incubation at the given temperature. The growth of the callus cultured in medium at pH 7.0 was 6.2-fold higher than in the same medium at standard pH of 5.8 (Zhou and Singh, 2007).

References

Ahuja, S., Kaack, B. and Roberts, J., 1998. Loss of fimbrial adhesion with the addition of *Vaccinum macrocarpon* to the growth medium of P-fimbriated *Escherichia coli*. *J Urol.*, 159: 559–562.

Averill, A.L., Averill, M.M. and DeMoranville, C.J., 1994. Alternative management strategies: Impact of late water on cranberry fruitworm and mites. *Cranberries*, 58(4): 4, 23–25.

Averill, A.L. and Sylvia, M.M., 1998. *Cranberry Insects of the Northeast.* East Wareham, MA: UMass Amherst Cranberry Sta. Ext. Publ. 112 p.

Averill, A.L., Sylvia, M.M., Kusek, C.C. and DeMoranville, C.J., 1997. Flooding in cranberry to minimize insecticide and fungicide inputs. *Amer. J. Alt. Agric.*, 12: 50–54.

Avorn, J., Monane, M., Gurwitz, J.H., Glynn, R.J., Choodnovskiy, I. and Lipsitz, L.A., 1994. Reduction of bacteriuria and pyuria after ingestion of cranberry juice. *J. Amer. Med. Assoc.*, 271: 751–754.

Beckwith, C.S., 1925. Control of the cranberry by submerging in water. *Agr. Exp. Sta. Bull. No.*, 441. New Brunswick, Rutgers Univ., NJ.

Birrenkott, B.A. and Stang, E.J., 1989. Pollination and pollen tube growth in relation to cranberry fruit development. *J. Amer. Soc. Hort. Sci.*, 114: 733–737.

Birrenkott, B.A. and Stang, E.J., 1990. Selective flower removal increases cranberry fruit set. *Hort. Sci.*, 25: 1226–1228.

Botelho, M.R. and Vanden Heuvel, J.E., 2006. Preliminary assessment of the impact of current flooding practices on nonstructural carbohydrate concentrations of cranberry. *Hort. Technol.*, 16: 277–285.

Burger, O., Weiss, E.I., Sharon, N., Tabak, M., Neeman, I. and Ofek, I., 2002. Inhibition of *Helicobacter pylori* adhesion to human gastric mucus by a high-molecular-weight constituent of cranberry juice. *Crit. Rev. Food Sci. Nut.*, 42(Suppl.): 279–284.

Cane, J.H. and Schiffhauer, D., 2003. Dose-response relationships between pollination and fruiting refine pollinator comparisons for cranberry (*Vaccinium macrocarpon* [Ericaceae]). *Amer. J. Bot.*, 90: 1425–1432.

Caruso, F.L., 2008a. Cranberry cultivars. In: *Cranberry Production: A Guide for Massachusetts*, (Eds.) H.A. Sandler, and C.J. DeMoranville. East Wareham, MA: UMass Cranberry Station, CP–08, pp. 72–84.

Caruso, F.L., 2008b. Disease management 2008. In: *Cranberry Chart Book: Management Guide for Massachusetts*, (Eds.) M.M. Sylvia, and N. Guerin. East Wareham, MA: UMass Cranberry Sta. Ext. Publ., p. 1–6.

Caruso, F.L. and Ramsdell, D.C., 1995. *Compendium of Blueberry and Cranberry Diseases*. St. Paul, MN: Amer. Phytopath. Soc. 87 p.

Cockfield, S.D. and Mahr, D.L., 1992. Flooding cranberry beds to control blackheaded fireworm (*Lepidoptera: Tortricidae*). *J. Econ. Entomol.*, 85: 2383–2388.

Cristoni, A. and Magistretti, M.J., 1987. Anti-ulcer and healing activities of *Vaccinium myrtillus* anthocyanosides. IL Farmaco. *Edizione Pratica*, 42: 29–43.

Cross, C.E. and Demoranville, I.E., 1978. Resanding of Massachusetts cranberry bogs (Part I). *Cranberries*, 45(2): 9–11.

Dana, M.N., 1990. Cranberry management. In: *Small Fruit Crop Management*, (Eds.) G.J. Galletta, and D.G. Himelrick. Englewood Cliffs, NJ, Prentice–Hall, pp. 334–362.

Davenport, J.R. and DeMoranville, C.J., 2004. Temperature influences nitrogen release rates in cranberry soils. *Hort. Sci.*, 39: 80–83.

Davenport, J.R., Pitts, M.T., Provance, W. and DeMoranville, C.J., 1997. Influence of soil iron and aerobic status on phosphorus availability in cranberry (*Vaccinium macrocarpon* Ait.) soils. *Acta Hort.*, 446: 369–379.

DeMoranville, C.J., 1992. Cranberry nutrients, phenology, and nitrogen-phosphorus-potassium fertilization. *Ph.D. Thesis*. Plant and Soil Sciences. Univ. Massachusetts, Amherst, 297 p.

DeMoranville, C.J., 2006. Cranberry best management practice adoption and conservation farm planning in Massachusetts. *Hort. Technol.*, 16: 393–397.

DeMoranville, C.J., 2008a. Nutrition management for producing bogs. In: *Cranberry Chart Book: Management Guide for Massachusetts*, (Eds.) M.M. Sylvia, and N. Guerin. East Wareham, MA. Access date: 2 Dec. 2008. http://www.umass.edu/cranberry/services/chartbooks.shtml: UMass Amherst Cranberry Sta. Ext. Publ., pp. 37–46.

DeMoranville, C.J., 2008b. Properties of bog soil. In: *Cranberry Production: A Guide for Massachusetts*, (Eds.) H. A. Sandler, and C. J. DeMoranville. East Wareham, MA: UMass Cranberry Station, CP–08, pp. 11–12.

DeMoranville, C.J. and Davenport, J.R., 1997. Phosphorus forms, rates, and timing in Massachusetts cranberry production. *Acta Hort.*, 446: 381–388.

DeMoranville, C.J., Davenport, J.R., Patten, K.D., Roper, T.R. Strik, B.C., Vorsa, N. and Poole, A.P., 1996a. Fruit mass development in three cranberry cultivars and five production regions. *J. Amer. Soc. Hort. Sci.*, 121: 680–685.

DeMoranville, C.J., Howes, B., Schlezinger, D. and White, D., 2009. Cranberry phosphorus management: How changes in practices can reduce output in drainage water. *Acta Hort.*, 810: 633–640.

DeMoranville, C.J., Sandler, H.A. and Bicki, T., 1996b. Best management practices guide for cranberry production in Massachusetts. UMass Amherst Cranberry Sta. Ext. Publ. http://www.umass.edu/cranberry/services/bmp/. Last accessed: Sept. 10, 2009.

DeMoranville, C.J., Sandler, H.A. and Bicki, T., 1996c. Sanding. Univ. Massachusetts Cranberry Sta. Ext. Publ. http://www.umass.edu/cranberry/services/bmp/. Last accessed: 24 Jan. 2006.

DeMoranville, C.J., Sandler, H.A. and Caruso, F.L., 2001. Planting new cranberry beds. UMass Amherst Cranberry Sta. Ext. Publ. http://www.umass.edu/cranberry/downloads/Planting per cent 20New per cent 20Cranberry per cent 20Beds.pdf. Last accessed: Sept. 10, 2009.

DeMoranville, C.J., Sandler, H.A., Shumaker, D.E., Averill, A.L., Caruso, F.L., Sylvia, M.M., and Pober, D.M., 2005. Fall flooding for management of cranberry fruitworm (*Acrobasis vaccinii*) and dewberry (*Rubus hispidus*) in Massachusetts cranberry production. *J. Crop Prot.*, 24: 999–1006.

Demoranville, I.E., 1984. Weeds of Massachusetts cranberry bogs, Part 1. East Wareham, MA: Univ. of MA. 22 p.

Deubert, K.H. and Caruso, F.L., 1989. Bogs and cranberry bogs in Southeastern Massachusetts. East Wareham, MA: Mass. *Agr. Expt. Sta. Res. Bull.*, 727: 31.

Eastwood, B., 1856. *The Cranberry and its Culture: A Complete Manual for the Cultivation of the Cranberry.* New York, NY: C.M. Saxton, Barker, and Co., 120 p.

Eck, P., 1990. *The American Cranberry.* New Brunswick, NJ: Rutgers Univ. Press., 420 p.

Fellers, C.R. and Esselen, W.B., 1955. Cranberries and cranberry products. East Wareham, MA: Univ. of Mass. *Coop. Ext. Bull.,* 481: 58.

Fitzpatrick, S.M., 2007. Survival of submerged larvae of cranberry girdler, *Chrysoteuchia topiaria*, in the laboratory. *Crop Prot.*, 26: 1810–1816.

Franklin, H.J., 1913. *The Cranberry Girdler.* East Wareham, MA: Mass. Agr. Exp. Sta., 27–28 p.

Fuleki, T. and Francis, F.J., 1968. Quantitative methods for anthocyanins: IV. Determination of individual anthocyanins in cranberry and cranberry products. *J. Food Sci.,* 33: 471–478.

Fuleki, T. and Francis, F.J., 1971. Quantitative methods for anthocyanins. VI. Flavonols and anthocyanins in cranberries. *J. Food Sci.,* 36: 1056–1060.

Gleason, H.A. and Cronquist, A., 1991. *Manual of Vascular Plants of Northeastern United States and Adjacent Canada*, 2nd edn. Bronx, NY: New York Botanical Garden, 910 p.

Greidanus, T. and Dana, M.N., 1972. Cranberry growth related to tissue concentration and soil test phosphorus. *J. Amer. Soc. Hort. Sci.,* 97: 326–328.

Greidanus, T., Peterson, L.A., Schrader, L.E. and Dana, M.N., 1972. Essentiality of ammonium for cranberry nutrition. *J. Amer. Soc. Hort. Sci.*, 97: 272–277.

Hanson, E., 1996. *Fertilizing Fruit Crops.* Michigan State University Ext. Publ. E–852.

Hanson, E. and Hancock, J., 1996. *Managing the Nutrition of Highbush Blueberries.* Michigan State University Ext. Publ. E–2011.

Harborne, J.B. and Grayer, R.J., 1988. The anthocyanins. In: *The Flavonoids: Advances in Research Since 1980*, (Ed.) B.H. Jeffrey. Chapman and Hall, New York, pp. 1–20.

Holton, T.A. and Cornish, E.C., 1995. *The Plant Cell*, 7: 1071.

Howell, A.B., Leahy, M., Kurowska, E. and Guthrie, N.I., 2001. *In vivo* evidence that cranberry proanthocyanidins inhibit adherence of p-fimbriated *E. coli* bacteria to uroepithelial cells. *Fed. Amer. Soc. Exp. Bio. J.*, 15: A284.

Hunsberger, L.K., DeMoranville, C.J., Autio, W. R. and Sandler, H.A., 2006. Uniformity of sand deposition on cranberry farms and implications for swamp dodder control. *Hort. Technol.,* 16: 488–492.

Kosola, K.R., Workmaster, B.A. and Spada, P.A., 2007. Inoculation of cranberry (*Vaccinium macrocarpon*) with the ericoid mycorrhizal fungus *Rhizoscyphus ericae* increases nitrogen flux. *New Phytol.,* 176: 184–196.

Kamei, H., Kojima, T., Hasegawa, M., Koide, T., Umeda, T., Yukawa, T. and Terabe, K., 1995. Suppression of tumor cell growth by anthocyanins *in vitro. Cancer Invest.,* 13: 590–594.

Koide, T., Kamei, H., Hashimoto, Y., Kojima, T. and Hasegawa, M., 1996. Antitumor effect of hydrolyzed anthocyanin from grape rinds and red rice. *Cancer Biother. Radiopharm.,* 11: 273–277.

Lampinen, B.D., DeMoranville, C.J. and Sandler, H.A., 2000. *Irrigation Management.* East Wareham, MA: UMass Extension, Best Management Practices Guide.

Lees, D.H. and Francis, F.J., 1971. Quantitative methods for anthocyanins: 6. Flavonols and anthocyanins in fresh cranberries. *J. Food Sci.*, 36: 1056.

Marucci, P.E. and Filmer, R.S., 1964. Preliminary cross-pollination tests on cranberries. In: *Proceed. Ann. Meeting Amer. Cranberry Growers Assoc.*, 94: 48–51.

McKay, D.L. and Blumberg, J.B., 2007. Cranberries (*Vaccinium macrocarpon*) and cardiovascular disease risk factors. *Nutr. Rev.*, 65(11): 490–502.

Neto, C.C., 2007. Cranberry and its phytochemicals: A review of *in vitro* anticancer studies. *J. Nut.*, 137: 186S–193S.

Oudemans, P.V., Caruso, F.L. and Stretch, A.W., 1998. Cranberry fruit rot in the Northeast: A complex disease. *Plant Dis.*, 82: 1176–1184.

Pilcher, L.S., 1985. Appraising sites for cranberry culture using seasonal temperature and daylength data. *M.S. Thesis*. Environmental Studies. Western Washington Univ., Bellingham, WA.

Prasain, J.K., 2008. Effect of cranberry juice concentrate on chemically-induced urinary bladder cancers. *Oncol Rep.*, 19(6):1565–70.

Reed, J., 2002. Cranberry flavonoids, atherosclerosis and cardiovascular health. *Crit. Rev. Food Sci. Nut.*, 42(Suppl.): 301–316.

Roper, T.R., 2009. Mineral nutrition of cranberry: What we know and what we thought we knew. *Acta Hort.*, 810: 613–525.

Rosen, C.J., Allan, D.L. and Luby, J.L., 1990. Nitrogen form and solution pH influence growth and nutrition of two *Vaccinium* clones. *J. Amer. Soc. Hort. Sci.*, 115: 83–89.

Sandler, H.A., 2003. Efficacy of nonchemical options for the control of dodder (*Cuscuta gronovii*) and several broadleaved species in commercial cranberry production in Southeastern Massachusetts. In: *Northeast Ecology and Evolution Conference Proceedings, Vol. 1.* New Brunswick, NJ, Rutgers Univ., pp. 23 .

Sandler, H.A., 2004. Factors influencing the colonization and establishment of plant species on cranberry bogs. *Ph.D. Diss.* Plant and Soil Sciences. Univ. Massachusetts, Amherst.

Sandler, H.A., 2008. Botany and basic farm features. In: *Cranberry Production: A Guide for Massachusetts* (Eds.) H.A. Sandler, and C.J. DeMoranville. East Wareham, MA: UMass Cranberry Station, CP–08, pp. 3–6.

Sandler, H.A. and DeMoranville, C.J., 2009. Economic analysis of nitrogen rate on vine production and fruit yield of pruned cranberry beds. *Hort. Technol.*, 19: 572–579.

Sandler, H.A., DeMoranville, C.J. and Autio, W.R., 2004. Economic comparison of initial vine density, nitrogen rate, and weed management strategy in commercial cranberry. *Hort. Technol.*, 14: 267–274.

Sandler, H.A., Else, M.J. and Sutherland, M., 1997. Application of sand for inhibition of swamp dodder (*Cuscuta gronovii*) seedling emergence and survival on cranberry (*Vaccinium macrocarpon*) bogs. *Weed Technol.*, 11: 318–323.

Sandler, H.A. and Mason, J., 2004. Efficacy of flooding for the control of dodder (*Cuscuta gronovii*) and several broadleaf species in commercial cranberry production in Southeastern Massachusetts: A two-year study. In: *Proc. Northeast Weed Sci. Soc.*, 58: 163.

Sapers, G.M. and Hargrave, D.L., 1987. Proportions of individual anthocyanins in fruits of cranberry cultivars. *J. Am. Soc. Hort. Sci.*, 112: 100–104.

Sapers, G.M., Graff, G.R., Phillips, J.G. and Deubert, K.H., 1986. Factors affecting the anthocyanin content of cranberry. *J. Amer. Soc. Hort. Sci.*, 111: 612–617.

Schwinn, K.E., Markham, K.R. and Given, N.K., 1993. Floral flavonoids and the potential for pelargonidin biosynthesis in commercial chrysanthemum cultivars. *Phytochem.*, 35: 145–150.

Shawa, A.Y., Shanks, C.H., Bristow, P.R., Shearer, M.N. and Poole, A.P., 1984. Cranberry production in the Pacific Northwest. Pullman, WA: Pac. *Northwest Coop. Ext Bull.*, PNW–247, 50 p.

Smith, J.D., 1993. Uptake and utilization of nitrogen sources by cranberry plants (*Vaccinium macrocarpon* Ait.). *Ph.D. Diss.*, Dept. of Horticulture. University of Wisconsin, Madison, WI.

Stackpoole, S., 2008. Nitrogen dynamics in cranberry beds. *Ph.D. Diss.* Dept. of Horticulture, University of Wisconsin-Madison, Madison, WI.

Stribley, D.P. and Read, D.J., 1974. The biology of mycorrhiza in the Ericaceae IV. The effect of mycorrhizal infection on uptake of N-15 from labeled soil by *Vaccinium macrocarpon* Ait. *New Phytol.*, 73: 1149–1155.

Strik, B.C. and Poole, A.P., 1992. Alternate-year pruning recommended for cranberry. *Hort Sci.*, 27: 1327.

Strik, B.C. and Poole, A.P., 1995. Does sand application to soil surface benefit cranberry production? *Hort Sci.*, 30: 47–49.

Suhayda, B., DeMoranville, C.J., Sandler, H.A., Autio, W.R. and Vanden Heuvel, J.E., 2009. Sanding and pruning differentially impact canopy characteristics, yield and economic returns in cranberry. *Hort Technol.*, 19: 796–802.

Tomlinson, B., 1937. Proper sanding of great importance in good bog management. *Cranberries*, 1(9): 4,8–11.

Upton, Roy (Ed.) 2002. Cranberry Fruit: Standards of analysis, quality control, and therapeutics. *American Herbal Pharmacopoeia*, Santa Cruz, CA. 30 pp.

Vanden Heuvel, J.E. and DeMoranville, C.J., 2009. Competition between vegetative and reproductive growth in cranberry. *Hort Sci.*, 44: 322–327.

Vitosh, M.L., Johnson, J.W. and Mengel, D.B., 1995. Tristate fertilizer recommendations for corn, soybeans, wheat, and alfalfa. Ohio State University Extension Pub. E–2567. www.ohioline.osu.edu/e2567/index.html. Last accessed: Sept. 10, 2009.

Vorsa, N. and Novy, R.G., 1994. DNA fingerprinting the American cranberry. *Cranberries*, 58(11): 10–15.

Wang, H.B., Nair, M.G., Strasburg, G.M., Chang, Y.C., Booren, A.M., Gray, J.I. and DeWitt, D.L., 1999. Antioxidant and anti-inflammatory activities of anthocyanins and their aglycon, cyanidin, from tart cherries. *J. Nat. Prod.*, 62: 294–296.

Weiss, E.I., Lev-Dor, R., Sharon, N. and Ofek, I., 2002. Inhibitory effect of high-molecular-weight constituent of cranberry on adhesion of oral bacteria. *Crit. Rev. Food Sci. Nut.*, 42(Suppl.): 285–292.

Yan, X., Murphy, B.T., Hammond, G.B., Vinson, J.A. and Neto, C.C., 2002. Antioxidant activities and antitumor screening of extracts from cranberry fruit (*Vaccinium macrocarpon*). *J. Agric. Food Chem.*, 50: 5844–5849.

Zhou, Y. and Singh, B.R., 2002. Red light stimulates flowering and anthocyanin biosynthesis in American cranberry. *Plant Growth Reg.*, 38: 165–171.

Zhou, Y. and Singh, B.R., 2004. Effect of light on anthocyanin levels in submerged, harvested cranberry fruit. *J. Biomed. Biotechnol.*, 5: 259–263.

Zhou, Y. and Singh, B.R., 2007. Characterization of environmental stress-regulated anthocyanin production and growth of cranberry callus. *J. Appl. Hort.*, 9(1): 17–21.

2011, Biodiversity in Horticultural Crops Vol. 3 *Pages* **74–84**
Editor: **Professor K.V. Peter**
Published by: **DAYA PUBLISHING HOUSE, NEW DELHI**

Chapter 5

Jamun

A.K. Singh* and Anju Bajpai
Central Institute of Subtropical Horticulture,
Rehmankhera, P.O. Kakori, Lucknow – 227 107, U.P.
**E-mail: aksinghcish@rediffmail.com*

Global food security has become increasingly dependent on only a handful of crops. Over 50 per cent of the global requirement for proteins and calories are met by just three crops- maize, wheat and rice. Only 150 crops are traded on a significant global scale. Yet, surveys indicate that there are over 7000 plant species across the world which are cultivated or harvested from the wild for food. These neglected and underutilized species play a crucial role in the food and nutrition security, income generation and food culture of the rural poor. Lack of attention in these crops has resulted in under-estimated and under-exploited potential value. It also places them in danger of continued genetic erosion and disappearance. This would further restrict development options for the poor.

The genus *Syzygium* comprises about 1000 species of evergreen trees and shrubs, most of them are tropical in origin. Jamun is found in Western Ghats and very extensively in the tropical region. Diversity is due to high heterozygosity and seed multiplication. Some of the species are described in Table 5.1.

Jamun is a common, large, evergreen and important fruit tree of Indian subcontinent. It falls in the broad basket of 'minor crops'. The use of the two terms neglected and underutilized pinpoints and highlights the less degree of attention paid by users and the level of research and conservation efforts spent on the fruit. The scientific name of Jamun is *Eugenia jambolana* or *Syzygium cumini* L. and it belongs to family Myrtaceae. Common names are java plum, black plum, jambul and Indian blackberry. Fruits are generally round in shape, deep purple or bluish in colour, having juicy, sweet pulp and a small stone.

Table 5.1: Species of *Syzygium*, Common Names and Distribution

Species	Common Name and Use	Distribution
S. aquem	Watery, Rose-apple, fruits edible	A small tree distributed in Assam and Meghalaya
S. amottlanum	Produces edible fruits	Western Ghats, The Nilgris, Palni and Anamalai hills
S. aromaticum	Clove, dried flower buds are of commercial importance	Evergreen trees cultivated in Tamil Nadu and Kerala
S. claviflorum	Fruits are acidic and edible	Andamans
S. cumini	Java plum, Jamun, Jambu	Throughout India
S. fruticosum	Wild jamun	Avenue tree
S. jambos	Rose-apple	Many parts of India
S. mappaceum	Grown as ornamental plant	Assam, Meghalaya, Arunachal Pradesh and Tamil Nadu
S. samarangense	Wax Jambu, fruits edible	Andamans and many parts of India
S. zeylanicum	Aromatic fruits are edible	Maharashtra, Karnataka, Orissa, Kerala and Andamans

Origin and Distribution

Jamun is one of the important but under exploited indigenous fruit crops of India. It originated from Indonesia and India; now grows abundantly in Southern Asia (Periyathambi, 2007). There are about 400 to 500 species of which a few provide edible fruits. Some species may be globally distributed, but tend to occupy special niches in the local ecology and in production and consumption systems. It is also found in Thailand, Philippines, Madagascar and a few other countries. The jamun is successfully introduced in many other tropical countries like West Indies, California, Algeria and Israel. In India, maximum number of jamun trees are found scattered throughout the tropical and subtropical regions. It also occurs in the lower ranges of the Himalayas up to an elevation of 1300 m. and in the Kumaon hills up to 1600 m. It is widely grown in the larger part of India from the Indo-Gangetic plains in the North to Tamil Nadu in the South (Chaturvedi, 1956, Singh *et al.*, 1967). While Jamun continues to be maintained by cultural preferences and traditional practices, they remain inadequately characterized and neglected by research and conservation. Like other underutilized crops, Jamun was once more widely grown but are falling into disuse for a number of reasons. Farmers and consumers are using these crops less because they are in some way not competitive with other crop species in the same agricultural environment. The decline of these crops may erode the genetic base and preventing the use of distinctive useful traits in crop adaptation and improvement. It is one of the most hardy fruit crops and can easily be grown in neglected and marshy areas where other fruits plants cannot be grown successfully.

Area and Production

Due to neglect and underestimation, Jamun is grown primarily in the centers of origin or centers of diversity by traditional farmers, where they are still important for the subsistence of local communities. It is not grown on a commercial scale but tends to occupy special stretches and pockets in the local ecology and village production system. As such they are common sites in roadside avenues and wind breaks. The production in India is mainly concentrated in the dry regions and the produce is collected by the villagers and sold in the local market. The production potential of Jamun depends on the demand for the products. Its cultivation can be introduced in arid and semiarid, resource-poor and wastelands areas where other fruit crops cannot be grown. Lack of commercial plantings is responsible for the produce to be largely collected and sold in village and nearby markets.

Biodiversity

In India, fruits like jamun (*Syzygium cuminii*) have a lot of diversity which are grown in a wide range of ecological situations, under tropics, subtropics and temperates. They are grouped as underutilized. Jamun fruits yield juice with excellent flavour, which can be converted into blended beverages and these could play an important role in meeting demand for nutritious, pleasantly flavoured and attractive natural food of high therapeutic value. Encouraging local people to produce these fruits helps in significantly, checking uncontrolled harvesting from wild and assisting in retention of various species in their native habitats where they perform the best.

It is well known that genetic resources are the building blocks of new varieties and are the common heritage of mankind. A few of the valuable genetic resources are eroding fast under varying pressures. Collection, conservation and utilization of Jamun and related species must, therefore, receive considerable attention. A large number of *spp.* of these fruits are conserved *in situ* at different locations. Due to its enormous size, large areas are required for their conservation. Erosion of genetic resources due to biotic and abiotic stresses, variability and poor viability of seed, vegetative propagation, urbanization and indiscriminate land utilization and resource crunch are a few of the constraints in effective conservation of these genetic resources.

The native diversity, in particular, occurs largely in small farms/orchards, in backyards, roadside avenues and marginal lands across the country, which has predominantly arisen from seedling origin. The choice of seedlings is for multipurpose trees which promote socioeconomic development, raising rural income and those that could be managed by small farming communities on a sustainable basis. Owing to cross-pollination and seed propagation, there is enormous variability in its fruit morphology, fruit quality, maturity and productivity. A number of seedling strains with lot of variation in respect of fruit shape and size, pulp colour, TSS, acidity and earliness particularly in Uttar Pradesh, Gujarat ad Maharashtra provide a good scope for selection of better varieties. Lancaster Percy and Bose, (1965) noted that F1 hybrid between the Alba variety of water apple *(S. javanicum)* and rose apple *(S. jambos)* had prolific bearing with large fruits than those of the parents. The fruits had fragrance and sweetness of rose apple. Survey in Pune and Ahmednagar districts of Maharashtra revealed wide variation in fruit weight (3.5 to 16.5 g), pulp content (54.29-85.71 per cent), TSS (4.5-17 per cent) and acidity (0.16 - 0.55 per cent) from which promising types No.15,4, 14 and 13 were identified (Keskar *et al.*, 1989a). Ashraf *et al.* (1987) reported that fruit shape in jamun varied from round to oblong and apex of fruits from flat to pointed. He also observed great variability in physico-chemical characteristic of fruits offering possibility of selecting a variety suitable for fresh market and processing. Small seed size and high pulp content with better chemical qualities are ideal characteristics. Singh *et al.* (1999) evaluated eight genotypes of jamun under Faizabad conditions and reported that oblong types had more fruit weight and relatively less seed weight. Among the locally available genotypes of jamun in West Bengal, Selection No.1 (oval shaped large fruit) and Selection No.2 (Cylindrical shaped medium sized fruit) are better on the basis of yield and fruit quality attributes (Kundu *et al.*, 2001). A survey was undertaken in Karnataka state to investigate the nature and extent of variability in jamun seedling progenies for morphological characters of tree. High degree of variability was observed for plant girth, leaf area, petiole length and leaf length to petiole length ratio (Prabhuraj *et al.*, 2002). A survey was carried out in north Goa to identify elite genotypes of jamun and individual fruit weight ranged 3.42 to 13.67 g, length 3.31 to 5.26 cm, girth 5.21 to 9.82 cm, pulp (per cent) 58.57 to 84.55, TSS 12.00 to 26.80

per cent, titratable acidity 0.59 to 1.63 per cent, total sugars 6.87 to 25.31 per cent and sugar: acid ratio from 15.39 to 27.92 (Devi *et al.*, 2002).

Many of these species are also now taken up for improvement through research and development efforts by the national programmes like National Network Project on Underutilized Fruits initiated by ICAR, New Delhi. Diversity in fruit attributes remains poorly represented in Jamun. Gaps for collecting the diversity within the region exist, both in terms of areas and species, and within species variability as such, in Jamun.Local knowledge on the use and conservation of native diversity is poorly recorded. Proper assessments through detailed genetic diversity studies have not been carried out. Such assessments would lead to effective utilization and conservation of *Syzygium* and Myrtaceae fruit species.

Ecogeographic Survey to Collect and Assess Genetic Diversity in Jamun spp.

The distribution of *Syzygium* spp. family Myrtaceae, in natural habitats points out that either the species diversity is widely distributed or is scattered in small pockets, and often several species may be truly endemic. A good deal of data compiled from national/regional herbaria will be helpful in providing information on the site, terrain, ecology, climate, phenology, uses, etc. There are about 400 to 500 species of which a few provide edible fruits (Chundawat, 1990). Genus *Syzygium*, consists of about 75 indigenous species, a few are of commercial significance. The *Syzygium cumini* Skeels is a large, evergreen tree, with dark purple date-like fruits with prominent elongated seeds. Other related species are: *S. jumbos* (rose-apple or safed jamun). It is commonly found in Assam, Bihar, West Bengal and Tamil Nadu. Its fruits are light yellow with fragrance and sweetness of rose-apple. Other species *S. fruiticosum* and *S. uniflora* bear small-sized edible fruits, whereas *S. densifolium* is free from termite attack. It is suitable for use as rootstock for *S. cumini*. It is a large, evergreen tree, producing dark purple, oval delicate fruits with prominent seeds. Flowers are borne in the axils of leaves of branchlet and are hermaphrodite. Some of the species are described below:

Syzygium jambos (Rose Apple or Sated Jamun)

The tree is ornamental. The fruits are light yellow- white in colour. They are rose scented and calyx persistent. The trees are medium sized and evergreen. The leaves have very small petiole. The seeds are polyembryonic and it is being grown in Assam, Bihar, Andhra Pradesh, Tamil Nadu, West Bengal, Coastal areas of Maharashtra and Gujarat.

Syzygium fruitecosum

The trees are suitable for windbreak. Fruits are edible and small in size.

Syzygium javanica (Water Apple)

This species is found in South India and West Bengal.

Syzygium densiflora

It is suitable for use as rootstock for *Syzygium cuminii*. It is resistant to attack of termites.

Syzygium uniflora (Surinam Cherry or Pitanga Cherry)

It is also a small tree and bears small-sized fruits having bright red colour and aromatic flavour. The tree is found in South India.

Syzygium zeylanica

The tree is small and bears edible fruits. It is found in Western Ghats of India. Jamun is an evergreen tree - attaining medium to large size. The trunk is thick and greyish white in colour. Leaves opposite, simple, mostly firm and glossy, elliptic, pinnately veined with lateral veins close together; flowers in a few flowered panicles. Flowers are hermaphrodite, light yellow in colour, borne in the axils of leaves on branchlet, calyx lobes 4, calyx tube not extending beyond summit of ovary, petals 4, white, spreading, stamens many, ovary inferior, 2-celled (Bailey and Bailey, 1978).

Study and Classification of Genetic Diversity through Use of Morphological and Molecular Methods

The use of modern techniques in assessing diversity to quantify genetic diversity in priority gene pools and classification of the available diversity is important. The use of biochemical and molecular techniques along with morphometric methods provides clues to demarcate *in situ* conservation sites, and for better understanding of the extent of diversity.

In Jamun, molecular analysis was initiated by RAPD primers. Good differentiation and diverse genetic bases of Indian accessions were demonstrated. As J-28, J-30 and J-33 are selections from Karnataka (Arvhavi), these show close affinities among each other. This demonstrates agroecological adaptation to be important for clustering. The dendogram analysis clearly grouped the accessions as per their different geographical regions. The seedling selection from the same geographical regions where they have been traditionally grown over the years normally group together.

Development of Seed Conservation Technology for Jamun

Jamun possesses recalcitrant seeds which are difficult to conserve. These are also vegetatively propagated and *in vitro* conservation techniques are important for this medicinally important plant.

Budding

In-situ budding was performed on young seedlings of jamun at Central Institute for Subtropical Horticulture, Lucknow with a success of 68 per cent in February (Singh *et al.*, 2007). In India, shield, patch and Forkert methods of budding are generally employed. Singh *et al.* (1979) at Basti (Uttar Pradesh) noted 70 per cent success by patch budding in March. Patch budding was successful when performed in June under Gujarat conditions (Chovatia and Singh, 2000).

Soft Wood Grafting

Soft wood grafting was performed on 8-10 months old seedlings using cleft method in every month of the year at Central Institute for Subtropical Horticulture, Lucknow and minimum days taken for bud sprout were recorded in February and March (16 days). Maximum success was recorded during March, April and July (80 per cent) of grafting (Singh *et al.*, 2007). This method was found to be better for *in situ* grafting under rain fed conditions. Madalageri *et al.* (1991) reported that soft wood grafting was successful method for multiplication of jamun under Karnataka conditions. Chovatia and Singh, (2000) recorded 41.67 per cent success through soft wood grafting during June under Gujarat conditions.

Inarching

Jamun can be propagated by inarching but it is not adopted commercially. In June-July, one-year-old rootstocks are inarched with the matching thickness of scion.

Veneer Grafting

Veneer grafting was performed on 8-10 months old rootstocks in every month of the year at Central Institute for Subtropical Horticulture, Lucknow and success rate was 60 per cent in February-March followed by 50 per cent in September (Singh *et al.*, 2007). Bose *et al.* (2001) recorded 31 per cent success in veneer grafting when one-year-old seedlings were used as rootstock. The shoots were taken from spring flush and the operation was done in July.

Airlayering

At Annamalai (A.P.), all the air layered shoots treated with IBA and NAA in Lanolin paste at 0.1 per cent rooted where as untreated layers failed to root (Shanmugavelu, 1967).

Cutting

Tip cuttings of *S. javanicum* rooted well under intermittent mist and treatment with IBA at 5000 ppm produced 100 per cent rooting. Shanmugavelu (1966) reported 45 per cent rooting with 100 ppm IBA and 40 per cent with 100 ppm IAA treatments.

Top Working

The old, unproductive trees or damaged young plants can be rejuvenated easily by top working, employing scion wood or buds from superior clone.

Micropropagation

Yadav *et al.* (1990) induced multiple shoots from nodal and shoot tip segments of 10-15 days old seedlings of *S. cuminii* on modified MS medium supplemented with BA singly and in combination with NAA or IBA. Excised shoots were placed for root induction on MS medium containing NAA or IBA and then transferred to MS basal medium to form complete plantlets. The regenerated plantlets were acclimatized and successfully transferred into the soil. Propagation from elite mother trees is very important. Roy *et al.* (1996 a) induced multiple shoots from nodal explants of l0-year-old elite trees of *S. cumini* on MS medium supplemented with 2.5 mg Kinetin/l. Explants taken from *in vitro* - proliferated shoots produced further multiple shoots when cultured on the same basal medium also containing 2.5 mg kinetin/l. Repeated subculture resulted in rapid shoot multiplication at an average of 10 shoots/subculture. Jain and Babbar (2000) obtained multiple shoots from the epicotyl segments bearing scaly leaves, excised from *in vitro*- grown seedlings of *S. cuminii*, on MS medium supplemented with different concentrations of BA. An average of 8.6 shoots/explant were produced in 60 days after inoculation, following transfer to fresh medium after 30 days. The shoots were excised and the residual explants were transferred to fresh medium, where they again developed shoots. A protocol was developed to raise plants of *S. cuminii* at any time of the year. Somatic embryogenesis is found successful for multiplication of jamun plants (Litz, 1984). A successful micropropagation protocol of *Syzygium cuminii* L. Skeels. was developed at CISH, Lucknow using nodal explants from mature tree.

Conservation and Research for Breeding

Success of the breeding programs of major crops depends by and large on the amount of genetic diversity collected during hundreds of germplasm collecting missions. Almost all kinds of fruits are grown in India due to prevalence of varied ecological conditions. Conservation of fruit germplasm is done for current and future uses and comprises entire diversity of primary, secondary and tertiary gene pool. Almost all underutilized fruits were earlier in a state of neglected condition but Pareek (1988) did a comprehensive- review of genetic resources activities of arid zone fruits. Jamun was not

covered under the programme. There are no standard varieties available in India, however a few of common types grown in north India are known as *Rajamun*. A type having large sized fruits is known as 'Paras' in Gujarat. A selection known as Narendra Jamun-6 was identified with desirable traits at Faizabad. However, it has not been popularised.

In recent years under National Network Project on Underutilized Fruits (NNPUF) the work on jamun is carried out in Central Institute for Subtropical Horticulture, Lucknow. The following varieties/accessions were identified and established in field gene bank. The description of varieties/accessions are as follows:

CISH J-37

A superior genotype selected from Central Institute for Subtropical Horticulture, Lucknow and established in field gene bank. Average fruit weight recorded 24.05g, length 3.90 cm and diameter 3.03, pulp 92.26 per cent, TSS 16.4 °Brix, ascorbic acid 49.88mg/100g and antioxidant 1467.9μg/100ml AEAC unit (Singh *et al.*, 2007).

CISH J-42

Seedless genotype selected at Central Institute for Subtropical Horticulture, Lucknow during 2008 and multiplied by vegetative propagation and established in field gene bank. Average fruit weight recorded 6.87g, length 2.57 cm, pulp, 97.9 per cent, TSS 14.67 °Brix and ascorbic acid 34.14mg/100g (Singh *et al.*, 2007).

Jamun GJ 40

It was collected from Ode village of Anand district of Gujarat. Peak period of flowering is March. It ripens in the 3rd week of May and recorded 22.00 g average fruit weight, 86.00 per cent pulp, 14.10 per cent TSS, 0.37 per cent acidity, 10.00 per cent total sugar and 43.10 mg/100g vitamin C.

Jamun GJ-2

It was collected from Ode village of Anand district of Gujarat. Peak period of flowering is March. It ripens in the 4th week of May and recorded 20.10g average fruit weight, 85.57 per cent pulp, 17.50 per cent TSS, 0.37 per cent acidity, 12.40 per cent total sugar and 46.10 mg/100g vitamin C.

Jamun G J-2

Rajamun

It bears large-sized (length 2.5-3.5cm), oblong, deep purple fruits having purple pink, juicy and sweet pulp and small seed. Fruit ripens in June-July and possesses small size stone.

Late Maturing Variety

Bears a small sized (length 1.5-2.0 and diameter 1-1.5cm), slightly round fruit, deep purple or blackish in colour at full ripe stage. The stone present in this cultivar is comparatively large in size. Fruits ripen in August.

A Seedling Selection 'Paras'

A seedling selection `Paras' yields sweet fruits. Considerable tree variation exists in the Pune and Ahmednagar districts of Maharashtra. Extracts of stems, leaves, buds and flowers possess moderate antibiotic activity against *Micrococcus pyogenesvar, aureus.*

Complementary Conservation for Meeting New Needs

As for those accessions which are collected and placed in *ex situ* collections, adequate measures are taken for their multiplication and evaluation. Still efforts need to be directed towards better maintenance of the resource base, both through *ex situ* and *in situ* conservation methods, to ensure the development and sustainable use by present and future generations. Ability of modern technologies to transform crops and other plants into diverse products, to extend their shelf life, to improve their transformation by agro-industries are also new opportunities to strengthen the marketing of these species and their products.

Improving Nutrition Levels

Jamun does fit into scheme of things as it is "species with underexploited potential for contributing to food security, nutrition, health, income generation and environmental services."It is nutritionally rich and is adapted to low input agriculture. The erosion of these species can have immediate consequences on the nutritional status and food security of the poor. Wild, semi-domesticated or fully cultivated, these species are important for the well being of local communities. Their enhanced use can bring about better nutrition and fight hidden hunger.

Improving Incomes

Growing market opportunities for Jamun may generate additional income to those poor farmers in less favored environments where these crops have comparative advantages over staple or major crops. It is a typical example of a highly nutritive species whose versatile applications are still largely untapped by the market. Unripe fruit is used for making vinegar. The juice is also criminating, diuretic and gives a soothing effect on human digestive system. Fully ripe fruits are eaten fresh and can be processed into beverages, jelly, jam, squash, wine, vinegar and pickles (Ochse *et al.*, 1961). Juice of ripe fruit is used for preparing sauces and beverages. It is also dried with salt and preserved as a digestive powder or churan. Jamun fruit reduces sugar in blood and is very good for the control of diabetes. Its seeds contain Glucoside, Jamboline and Ellagic acid, which have the ability to check conversion of starch into sugar in case of excess production of glucose. Jamun seeds are also used as a remedy for diabetes. The seed and bark have several applications in Ayurveda, Unani and Chinese systems of medicine. The seed is also rich in protein and carbohydrates. It also contains traces of calcium. These are, therefore, widely used as cattle feed, a medicine against diabetes and antidote in case of soft-food poisoning.

Improvement of Environmental Health

Climate change and the degradation of land and water resources have led to a growing interest for crops and species, adapted to difficult environments. Many neglected and underutilized species occur wild or under cultivation in these areas. In many instances, these species are the only crops that can cope with such harsh environments unfit for other crops, where they can provide sustainable productions. In this way, they contribute significantly to maintain diversity rich and hence more stable agro-ecosystems. Generally jamun trees are not manured because they can stand a good deal of neglect. However, an annual dose of about 20 kg of FYM during the per-bearing period and 50-80 kg/

tree to bearing trees is considered beneficial. On very rich soils, the tree has a tendency to put on more vegetative growth with the result that the fruiting is delayed. Under such conditions, the trees should not be manured and irrigation should also be given sparingly and withholding in September to October and again in February to March. This helps in fruit bud formation and in blossoming and fruit setting.

Developing Cooperative Links with the ICUC and other Organisations

International Centre for Underutilized Crops (ICUC) is developing a research and development network on tropical fruits in Asia with genetic resources as one element. NNPUF undertaken by ICAR is an excellent effort to collect, conserve, evaluate and document diversity from various centers across the country. The status of collecting and conserving Jamun in various centers under NNPUF is discussed below:

Table 5.2: Status Information of Jamun Collection at Different Centres of NNPUF

Sl.No	Name of Centre	Source	Total Collection
1.	CISH, Lucknow	A total of 43 superior genotypes was identified *in situ* on the basis of quality attributes from Lucknow, Barabanki, Unnao, Raebareilly, Faizabad, Saharanpur, Muzaffarnagar, Badot, Gautam Budh Nagar and Mathura districts of U.P., Jhajjar, Sonipat, Karnal and Faridabad districts of Haryana, Vadodhara, Godhra and Anand districts of Gujarat, Vengurla and Belgaum, Gokak and Arbhavi areas of Karnataka.	43
2.	CHES, Chettalli	Bahaduli, No.85, No.44, No.58, Gokak-1, Gokak-2, Gokak-3, Gokak-4, Bhubaneshwar, Vengurla, Krishnagiri, Sira, Trichy, twenty five accessions from Kerala and three accessions from Hosur.	41
3.	CHES, Godhra	Genotypes like Konkan Bahadaoli, Gokak 1, Gokak 2 and Gokak 3 were collected from FRS Vengurla, Maharashtra, Seedless and Seeded jamun from Salem, Tamil Nadu. Twenty five plants of jamun were collected from CISH, Lucknow.	66

Prospects for Jamun

The extreme richness of genetic resources in Jamun along with other fruits can be used to modify human food structure, improving health, even longevity of human beings. It is, therefore, necessary to make more efforts in collecting and preservation of Jamun germplasm as well as undertake studies on their uses. The loss of biodiversity poses serious threat to agriculture and livelihoods of millions of people. Conserving biodiversity and using it wisely are global imperatives. Biodiversity provides the foundation for our agricultural systems.The potential of the genetic resources will be fully tapped in this way. In addition, it is also necessary to intensify international cooperation and exchange of information and germplasm. There is need to integrate/coordinate national and regional efforts to promote and strengthen the PGR activities in underutilized fruits like Jamun.A bright future is expected in all these activities.

References

Ashraf, S.M., 1987. Studies on postharvest technology of jamun fruits. *Ph.D. Thesis*, Department of Horticulture, NDUA&T, Kumarganj, Faizabad.

Bailey, L.H. and Baily, E.J., 1978. *Hortus Third: A Condensed Dictionary of Plants Cultivated in the United States and Canada*. MacMillan Publishing Co., New York, pp. 457–458.

Bose, T.K., Mitra, S. K. and Sanyal, D., 2001. *Fruits: Tropical and Subtropical,* Vol. II. Naya Udyog, Calcutta, India, pp. 643–656.

Chaturvedi, M.D., 1956. Jamun is another of our prized trees. *Indian Fanning,* 5: 17–19.

Chovatia, R.S. and Singh, S. P., 2000. Effect of time on budding and grafting success in jamun *(Syzygium cuminii* Skeels). *Indian J. Hort.*, 57(3): 255–258.

Chundawat, B.S., 1990. *Arid Fruit Culture*. Oxford and IBH Publishing Co. Pvt. Ltd., New Delhi, pp. 165–171.

Devi, S.P., Thangam, M., Desai, A.R. and Adsule, P.G., 2002. Studies on variability in physico-chemical characters of different jamun (*Syzygium cuminii* Skeels) accessions from Goa. *Indian J. Hort.,* 59(2): 153–156.

Jain, J. and Babbar, S.B., 2000. *Plant Cell Respiration,* 19: 519–524.

Keskar, B.G., Karale, A.R., Dhawale, B.C. and Chaudhary, K.G., 1989a. Improvement of Jamun (*Syzygium cuminii* Skeels*)* by selection. *Maharashtra J. Hort.*, 4: 117–120.

Kundu, S., Ghosh, D.K. and Maiti, S.C., 2001. Evaluation of some local types of jamun (*Syzygium cuminii* Skeels) of west Bengal. *Environment and Ecology,* 19(4): 872–874.

Lancaster Percy, S. and Bose, T.K., 1965. Studies on hybrids of *Syzygium cuminii* Skeels. *Indian J. Hort.,* 22: 87–88.

Litz, R.E., 1984. *Hort Science,* 19: 720–722.

Madalageri, M.B., Patil, V.S. and Nalawadi, U.G., 1991. Propagation of jamun (*Syzygium cuminii* Skeels) by soft wood wedge grafting. *My Forest,* 27(2): 176–178.

Ochse, J.J., Soule, M.J. Jr., Dij, M.J., Kman and Wehlberg, C., 1961. *Tropical and Subtropical Agriculture.* McMillan, New York.

Pareek, D.P., 1988. Present status and future needs for genetic resources in arid zone fruits. In: *Plant Genetic Resources: An Indian Perspective, Symposium Proceedings*. National Bureau of Plant Genetic Resources, New Delhi, p. 320–334.

Periyathambi, R., 2007. *Jamun: The Potential Untapped*. Kerala Calling, February, pp. 30–32.

Prabhuraj, H.S., Swamy, G.S.K., Athani, S.I., Patil, B.R., Hulamani, N.C. and Patil, P.B., 2002. Variability in morphological characteristics of Jamun (*Syzygium cuminii* Skeels) trees. *My Forest,* 38(2): 187–189.

Roy, S.K., 1996a. *Bangaladesh Journal of Botany,* 25: 159–164.

Shanmugavelu, K.G., 1966. Studies on the effect of plant growth regulators on some forest plant species. *Madras Agric. J.*, 53: 470–2.

Shanmugavelu, K.G., 1967. A note on airlayering of Eugenia (*Syzygium cumini*) and neem (*Azardirecta indica*) *South Indian Hort.*, 15: 70–1.

Singh, A.K., Bajpai, Anju and Shukla, S.K., 2007. *Annual Report*. National Network Project on Underutilized Fruits.

Singh, I.S., Srivastava, A.K. and Singh, V., 1999. Improvement of some underutilised fruits through selection. *J. Appl. Hort.*, 1(1): 34–37.

Singh, R.K., Pandeh, I.C. and Prasad, R.S., 1979. Propagation of jamun by Budding. *Punjab. Hort. J.*, 19: 74–75.

Singh, S.K., Krishinamurthy, S. and Katyal, S.L., 1967. *Fruits Culture in India*. ICAR, New Delhi, pp. 255–259.

Yadav, U., Lal, M. and Jalswal, V.S., 1990. *Plant Cell Tissue Organ Culture*, 21: 87–92.

2011, Biodiversity in Horticultural Crops Vol. 3 *Pages* ***85–119***
Editor: **Professor K.V. Peter**
Published by: **DAYA PUBLISHING HOUSE, NEW DELHI**

Chapter 6

Litchi (*Litchi chinensis* Sonn.)

Rajesh Kumar
National Research Centre for Litchi,
Muzaffarpur – 842 002, Bihar
E-mail: rajeshkr_5@yahoo.com

Litchies are cultivated for their very attractive and popular fruits which have a long history of acceptance in many parts of Southeast Asia. The demand is for large bright red/pink fruits with small seeds, crisp and sweet aril. The important litchi growing countries are China, Brazil, India, Malaysia, Thailand, Vietnam, Myanmar, Mauritius, South Africa, Australia, New Zealand, Madagascar and Taiwan. The pearly white, translucent fleshy succulent aril or edible pulp of litchi fruit is favourite of millions in these countries due to its attractive appearance, delicate flavour, pleasant fragrance and distinctive taste. The other way of utilizing fruits during off season is the preserved or canned fruits. China, Taiwan, Thailand and India have substantial canning industries. A large portion of the crop in China is also used as dry nuts which are not very common in other countries. However, at present in India, 60 per cent of the crop is utilized in the domestic markets in non traditional areas, 20 per cent in the production catchments area and rest for processing and value added products like concentrates, juice and squash. The other area for utilization may be the alcoholic beverages. The food value of fruits lies in its sugar content.It also contains protein, fat, minerals (Ca and P) and is a reasonable source of vitamins C, A, B and B_2.

Litchi is an important commercial fruit but is meagerly exploited for its potential outcome. The ever increasing demand by rapidly growing population has forced the situation to exploit its potential through systemic research and development. Though it has established its significance for the human beings due to essential characters like good taste, nutrition, storability, blendability, palatability, health value and preservability. Its commercial significance stands because it bears fruits having attractive skin colour, excellent taste with flavour and treated as one of the finest fruits with high nutritional value popularly known as "Queen" of the fruits and demarcated as a crop having high export potential. It is an evergreen subtropical fruit crop and is highly specific to climatic requirements and probably due to this reason, its cultivation is restricted to a few countries in the world and a few states in India. Litchi provides livelihood to millions of people as it provides both on-farm and off-farm employment and income. The area under cultivation is expanding with many new orchards in China, Viet Nam and India. The average yields are below 5 t/ha, where as some better orchards produce three times this. New plantings are ensuring a continuous growth of litchi in the world. Though with slow pace, area, production and productivity of litchi are increasing in the country.

Most of the genetic diversity is found in China and most of the production comes from South Asian countries. The genetic base in major producing countries is quite narrow and if the targeted goals are to be achieved, diverse genotypes are to be introduced from China, its neighbouring areas as well as from Australia and need to be exploited. A detailed assessment of biodiversity of litchi in different growing areas, collection, conservation and utilization are of prime concern to bring about desired significant accomplishments through breeding.

Origin, History and Spread

Origin

Litchi originated in the subtropical low elevations of Southern China, first in the north tropical rainforests and mountain forests of Southern China, where it is a dominant tree species. However, there are different opinions about the origin of litchi. According to De Condolle (1909) litchi was only known by the Chinese in the third Century. Popenoe (1920) mentioned in Blume Cochin that China and Philippine islands are the land of origin of litchi. Toa (1955) viewed that litchi originated in Guangtong and Fukien province of China where as Ochse *et al.* (1961) opined that litchi is grown in China from the last 40 centuries and therefore it is native of only this region. The other pocket of wild litchi trees was located in the southern sub tropical low elevations of Southern China in the provinces of Guangdong/Kwangtung and Fujian/Fukien region primarily along rivers and near the seacoast. After intensive survey of literature and consultation with experts, Arora *et al.* (1996) concluded that the *Litchi chinensis* originated in northern Vietnam, parts of Malaysia and southern tropics of South China. Wild species of litchi with variable fruit types are still being grown in southern Hainan, South western Guangdong, Southern Yunnan and South eastern part of Guangxi Zhuang. Still the wild litchi forest in about 45 ha exists in Bawang Ling ridges in Changjiang country of Hainan province. Above all, the evidences and available literature almost confirm the fact that the original abode of litchi is in South China and adjoining areas. These lowland litchies are the most likely to be a result of fruit/seeds which were washed down by rivers from the mountains or transported by birds and then deposited naturally in great conditions for litchi growth. Hence, the provinces of Kwangtung and Fukien with the present available evidences are thought to be the main centre of origin. The main centre of origin of litchi might have been between 23° and 27° North latitude in the subtropical parts of southern China, northern Vietnam and Malayasia (Ghosh and Mitra, 2000). The species appears to be no longer in the wild stage (Groff, 1943). However, wild species of litchi are reported in areas of Southern Hainan, South-western Guangdong, southern Yunan and southeastern part of Gunagxi Zhuang region in China (Hong *et al.*, 1996). It is reported to be in cultivation since 1500 BC by the people of Malayan descent (Chandler, 1958) and is subjected to intense selection. In southern Guangdong province of China, litchi is known to be grown for thousands of years. There are still villages in Southern China today with litchi trees which are over 1000 years old! There is some evidence to suggest that wild litchi may have also originated in northern mountain regions of Vietnam.

History

Litchi, the important Sapindaceae crop spread from China across the subtropical countries since 17th century. There is a long history of litchi cultivation in China but are relatively new in most of the extensively growing countries due to slow pace of area expansion and spread. The name Litchi or li-zhi in Chinese Pinyin, coined only at its centre of origin, meaning "to be detached from the branch" is so-called because once the fruit is picked it deteriorates quickly, first appeared in the text in 200 BC (Wu, 1998). History envisages that the litchi cultivation, spread and its likings have remained as a part of royal confine since ancient times. According to Ge Hong (281-314 AD), the litchi fruits along with some other tropical and sub-tropical fruits were presented to the then Emperor Kao Tsu, by the landlords in the Southern China as early as 200 BC. The emperor Wu Ti (140-87 BC) attempted many times to establish the litchi plants at Changan but could not succeed, may be due to the area falling under temperate zone. The first reference on litchi was published in the literature of the Han dynasty between 140 to 86 BC. Then the several monographs have provided the details on cultivars, tree management, pre- and post-production as well as utilization aspects. However, a monograph on litchi published by Ts'ai Hsiang in 1059 AD, is the first systematic publication and it was mentioned specifically, while compiling and publishing the first detailed description of cultural aspects of litchi cultivars in English by Groff (1921). According to him, the earliest known record of cultivars was made by Guo Yigong in his Guangzhi Register published 1800 years ago. There are still villages in Southern China with litchi trees over 1000 years old! There is some evidence to suggest that wild litchi may have also originated in northern mountain regions of Vietnam. The book written by Singh and Singh (1954) is the first Indian publication towards the identification and documentation of litchi cultivars in India. The earlier propagation was done on through seeds up to 10th century and later as records say that only in 16th century the other improved methods of litchi plant propagation like layering, grafting and cutting came into use.

In China, numerous wild litchi trees are found in moist forests in Hainan Island from low elevation up to 600 to 1,000 m, and below 500 m in hilly areas of the Leizhou Peninsular, west Guangdong and east Guangxi. Litchi is usually the dominant species in the upper layer of the forest. Wild trees can also be found in parts of Viet Nam, north of Hanoi, although there are a fewer pockets of natural rainforests than in China. Wild specimens are similar in general appearance to some cultivars grown in China and Viet Nam. The fruits are edible, but the flesh or aril is relatively thin and sour, and not commercial. The wild types evolved in two directions, the skin segments becoming protruded and long as in "Tai So" and "Kwai May", or "China (Bombai)" or flattened as in "Sum Yee Hong", "Souey Tung", "No Mai Chee" and "Wai Chee" or in "Shahi" or "Rose scented" and "Bedana" of Indian origin.

Spread

The spread and commercialization of litchi species around the world were slow mainly due to the specific climatic requirements and poor cropping of trees in many areas, as well as the short life of the seeds. Despite the fact, the excellent fruit characteristics and taste, the species and its commercial cultivation remained confined to southern China and possibly northern Vietnam until the 17th century and only during 17th to 18th century its spread took place to other parts of the world (Tindall, 1994; Hai and Dung, 2002; Singh and Babita, 2002). It was taken to most of the tropical and sub-tropical world in the last 400 years. Apart from China, the fruits became very popular in South and Southeast Asia, but less well known in Africa, the Middle East and America. The route of litchi passed from China to other parts of the world as documented by Galan Sauco and Menini (1989). According to them, litchi was first introduced to India to its eastern part in the end of 17th century from Burma and later after 100 years, it again came to Bengal during the end of 18th century (Goto, 1960; Knight, 1980; Liang, 1981; Pandey and Sharma, 1989). This is the period when litchi was introduced in most of countries in South East Asia like Bangladesh, Nepal, Thailand, Indonesia and other countries through Burma (Myanmar) and India and became commercial occupying around 3000 ha in Nepal, 12,000 ha in Bangladesh and 23,000 ha in Thailand. In the Encyclopedia Britanica, it is mentioned that litchi was the favourite fruit of the Cantonese since ancient times and was introduced to Jamaica in 1775. In Hawaii, litchi was introduced in 1873, in West Indies during 1775 while in Natal Province of South Africa, it was introduced during 1869 (Marloth, 1947). In United States of America, the litchi was introduced from India (Saharanpur area) in 1883 and first planted at Florida. Subsequently it reached California during 1897. It was then introduced to the Reunion Island in 1764 by Joseph-Francois Charpentier de Cossigny de Palma. It was later introduced to Madagascar which became a major producer. Production goes back at least 2,000 years in southern China, but is relatively new in rest of Asia and the Pacific. It is thought that Chinese immigrants introduced litchi seed to north Queensland during the gold rush of the late 1800s. The record of cultivation of litchi in Mexico was traced during 1900s, which was considered introduced from China. Litchi reached Israel in the 1930s from South Africa (Mauritius), California (Floridian) and India (Bengali) (Goren *et al.*, 2001). The commercial production of litchi in Australia expanded only after 1970s.

The Genus *Litchi* and its Classification (Botany and Classification)

Litchi belongs to family Sapindaceae, (which includes mostly trees, shrubs and tendril-bearing vines) comprising about 140 genera and 1,500 species, which are native to Asia (>90 per cent), South America, Africa and Australia (Bailey, 1949; Leenhouts, 1971). Most of the species are tropical and sub-tropical fruits out of which only a few members are of much horticultural interest. This particular family is also known as soapberry (*Sapindus saponaria*) family due to presence of Saponin (used as soap substitute in the tropics) in the fruit (Nakasone and Paull,1998). The Sapindaceae was originally described by Carnbessedes in 1828 but the authentic and accepted official classification was made by Rodlkofer (1932) which was mainly based on leaf, flower, pollen, aril and fruit morphological characters. Litchi is only the crop of major significance in this family. The litchi can be systematically classified as follows :

Kingdom	:	Plantae
Phyllum/division	:	Magnoliophyta
Class	:	Magnoliopsida
Subclass	:	Eurocid-II
Order	:	Sapindales
Family	:	Sapindaceae
Subfamily	:	Sapindoideae
Tribe	:	Nephelieae
Genus	:	*Litchi*
Species	:	*chinensis*

Its standard and accepted binomial nomenclature is *Litchi chinensis* (Gaertn.) Sonn. (Bailey,1950). The other synonyms suggested by different botanists are *Euphoria litchi* Desf., *E. sinensis* Gmel.,

Figure 6.1: Litchi : Cv. Shahi

Figure 6.1–Contd..

Figure 6.2: Litchi : Cv. China

Figure 6.3: Late Bedana

Figure 6.4: Swarna Roopa

Figure 6.5: Kasba

Figure 6.6: Purbi

E. punicea Lamk., *E. litchi* Juss., *Dimocarpus lychi* Lour., *D. litchi* Willd., *Nephalium dimocarpus* Hf. and T., *N. duriocarpus* T., *N. litchi* Camb., *Sapindus edulis* Ait., *Scytalia chinensis* Geortn., *S. litchi* Roxb., *and S. locacon* Roxb. A new name *Litchi litchi* Britton was reported by Britton and Wilson (1926). Hooker (1875) reported 12 species of *Nephalium* growing in India. These are *Nephalium mutabile* Blume., *N. lappaceum* Linn., *N. glabrum* Nornh., *N. rubescens* Hiern., *N. costatum* Hiern., *N. longana* Camb., *N. malaiense* Griff., *N. eriopetalam* Miq., *N. gardneri* Thwaites *(syn. Euphoria gardneri* Bedd.*), N. stipulaceum* Bedd. (syn. *Otonephalium stpulaceum* Radlk.) (Groff, 1921) (c.f. The Litchi- Pandey and Sharma). Besides litchi,the other important members which belong to this family are longan (*Dimocarpus longan* syn. *Euphoria longana* Lam.), rambutan (*Nephelium lappaceum* L.) and Pulasan (*Nephalium mutabile* Blume.) The genus *Litchi* comprises only one species *i.e. L. chinensis*, consisting of three subspecies (Leenhoults, 1978). *Litchi chinensis* ssp., *chinensis, Litchi chinensis* ssp., *phillipinensis* (Radlk.) Leenh., from the Phillipines and New Guinea and *Litchi chinensis* ssp., *javanensis*, the cultivated subspecies from the southern Indo-Chinese region and Java are apparently adapted to a wet tropical climate (Leenhouts, 1978).

Although litchi is under cultivation for over 2000 years, the cytology is studied very recently. Liu (1954) suggested that the species may have derived from more than one wild progenitor. Haploid chromosome numbers of 14, 15, 16 and rarely 17 were found. It was considered that the progenitor, so called 'Mountain litchi' is clearly different from the commonly grown. According to Simmonds (1976) and Lu *et al*. (1987) the chromosome number of *Litchi chinensis* Sonn., is 2n=2x=30.

The vernacular names of litchi are linchi in Tahiland, letsias in the Philippines, litsi in Indonesia, laici in Malaysia, litchia in Portugal, litchi de chine in France and vai and caу vai in Viet Nam (Tindall, 1994). Yule *et al.* (1903) stated that the pronunciation of north China is Lee-chee while in the southern part of the country, it is ly-chee. In India and also in English, it is pronounced mainly as litchi and somewhere as lychee.

Morphology, Floral Biology and Fruit Set

Tree

Litchi trees are polygamous. It is a medium-sized evergreen tree, reaching 15–20 m tall. In general trees may grow to 12-15 m in height and have a beautiful, dense, rounded, symmetrical canopy extending nearly to the ground. Trees are very attractive with dark green foliage and reddish-coloured fruit. Typically major limbs of non trained trees begin within 0.9-1.0 m of the ground. The plant is a dense, symmetrical, oval evergreen tree with a dark brown, short and thick trunk. Growth habit differs considerably among cultivars: Trees may have erect or drooping branches and grow 6m to more than 12 m in height (Galan Sauco, 1989). Litchi trees can live for many years. Banta (1952) reported that two trees in China were 1,200 years old, the largest being 3.2 m in diameter.

Roots

The root system grows profusely in the upper 0-60cm strata of the soil. The horizontal root spread depended on type of soil. The greatest concentration of roots was in the 100-200cm zone (64.82 per cent) followed by 0-100 cm zone (29.67 per cent) and 200-300cm zone (5.50 per cent) from the tree trunk. Vertically, 43 per cent of the total roots and 52 per cent of the fibrous roots lesser than one mm thick were located in 25-50cm soil depth (Hasan and Chattopadhyay, 1991; Roy *et al.*, 1987). Root study in litchi by core sampling technique at subtropical Queensland, Australia showed that 80 per cent of the feeder roots were located to be varying within 0 to 60cm soil depth. The depth of rooting was the greatest in fine textured soils. Air layered trees have no tap root system, most of the feeder roots were within upper 45 cm and only a few roots below 90 cm level (Nakata and Suehisa, 1969).

Leaves

Leaves are compound, mostly paripinnate with 2 to 8 (in rare cases 9 to10), lateral leaflets (leaflet 5–10 cm long); the terminal leaflet is absent. Leaves appear as successive seasonal vegetative flush, may have an emerging reddish, pinkish, light bronze, light coppery, glossy light green in colour, when young, becoming shiny and bright green when mature. Successive flushes are separated by quiescent periods, during which the expanded leaves darken and thicken as it is observed that before turning green as they expand to full size. A new flush begins only when there is either climatic fluctuation, pruning training operation or the previous flush has matured (Menzel *et al.*, 2000; Olesen *et al.*, 2002). Leaves (compound leaf, 15–25 cm long) are arranged alternately, exstipulate, petiolate (petiole about 3-4cm long). Leaflets opposite or nearly so (*i.e.* in a few cases alternate or some leaflets pairs are alternate and the rest opposite), mostly 6 in 3 pairs, slightly stalked 0.3 to 0.8 cm long, exstipulate, lanceolate to oblong lanceolate, entire or slightly wavy, acute or acuminate, dark green and quite glabrous, average size 13.50 cm long to 4.15 cm wide.

Stem

Litchi has spreading branches and weak stem. Branches emerging from main trunk remain smooth with wider croches whereas sub branches sometimes have narrow croch angle. Vegetative shoot giving rise to branches may arise from terminal portion or from the tip/top point of invigoration.

Terminal twig first begins to give out leaf out growth, which later may give rise to reproductive bud at the terminal tip. The branches, if not supported, rarely grow straight and if left unpruned, make the spheroid canopy. The stem is found growing usually spiraled. Bark is grayish or grayish brown and rough.

Inflorescence

The inflorescence is determinant, small 2 to 3 mm, white to greenish-yellow flowers are produced on the current season's wood in terminal clusters (mainly branched panicles, 5-30 cm in length). The time of floral initiation varies with the genotype and the environmental condition. In the northern part of hemisphere, it usually occurs between November and February, while in southern hemisphere in between June to September. In different litchi growing parts of India, the flower bud differentiation start in December and gets completed in January (Shukla and Bajpai, 1974). Prior to floral bud initiation, the twig first begins to give out vegetative growth and then just at the terminal of this vegetative shoot, a reproductive bud appears (Singh and Singh, 1954). Sometimes, mixed inflorescences *i.e.* inflorescences having primary branches invariably in the axil of leaf are also found. The compound racemose inflorescence is composed of several multiple branched terminal panicle where flowers occur in cymes (Pandey and Sharma, 1989). The individual flowers are borne in profusion on the axis of the 3rd order, rarely on 4th order and occasionally on the main axis of the inflorescence (Robbertse *et al.*, 1995). The constitution of cymes with regard to sexuality of flowers often varies with the flowering season. Generally variations are reported during the transitional stage of flowering *i.e.*, when both female and male flowers occur together. Banerjee and Chaudhary (1944) reported occurrence of six types of cymes in litchi as:

1. Cymes composed entirely of male flowers,
2. Cymes composed entirely of female flowers,
3. Cymes terminating into male flower with lateral female flowers,
4. Cymes terminating into female flower with lateral male flowers,
5. Cymes terminating into male flower with lateral flowers of different sex,
6. Cymes terminating into female flower with lateral flowers of different sex.

Types of Flower

Flowers are small, greenish-white or yellowish-white, produced in panicles up to 30 cm long, and are borne on a large thyrse (a many-flowered inflorescence) which emerges at the ends of branches anytime from late December to April (more commonly February and March) in India, China, Florida and Thailand. Butcher (1957a) distinguished three types of flowers which bloom in stages: (i) Male or staminate flowers (M_1) with no functional ovaries, bloom first; (ii) female or functionally pistillate flowers (F) with anthers which do not dehisce, comprise the second stage; and (iii) imperfect hermaphrodite flowers (M_2), also lacking functional ovaries, bloom last. Each phase of bloom consists of flowers of the same type. There are three flower types: two male types (called M1 and M2) and one female (called F) (Stern and Gazit, 1996). In general, the M1 flowers open first, female flowers open second, and M2 flowers open third. The small, yellow-green or brownish yellow, apetalous with small valvate sepals, a fleshy disc and 6 to 8 stamens, litchi flowers are 3.0-6.0 mm in length when fully expanded and borne in profusion on leafless or leafy well branched terminal panicles. The flowers have a cup shaped calyx with 4-5 short sepals but no petals, about eight stamens, a two-lobed stigma, an ovary on a short stalk, and one ovule in each of its two or three sections (carpels) (Galan Sauco, 1989, Menzel, 1991).

Floral Biology and Fruit Set

There are three sex types found in litchi as male, hermaphrodite and hermaphrodite with only functional male (rudimentary pistil), which can be placed on the chronological order of development as Type I, Type II and Type III.

(*a*) Type I is morphologically and functionally male processing six to eight stamens producing an abundance of pollen and lacking ovule differentiation.

(*b*) Type II flower is morphologically a hermaphrodite, functioning mostly as female with a well developed two lobed pistil and a two–lobed stigma. The five to eight stamens do not dehisce normally and have little viable pollen.

(*c*) Type III flowers function as male, although it is morphologically a hermaphrodite with a rudimentary pistil lacking style and stigma. The six to eight stamens produce numerous viable pollen grains. Almost two third of the flowers on a panicle are composed of type III and only about 20 per cent of the flowers are potential to bear fruits, the ratio of functional females (Type III) to functionally staminate flowers (Type I and III) varies with cultivar and season/climate with major pollinators (Alexander *et.al.*, 1982)

Litchi floral anthesis occurs in cycles of sex types (Mustard,1954), with a period of about 10 days during which only functional period of 2-3 days when both staminate and female (Type II) flowers open and then 2 days of only female flowers. This is followed by 2-3 days of transmissible types and finally male flowers for a period of 7-10 days. The number of days to complete each of the cycles appears to differ somewhat among inflorescences between cultivars and with environmental conditions. Inflorescences produced around the same time generally produce the same sex types. In the inflorescence, it is found to begin floral anthesis at any sexual phase of the cycle, going through the above sequence. The expression of sex type may be to a higher percentage of type II flowers. The sex ratio in mature tree blooms differs widely among cultivars and the fluctuation in the same cultivar is governed by many internal and external factors. The three distinct flowering phases/waves can usually be observed in every flowering inflorescence in litchi. There is hardly any overlap among these waves in the same inflorescence. However, synchronization is not perfect and consistent (sometimes some irregularities also observed) among inflorescences on the same tree and even less so among the inflorescences of different trees of same cultivar. The M1 and M2 types flowers only produce viable pollens. Pollination with fertile pollen is essential for fruit set. The litchi flowering pattern tends to promote cross pollination. The normality of reproductive organs and pollen viability are influenced by many factors like genetic, nutritional and environmental. Female flower infertility was also observed in early flowering and also the prolonged low temperatures during flower development may be the factor responsible for this infertility (Batten and McConchie, 1992). Since litchi flowers are functionally unisexual, pollination of female flowers occurs only by close- or cross pollination. Litchi pollination by air borne pollen is apparently not fully effective (Pandey and Yadav, 1970). Litchi flowers are adapted for insect pollination (entomophilous)(McGregor, 1976). The flowers produce large quantity of nector, which attracts a large number of insect pollinators. Pollen does not serve as a major attractant for most litchi pollinators. Bees are the most effective insect pollinators (McGregor, 1976, Free, 1993). However, the partial overlap between the female bloom and the two male (M1 and M2) blooms enables close pollination among flowers on the same tree, or among trees of the same cultivars, thereby providing an opportunity for self fertilization (Joubert, 1986; Stern and Gazit, 1996). Litchi is self fertile, since monocultivar litchi plots are capable of bearing good yields and self-compatibility in litchi is confirmed in various litchi growing countries by hand pollination experiments (Batten and McConchie, 1992:

Fivaz and Robbertse, 1995 and Stern and Gazit, 1998) and self fertility also by isozyme analysis (Stern *et al.*, 1993b; Degani *et al.*, 1995b). Depending on phase of flowering (different types), almost all fruits on trees close to second cultivar are hybrids and the various studies revealed the fact that there is beneficial effect of cross-pollination (Degani *et al.*, 1995b). Successful pollination and fertilization are prerequisites for the set of a normal litchi fruit. Non pollinated flowers have abscised after 15-21 days (Stern, 1992). The main factors affecting initial fruit set are : 1. the presence of fertile female flowers, 2. adequate pollination with effective pollen, 3. environmental conditions favourable to pollination, 4. the fertilization process and 5. fruit development. The presence of flowers with at least one normal ovule is prerequisite for normal fruit set. Litchi usually produces a large number of female flowers also, but most of them tend to be defective (Stern *et al.*, 1996b, 1997a). During the fertilization process, the pollen grains deposited on the receptive stigma germinate readily at a wide range of temperatures (15°C to 33°C), then pollen tubes reach the ovaries of all pollinated flowers in about 24h to 48h but the penetration in ovules also varies with environmental condition. Short exposure during the fertilization process (6h after pollination) to high temperature >30°C irreversibly arrested pollen tube growth (Stern, 1992). The fertilization occurs 2 to 3 days after pollination and there are also reports of double fertilization contributing to initial fruit set (McConchie and Batten, 1989; Wang and Qui, 1997). On the whole, fruit setting in litchi is a function of interaction of number of interdependent factors like genotype, age of the tree, inherent nutritional and bio-chemical status of tree, flower bud differentiation time, panicle development, flowering and sex ratio. These factors are greatly influenced by the site and location of the orchard, environmental conditions of the particular year and the orchard management practices. Even and otherwise low and irregular bearing is a major problem in commercial production of litchi (Mustard *et al.*, 1956). Sometimes even after very profuse flowering, poor fruit set and fruit yield are observed due to many factors like low female ratio, heavy flower and fruitlet drop etc.

Fruits

Fruits are borne in loose clusters numbering 3 to 50 fruits and are round to oval and 25 to 38 mm in diameter. Botanically the fruit is a drupe, entire or deeply divided into 1-3 caucus, indehiscent of variable shape (globose, oblong, conical, ovate, oblique or lop sides). Epicarp - The outside is covered by a pink-red, roughly-textured rind with different shades of red and yellow, tubercled, leathery becoming brittle afterwards that is inedible but easily removed. Tubercles–angular becoming flattened when ripe. Mesocarp - The inside consists of a layer of sweet, translucent white flesh (the aril), which is edible consisting of a highly developed (aril)and enveloping seed. The skin (pericarp) ranges from yellow to pinkish or red and is leathery, with small, short, conical or rounded protuberances. The edible portion of the fruit (pulp) is called an aril that is succulent, whitish, translucent, with excellent subacid flavor. Fruits contain one shiny, dark brown seed, usually relatively large, but it may be small and shriveled (called chicken tongues) in some varieties. Fruit must be ripened on the tree for the best flavor. The fruit is a drupe, 3–4 cm long and 3 cm in diameter. The round fruit is pendant in a loose cluster of several dozen fruits up to 30 (Cobin, 1952) or may be more. The leathery skin is covered with sharp-tipped tubercles. The seeds are dark brown (Menzel, 1991). The fruit must ripen on the tree, then is harvested over a 6-weeks period. The shelf life of the fresh fruit is only 10 to 14 days (Menzel, 1991, Palmer, 1956).

Seeds

The seeds of litchi (*Litchi chinensis* Sonn.) are dark brown, covered with white fleshy juicy and transluscent aril. The seed, which lies in the center-portion of the fruit, is a single glossy dark brown nut like structure measuring 2 cm long and 1–1.5 cm in diameter. The seed, similar to a buckeye seed, is slightly poisonous and should not be eaten. Normally large size viable seeds are formed but chicken tongue/shriveled seeds are also formed. Generally, the seeds are cylindrical, compressed, plaino-convex or concavo-convex, exalbuminous, embeded in the pulpy aril, chocolate in colour. In seedless varieties, the seed is very much sunken, canine tooth like or chicken tongue as called by the Chinese. These varieties most probably seem self or cross sterile, and seeds contained by them are probably unfertilized ones. This fact requires further experimental evidence. Seed germination and seedling vigour are observed to be higher in large seeds. Germination at 25-30°C takes about 16 days. Litchi seeds are recalcitrant and should not be allowed to dry. When sown immediately after extraction, their germination was 100 per cent which reduced even after one day storage (Ray and Sharma,1985). Seeds stored in air had poorer germinability but 15 per cent of those stored in water developed splits in testa. Strong positive correlation between seed moisture content and germination was observed and 20 per cent moisture was considered critical. Seed can be preserved in fruit for short periods. Seed viability was preserved for 24 days by storing fruits in sealed polythene bags after treatment with benomyl (0.05 per cent) and 6 per cent wax emulsion (Roy and Sharma,1987).

Ecological Adaptations

Climate

Litchi chinensis is one of the most environmentally sensitive of the tropical tree crops. It is adapted to the tropics and warm subtropics, producing the best in regions with winters that are short, dry and cool but frost free, and summers that are long and hot with high relative humidity at the time of maturity. Good protection from wind is essential for cropping. Year-to-year variations in weather precipitate crop failures, for example, through untimely rain promoting flushing at the expense of floral development, or through poor fruit set following damp weather during bloom. Occurrence of high temperature and water stress during fruiting period, especially in the lower elevation results in excessive drop and cracking of fruits. Litchi, a subtropical fruit crop can be found growing an elevation ranging 350-1100 m above mean sea level. The litchi usually likes low elevations but can be grown up to an altitude of 800 m with varying degrees of success but in general not found growing above 1000 m. The altitudinal range of litchi growing region/areas are recorded as 8.7 meters at Canton, China (Groff, 1943) to 682.0 meters at Dehradun, India. Likewise the latitudinal range varies from 26°38′ S to 32°40′ N from equator. Production is the highest in Asia pacific region at 24°N latitude and 121°E longitude and 300 meter altitude (Batten,1982).

Commercial cultivars are geographically adapted and number over 100 world wide (Yokoyama *et al.*, 1991).Litchi thrives well in subtropical climate (Chia *et al.*, 1997). The climatic factors are important determinants in litchi fruit production like growing season thermal time (temperature range), moderate winter, actual rainfall with its specific annual distribution pattern and several of the interaction variables. Litchi has specific climatic requirements for successful cultivation and a set of environmental variables and their interactions determine the success of this crop. Poor production is experienced in many litchi producing countries because winters are not cool or dry enough to induce growth-check prior to flowering as litchi is native to Kwangtung and Fukien provinces of China where the winters are dry and the summers hot and very humid. This is the climate they require to set fruit readily. Litchi does need a seasonal variation in temperature to produce fruit. The chilling requirement is about 100 to 200 hours between 0° and 7.2° C. Young litchi trees are severely damaged or even killed at -2 to 3°C, with dormant and mature trees having more tolerance to low temperatures. Periodic cold is required for flowering, although in Guangzhou, air temperatures rarely drop below 3°C. When trees are deeply rooted, low air temperature is more important than soil moisture (Menzel *et al.*, 1989). There is considerable cultivar difference in response to temperature (Menzel *et al.*, 1989) while evaluating seven cultivars, it showed panicle emergence after 6 weeks at 15/10°C (Day/night) or at 20/15°C, however, there are differences in the degrees of flowering between the cultivars at 20/15 °C and no flowering at 25/20°C or 30/25°C. Variations in the times of panicle emergence, panicle development, and anthesis among the cultivars in relation to seasonal progressions in temperature affected the number of fruit set. The number of fruit set/panicle increased as the number of female flowers/panicle increased. Higher numbers of female flowers were associated with maximum temperatures during flower development of 18°C, with lower numbers at 23°C. Higher maximum temperatures during anthesis (30°C v. 24°C) increased the proportion of female flowers setting fruit (Menzel and Simpson, 1992). Flowering is irregular at high temperature. Pollination is optimum at about 19-22°C and occurs in 5-7 days. Winter/spring temperature extremes affecting bloom phenology and unsettled weather limiting bee flight during bloom are identified as other causes of reproductive failure (Batten, 1986). It varies greatly within panicles (McConchie and Batten, 1991), and ranges from 1-50 per cent of the female flowers produced (Galan Sauco, 1989). Reproductive failure is common and not always explained. Fruit set in litchi is also very much climate dependent and profoundly affected by temperature and then only by other climatic factors. Extended period of high temperatures above 30°C during anthesis can also reduce fruit set. Variations in the times of panicle emergence, panicle development, and anthesis among the cultivars in relation to seasonal progressions in temperature affected the fruit set (Menzel and Simpson, 1992). High temperature during fruit development is implicated in skin browning and skin cracking, especially when accompanied by low humidity. High temperature (>38°C) in April–May often accelerates fruit development at the expense of fruit weight. Litchi is a day-neutral plant and floral initiation is directly associated with the starch content of leaves and stems. In litchi cv. Nu Mai Zi, at light intensities >60klx, photosynthesis began to be inhibited. Leaf photosynthate accumulation was in the order of autumn>summer>winter (Deng and Ni,1994). High rainfall and humidity induce good growth in litchi. In China and Thailand, litchi is also grown on banks with high water-tables beneath them. Ample annual rainfall is considered around 1250 to 2000 mm for litchi. When rainfall is lower, irrigation is essential. A dry autumn and winter are important to prevent litchi vegetative growth and making essential condition for good flowering (Nakata and Watanaba, 1966) in South China. A certain degree of water stress is needed for flower initiation. It seems that low humidity is detrimental to litchi fruit set (Batten,1986a). A major problem of litchi culture in Hawaii, is that rainfall comes in late autumn and winter, making it difficult to control vegetative flushing. High temperature (38°C) and relative humidity below 60 per cent resulted severe skin cracking of fruits in India (Sharma and Ray,1978b). In areas of excessive rainfall and high humidity, even during the time of fruit maturity, commercial litchi growing may not be possible. Fruit set in litchi is climate dependent and profoundly affected by humidity in conjunction with other climatic factors. Overall, the optimum rains are conducive for production of quality litchi fruits, but sometimes early and continuous monsoons affect the quality due to severe attack of fruit borer and some rots. Like other important fruit crops, the initial establishment, proper growth and performance of litchi are found affected by nature of winds. The high speed windstorm or cyclones during fruit bearing period, more particularly at harvesting time may cause serious damage. It is found growing mainly in areas/places having mild tropical and subtropical climate but young plants are extremely sensitive to cold and require frost protection. Moreover, it needs to be protected from frost until it is well established (about 3 years). Mature trees tolerate a few degrees of frost. In India, the production tract is restricted in the north to north-eastern regions experiencing mild climate (in the foothills of Himalayas from Jammu and Kashmir to Tripura). Fog period extending upto 3-15 hours or more at a stretch affects flower bud development before flowering and expression of flowering sex. Fog definitely curtails light supply and reduces photosynthesis.

Soil

The soil in which litchi is grown is of vital importance, since a healthy root system depends entirely on good soil condition and interestingly, litchi grows under a wide variety of soils including alluvial soils, loams, heavy clays, organic soil, calcareous soil (with 30 per cent lime) and rock piles (Singh and Jawands, 1962; Nijjar, 1972; Chapman, 1984 and Pandey and Sharma, 1989). In China, it is cultivated in sandy or clayey loam, "river mud", moist sandy clay, and even heavy clay. The pH should be between 6 and 7. If the soil is deficient in lime, this must be added. In litchi, growth and its productivity are determined more by subsoils than by the kind and quality of surface soil. Ideal subsoil should be well drained and loose enough to permit root development. A sandy loam subsoil is preferred to heavy clay or compacted soil. The soil requirements for litchi cultivation are very fastidious in India (Katyal and Chugh, 1961) but in other countries, it is described as not very demanding and specific as because litchi could grow well on soils which were only 40 cm deep, even rocky, calcarious less fertile soils in Florida (Batten, 1982; Kambell and Knight, 1983; Sauco and Menini, 1989). The Chinese often plant litchi on the banks of ponds and streams *i.e.* in low, wet land to form raised beds with specified distance planting to have perfect drainage but the soil is always moist. Though the litchi has a high water requirement and can stand occasionally brief flooding but it cannot stand prolonged water-logging. The water table should be at least 1.2-1.8 m below the surface and the underground water should be moving in as much as stagnant water induces root rot. It will not thrive under saline conditions. The pH of litchi soils in China is usually 5.5. In India, the best litchies are grown (in the state of Bihar) on calcareous soils having sandy loam texture and are containing 24 to 30 per cent free calcium carbonate having the pH range of 7.5 to 8.5, though in other places in the country too, it is found successfully growing in the laterite soil having pH 5.0 to 5.5 (acidic condition) preferably with external lime application to bring the soil pH to 6.0-7.0. Calcareous soil appears to be more salubrious for litchi. It grows and yields well in calcareous soils in India, California, Florida and China (Ray, 1981; Nijjar 1972 and Chen, 1949). Its cultivation is successful even in low to medium fertile soils but appropriate nutrient management is required in early stages of growth. In alkaline soils, application of micronutrients (Zn, B, Mg) is essential (Tindall, 1994). The type of soils greatly influences quality of fruits. However, high organic matter (more than 1 per cent) and soil with poor drainage are not ideal for this crop as these conditions promote more vegetative phase and do not favour the growth of mycorrhizal fungi (Kadam and Despandey, 2005). Poorly drained soil or soil with impermeable layers is not suitable for litchi. Although, gravelly or rocky soils drain well but do not supply enough water to the trees, because of poor water holding capacity, require good irrigation practices such as wetting the soil more frequently with small quantities of water to make these soils more suitable. Litchi makes better growth if mycorrhizal fungi grow on its roots. The mycorrhiza was named as *Rhizophagus litchi* that belongs to the VAM of Phycomycetous endophytes. More recently this VAM fungus is renamed as *Glomus litchi*. It is a symbiotic association of fungi with the roots of vascular plants which act as an additional catchment and absorbing surface in the soil through interconnected network of external mycorrhizal hyphae and there by improve the water and nutrient uptake efficiency of the host plant. In India, new plants should be grown in soil taken from the vicinity of old trees to introduce the mycorrhiza (Chadha, 1968). In litchi, the establishment and growth in early stages are better in loamy, alluvial, well drained, aerated and deep soils even moderately fertile soil having sufficient organic matter, pH range from 5.5 to 7.5 and water table below 180 cm also proved better for litchi cultivation. The extremely sandy, shallow, rocky, waterlogged, heavy textured, highly alkaline or saline soils are not suitable for litchi cultivation as these problem soils require suitable reclamation with chemical and organic amendments. Litchi makes better growth if mycorrhizal fungi grow on its roots.

Litchi cultivation is found globally under wide deviations from its exacting soil and climatic requirements (Dutt, 1962) resulting erratic behavioral performance.

Varieties/Clones

Every country has its own uniqueness in the cultivars of prevalence. The cultivars in different countries vary with the region from where they are introduced and the extent of hybridization. There are hundreds of clones of litchi in field gene banks in different research institutions of world. Selection and hybridization work in litchi growing countries of the world mainly in mainland China have yielded a few clones under each category with distinct and unique features. Apart from this, a lot of expeditions are made mainly in centre of origin and neighbouring countries and a large number of useful collections are being maintained. Owing to poor response of breeding in litchi, major attempts for clonal selection are made mainly at the place of origin *i.e.* China (Arora *et al.*, 1996), while considerable efforts are also made in other countries like Miami in Florida (Knight, 1963), Hawaii (Storey *et al.*, 1953), Queensland (Cull and Lindsay, 1996) and at Saharanpur in India (Lal and Nirwan, 1980). These resulted in a few new varieties *e.g.* Groff–a seedling selection from Hak Yip, Brewster–a seedling selection from Chen Family Purple (Groff, 1948), Saharanpur selection - a chance seedling selection (Uttarakhand), Swarn Roopa- a clonal selection from local collections. Attempts are also made to identify some important genotypes/ideotypes (Singh, 1998) as detailed in the Tables 6.1 and 6.2.

Table 6.1: Ideotype of Litchi

Sl.No.	*Particulars*	*Traits/Characteristics*
1.	Plant traits	Dwarf
		Regular bearing
		Precocious
		Wider adaptability
		Early and late maturity
2.	Fruit yield	High
3.	Fruit quality	Nature of attachment of seed with flesh
		Skin surface
		Skin colour
		Aril fragrance
4.	Seedlessness	Chicken tongued seeds
5.	Resistance to physiological disorders	Resistance to cracking and sunburn

Varieties Grown in India

Litchi varieties grown in India are highly variable under different climate and soil conditions. When distinguishing the cultivar, the shape of skin segments and protuberances are the reliable and stable genetic characteristics. Fruit size, shape and taste are also variables but are influenced by other than genetic factors. Indian cultivars vary greatly in vegetative flushing pattern, flush colour and flowering ability. Based on these characteristics, cultivars can be classified in five groups give in Table 6.3.

Table 6.2: Identification of Litchi Cultivars Based on Ideotype

Sl.No.	*Particulars*	*Cultivars/Varieties*
1.	Plant and yield characters	
	☆ Wider adaptability and high yielding	☆ Hak Yip, Calcuttia, Bengal
	☆ Early to late maturity and high yielding	☆ Early Bedana, Late Bedana, No Mai Tsz, Brewster
	☆ Large fruit types	☆ Muzaffarpur, Rose Scented, Bombai, Calcuttia
	☆ Smooth skin and red colour	☆ Salatheil, No Mai Tsz
	☆ Good quality fruits	☆ No Mai Tsz, Calcuttia, Gulabi, Muzaffarpur
	☆ High quality of seedling fruits	☆ Early Bedana, Late Bedana, No Mai Tsz, Kwai Mai, Brewster, Fay Zee Siu, Salatheil
2.	For processed product	
	☆ Dry fruit	☆ No Mai Tsz, Hak Yip, Brewster, Salatheil, Kwai May Red
	☆ Canning	☆ Early Large Red, Early Bedana, Rose Scented, Shahi
3.	Tolerant to abiotic stresses	
	☆ Tolerant to hot and dry winds	☆ Calcuttia
	☆ Resistance to fruit cracking	☆ Swarna Roopa

c.f. : Litchi (Rai *et al.,* 2001).

Table 6.3: Key to Important Cultivars of Litchi

Groups	*Characteristics*	*Cultivars*
A	Flush pink, leaf boat-shaped, dark green, panicle long, fruits oblong with round apex.	
	☆ Colour of fruit deep pink	Shahi, Tirkolia
	☆ Rose flavour	Shahi, Rose Scented
	☆ Colour of fruit light and greenish	Green
	☆ High cracking and big seed	Ajhauli, Deshi
	☆ Late in maturity	Dehradun
B	Deep pink flush, leaf with twist along the length, curved upward from the midrib and down along their length, panicle long, fruit oblong with pointed apex.	
	☆ Fruit pink colour	China, Surguja 1
	☆ Fruit deep pink	Purbi, Mandraji
	☆ Fruits in bunches	Bombaia, Calcuttia
	☆ Early maturity	CHES-2
C	Dark pink flush, oval shaped leaves, compact and small panicles, fruit round, smooth, chicken tongue seed	
	☆ Early maturing	Early Bedana, Seedless
	☆ Late maturing	Late Bedana, Seedless
	☆ Deep pink colour, mid season maturity	Swarna Roopa
D	Deep pink flush, boat-shaped, dark green long leaves, panicle long, largest size fruit, deep in colour	Kasba
E	Small elongated leaves, light green in colour, panicle compact, fruit medium in size, very late maturity.	
	☆ Pulp sweet and excellent flavour	Longia.
	☆ Pulp sour	Kaselia, Khatti

Although availability of litchi fruit has short duration not exceeding 30-40 days in India, cultivars can be distinguished based on season and maturity, provided they are grown at same location. Accordingly, cultivars can be grouped into three categories namely, early, medium and late season. Fruit maturity also varies slightly and relatively from year to year depending upon prevailing weather conditions. The maturity period of the fruit also varies depending upon agro-climatic regions. Some of the varieties in West Bengal come to maturity earlier, while the same varieties can be late in Uttar Pradesh. Due to the production of litchi in varying agro-climatic conditions, maturity, fruit colour, shape and size are reported to be varying. Thus, there has been much confusion in the names of varieties. As a result, the same variety is called by different names at different location. Distribution of varieties grown in different litchi growing states of India has been given in the Table 6.4.

Table 6.4: Varietal Distributions of Litchi in Different States in India

States	*Varieties*
Bihar/Jharkhand	Deshi, Ajhauli, Green, Purbi, China, Kasba, Bedana, Dehrarose, Shahi, Mandraji, Longia, Trikolia, Kaselia and Swarnrupa.
Uttar Pradesh/Uttrakhand/ Himachal Pradesh	Early Large Red, Bedana, Late Large Red,, Rose Scented, Calcuttia, Extra Early, Gulabi, Pickling, Khatti, Dehra Dun, Piyazi.
West Bengal/Assam	Bombai, Ellaichi Early, China, Deshi, Purbi and Kasba.
Haryana/Punjab	Calcuttia, Early Seedless, Late Seedless, Seedless-1, Seedless-2
Chhattisgarh	Sarguja-1, Sarguja-2

Table 6.5: Recommended Cultivars of India with their Salient Characters

Cultivars/Varieties	*Salient Characters*
India (Pandey and Sharma, 1989; Singh, 1968; Jawanda and Singh,1977; Menzel and Simpson, 1986)	
☆ Shahi	Early maturing, oval to oblong shaped fruits with crimson red tubercles on uranium green skin background, fruit weighing about 22-25g, edible aril with rosy aroma, excellent quality, regular bearer, heavy yielder, highly commercial cultivar with export potential (Bihar)
☆ China	Late season maturity, oblong to conical shaped fruits with dark rosy(pink) coloured skin, fruit weighing about 20-25g, aril sweet juicy aroma with excellent quality, alternate bearer, heavy yielder, highly commercial cultivar
☆ Early Bedana	Synonym as Early seedless, early maturity, oval to heart shaped fruits having rough surface with uranium green skin covered with carmine red tubercles, small seeded fruit, average weight 18-20g, medium yielder, commercial cultivar with high marketability value.
☆ Late Bedana	Synonym as Late seedless, Late-season maturing, fuits heart to conical shape, vermilion to carmine red colour with blackish brown tubercles, fruit weighing about 20-22g, medium yielder, highly commercial cultivar having export potential.
☆ Rose Scented	Early maturing, synonym as Shahi, crimson red coloured fruit, edible aril with rosy aroma, excellent quality, regular bearer, heavy yielder, highly commercial cultivar with export quality fruits.
☆ Dehradun	Synonym as Dehra Rose, Mid-early maturing, fruits obliquely heart to conical shape, bright rose pink coloured skin with attractive fruits weighing about 16-18g, high yielder, commercial cultivar
☆ Bombai	Early maturity, obliquely heart shaped with attractive carmine red with uranium green skin background fruits weighing about 15-20g, high yielder, prone to fruit cracking, commercial cultivar
☆ Kasba	Mid-season maturity, local selection, deep red coloured skin, fruits oval-round in shape large sized fruits weighing about 22-25g, high yielder, commercial cultivar

Contd...

Table 6.5–Contd...

Cultivars/Varieties	*Salient Characters*
☆ Longia	Late season maturity, medium sized oval to round in shape, red skin fruits weighing 18-20g, quality fruits with rosy aroma, medium yielder, commercial cultivar
☆ Trikolia	Mid-late maturity, red fruit, oval to round shape, average wt. 20g, medium yielder, commercial cultivar
☆ Ajhauli	Earliest maturity, oval to round shaped, red coloured skin fruits weighing about 15-18g, high yielder with quality fruits, commercial cultivar
☆ Mandaraji	Mid-late maturity, scarlet-red to dark-red skin, fragrant fruits weighing about 15-20g, high yielder, commercial cultivar
☆ Swarna Roopa	Late-season maturity, dark-red skin, small seeded fruit, average wt. 12-15g, medium yielder, cracking resistant cultivar
☆ Calcuttia	Late season maturity, oblong to round shaped fruits with dark pink coloured skin, fruit weighing about 18-20g, alternate bearer, heavy yielder, highly commercial cultivar
☆ Purbi	Synonym as Bengal (W.B. and Australia), large size fruit, skin with bright-red tubercles on uranium green background colour and egg-round to lopsided heart shaped weighing about 23-27g. Commercial cultivar.
☆ Kasailiya	Synonym as Khatti, oval to round, dull red (yellowish tinged) fruits weighing about 18-20g, pickling type, less commercial
☆ Ellaichi	Mid-season cultivar with fruits having conical, marigold-orange red in colour weighing about 12-15g, comparatively small seeded, good quality medium yielder, good marketing potential
☆ Green	Early maturity, oval to round shaped with florescent green to reddish green skin at maturity, fruits weighing about 15-18g, medium yielder, less marketability
☆ Gulabi	Late season maturity, oval to oblong, medium in size with carmine red backgorund with mandarin red tubercles, average fruit weight 20g, less commercial.
☆ Saharanpur selection	Late season maturity, oval to oblong, medium in size with dark red skin fruit weighing about 18-20g, commercial cultivar.
☆ Sabour Madhu	Late season maturity, reported as hybrid (Purbi x Bedana), fruits oval to oblong, medium in size with red skin, average fruit weight 20g, non-commercial.

Exotic Varieties

Table 6.6: Recommended Cultivars of Different Litchi Growing Countries with their Salient Characters

Cultivars/Varieties	*Salient Characters*
	China (Arora *et al.*, 1999; Stephens, 1955)
☆ Sanguehong	Early maturing, pink red fruit weighing about 37-42g, heavy yielder, commercial cultivar
☆ Baitangying	Early maturing, red skin fruit weighing about 24.8g, heavy yielder, commercial cultivar
☆ Yuanzhi	Mid season maturity, light pink red fruit average weight 24g, heavy yielder, commercial cultivar
☆ Feizixiao	Mid-early maturing, greenish red fruit weighing about 30g, heavy yielder, commercial cultivar for export too.
☆ Heije	Mid-early maturing, red fruit weighing about 18-28g, heavy yielder, commercial cultivar
☆ Guiwei	Mid-early maturing, deep red fruit weighing about 16-18g, high yielder, commercial cultivar

Contd...

Table 6.6–Contd...

Cultivars/Varieties	Salient Characters
☆ Huaizhi	Late maturity, red fruit, average wt. 22-25g, high yielder, stress resistant, suitable for table and processing
☆ Zeng Cheng Gua Li	Late maturity, deep red fruit, oval-round fruit, average wt. 25g, high yielder
☆ E Dan Li	Mid season maturity, extra large red skin fruit, average wt. 52.9g, medium yielder, commercial cultivar
☆ Qing Pi Tian	Mid-late maturity, dull red fruit, oval to chordate shape, average wt. 25g, high yielder, suitable for canning/processing
☆ Sun Yit Hoong	Earliest maturity, red fruit, medium yielder, average fruit weight 20-22g, less commercial
☆ Nuo Mi Ci	Mid-late maturity, scarlet-red skin fragrant small seeded fruit, average wt. 25g, high yielder, commercial cultivar
☆ Baila	Mid season maturity, pink-red skin fragrant small seeded fruit, average wt. 24g, heavy yielder, commercial cultivar
	Thailand (Anupunt and Sukhvibul, 2005)
☆ Hong Huai	Subtropical group, upland adaptability, bright red skin fruit, large egg shaped, average weight 20-30g, commercial cultivar.
☆ Kim Cheng	Subtropical group, upland adaptability, bright red skin oval shaped fruit, commercial cultivar
☆ Haeo	Subtropical group, upland adaptability, mid season maturity, blotchy yellow to deep red skin fruit, heart shaped, average fruit weight 20-28g, commercial cultivar
☆ Chakrapat	Subtropical group, upland adaptability, late maturity, bright red skin fruit, large size, heart shaped, average fruit weight 20-24g, high commercial value cultivar
☆ Khom	Tropical group, lowland adaptability, early maturity, blotchy yellow to purplish red skin fruit, almost round shape, average fruit weight 11-20g, commercial cultivar
☆ Krathon Thong Phrarong	Tropical group, lowland adaptability, large dark red skin fruit, almost heart shaped, average fruit weight 20-25g, commercial cultivar
☆ Khiao Wan	Tropical group, lowland adaptability, dark red skin fruit, almost round shaped, average fruit weight 11.9g, commercial cultivar
☆ Samphao Kaeo	Tropical group, lowland adaptability, dark red skin fruit, almost heart shaped, average fruit weight 15-20g, commercial cultivar
	Philippines (Boucato, 1995; Boucato and Ringor, 1998)
☆ UPLB Red	Low land adaptability, dark red fruit weighing 14g, commercial cultivar
☆ Sinco	Upland (hilly) adaptability, medium sized fruit with red skin, average weight 20g, medium yielder commercial cultivar
☆ Olan	Low land (warmer area) adaptability, dark red skin with oval shape weighing about 26g, medium yielder commercial cultivar
☆ Mauritius	Upland(hilly) region adaptability, bright red skin fruit, large egg shaped, average weight 20-30g, commercial cultivar

Contd...

Table 6.6–Contd...

Cultivars/Varieties	*Salient Characters*
	Queensland (Australia) (Winston and Farrel, 1993)
☆ Tai So	Upland(hilly) region adaptability, bright red skin fruit, large egg shaped, average weight 20-30g, commercial cultivar
☆ Haak Yip	Mid season maturity, heart shaped with blotchy yellow to deep red fruits at maturity weighing about 21g, high yielder commercial cultivar
☆ Brewster	Mid late maturity, medium sized oval-round shaped, bright red coloured thick skin fruits weighing 20g, highly commercial cultivar
☆ Bengal	Mid late maturity, attractive bright red coloured thick skin fruits weighing 23-27g, highly commercial cultivar
☆ Sovey Tung	Late maturity, uneven heart shaped to oval, dark red skin fruit weighing 20 g, medium yielder, average quality fruit, less commercial
☆ Chacapat	Subtropical group, upland adaptability, late maturity, bright red skin fruit, large size, heart shaped, average fruit weight 20-24g, high commercial value cultivar
☆ Kiamana	Mid-season maturity, bigger sized heart shaped with purple-red skin, fruits weighing 25g, high yielding commercial cultivar.
☆ Salathiel	Early to mid maturity, egg to ball shaped fruit with blotchy yellow to deep red at maturity, average weight 15-18g, most fruits have chicken tongue seeds, commercial cultivar
	Florida and Hawaii (USA) (Chen, 1949; Li and Chu,1948;, Hamilton and Yee, 1970; Menzel and Simpson, 1986; Knight, 1963; Knight Jr., 2000)
☆ Brewster	Early maturity, medium to large size fruit weighing 20-26g, heart shaped with bright pink-red skin, high yielder, commercial cultivar
☆ Mauritius	Subtropical group, upland adaptability, bright red skin fruit, large egg shaped, average weight 20-30g, commercial cultivar
☆ Emperor	Subtropical group, upland adaptability, bright red skin fruit, round shaped, average weight 20g, medium yielder, less commercial cultivar
☆ Haak Yip	Mid season maturity, heart shaped with blotchy yellow to deep red fruits at maturity weighing about 21g, high yielder commercial cultivar
☆ Kaimana	Mid-season maturity, bigger sized heart shaped with purple-red skin, fruits weighing 25g, high yielding commercial cultivar.
☆ Kwai May Pink	Mid-season maturity, medium sized yellow-pink to orange pink fruits weighing 18-22g, high yielding commercial cultivar with export value.
☆ Kwai Mi	Mid-early maturing, deep red fruit weighing about 16-18g, high yielder, commercial cultivar
	South Africa (Penter *et al.*, 2000; Yamanishi and Kavati, 2000)
☆ Red Mclean's	Late maturity, attractive bright red coloured thick skin fruits weighing 22 g, medium yielder commercial cultivar
☆ Mauritius	Subtropical group, upland adaptability, bright red skin fruit, large egg shaped, average weight 20-30g, commercial cultivar

Contd...

Table 6.6–Contd...

Cultivars/Varieties	*Salient Characters*
	Israel (Goren and Gazit, 2000)
☆ Mauritius	Subtropical group, bright red skin fruit, large egg shaped, average weight 20-30g, commercial cultivar
☆ Nuo Mi Ci	The heart shaped, bright red skin fruits weighing 22g, small seeded, medium yielder, highly commercial cultivar
☆ Huai Zhi	Late maturity, fruits oval to round shaped, blotchy yellow to deep red skin at maturity, average fruit weight 20 g, small (chicken toungue) seeds, medium yielder and commercial cultivar.
	Canary Island (Cull and Lindsay, 1995; Galan *et al.*, 2000)
☆ Yu Her Pao	Early maturing, precocious, oval to round fruits with dark red tubercles on yellowish green skin background, fruit weighing about 18g, good quality, regular bearer, heavy yielder, highly commercial cultivar
☆ Early large red	Early maturing, precocious, oval to heart-shaped fruits with dark red skin, fruit weighing about 22g, good quality, regular bearer, high yielder, highly commercial cultivar
☆ Lenz	Late season maturity, oblong to round shaped fruits with dark red coloured skin, fruit weighing about 20g, alternate bearer, high yielder, highly commercial cultivar
☆ Calcuttia late	Late season maturity, oblong to round shaped fruits with dark red coloured skin, fruit weighing about 20g, alternate bearer, high yielder, highly commercial cultivar
☆ Brewster	Mid late maturity, medium sized oval-round shaped, bright red coloured thick skin fruits weighing 20g, highly commercial cultivar
	Pakistan (Panhwar,2000)
☆ Bengal	Late season maturity, attractive bright red coloured thick skin fruits weighing 25g, medium yielder, highly commercial cultivar
☆ Dehradun	Late season maturity, oblique heart to round shaped, attractive bright rose-pink coloured thick skin fruits weighing 20 g, good yielder, highly commercial cultivar
☆ Bedana	Late maturity, fruits heart shaped, orange-red skin at maturity, average fruit weight 25 g, small (chicken tongue) seeds, medium yielder, commercial cultivar.
	Bangladesh
☆ Muzaffarpuri	Early maturity, pink colour fruits, average weight 20g, commercial
☆ Bombai	Mid season maturity, carmine red colour fruits, average weight 18-20g, commercial cultivar and widely cultivated
☆ Bedana	Late maturity, bright red skin fruits, average weight 25-28g, highly commercial.
☆ China-3	Late maturity, fruit skin with patches of red, orange and green colour, average fruit weight 25g, best for commercial cultivation

Biodiversity of Litchi in Major Growing Countries

Biodiversity in litchi can be referred at three levels namely species level, genetic level and ecosystem level. The diversity of litchi cultivars results from natural selection (natural selection, genetic drift and mutation) and artificial process from generation to generation to adapt to human requirements under effects of social-economic/cultural and environment/biological factors. As litchi originated in Southern

China and is in cultivation since last 2000 years. It was introduced widely as a cultivated plant in the tropical and subtropical world (Campbell and Knight, 1990). Depending on the environment of certain regions, cultivation practices and uses, litchi cultivars developed differently. The origins had been diversified or lost and several litchi cultivars were renamed or mislabeled, resulting in confusion in the nomenclature of litchi cultivars (Aradhya *et al.*, 1995; Degani *et al.*, 1995). Apparently the same variety/cultivar is described under different names and in some locations/regions different cultivars are given the same name.

As mentioned earlier, the genus *Litchi* comprises of only one species *i.e. L. chinensis*, consisting of three subspecies (Leenhoults,1978). *Litchi chinensis* ssp., *chinensis*, from China the cultivated subspecies from the southern Indo-Chinense region adapted to sub-tropical climate; *Litchi chinensis* ssp., *phillipinensis* (Radlk.)Leenh., from the Phillipines and New Guinea and *Litchi chinensis* ssp., *javanensis*, from North Vietnam, Western Java and Cambodia and adapted to wet tropical climate. (Leenhouts,1978).

The various classification or attempts to group the cultivars/varieties are made in almost all the important litchi growing countries based on the agro-morphological factors like origin and source of material, morphology, environmental/ecological adaptation, and utilization. The programmes of work on litchi biodiversity require taxonomic capacity building to achieve the conservation objectives and status along with the care and maintenance of present state of the genetic resources conservation and the risk of future genetic erosion. Efforts are on in this aspect, as according to Galan Sauco (1995), FAO in 1993 established a mechanism for strengthening collaboration in fruit plant genetic resources and production called MESFIN (Mediterranean Selected Fruit Intercountry Network) which included litchi. Conservation status in litchi is now being updated and computerized in many countries.

Germplasm Collection, Evaluation, Conservation, Utilization and Exchange

According to Menzel (1992), the genetic diversity within litchi germplasm is limited, since most of the cultivars are selected from South China conditions. There is very little or negligible efforts made to collect the germplasm from original place *i.e.* China, the mainland. Systematic collection of germplasm of litchi is in advance stage in some important litchi growing countries. The specific criteria for collection of germplasm are only to widen the genetic variability, but for conservation point of view the important basis should be 1. Genic variation 2. Gene-ecological variation 3. Population size, 4. Legal conservation status 5. The Socio-economic contest and 6. Commercial importance. According to the situations, the various conservation methods are applied like *in-situ* conservation, *ex-situ* conservation, *in-vitro* conservation (such as embryo-shoot tip culture) and cryopreservation. For advanced conservation strategies, support from National and International programmes and International Plant Genetic Resource Institute (IPGRI)(Now Biodiversity International) may be sought upon. As per the information available, the population structure of litchi in its major producing countries is outlined as follows:

China

In China, the existing elite and the oldest germplasm has a history of over 400 years and the trees are still thriving very well and bear heavily. The litchi variable germplasm is mainly distributed in Guangdong, Guangxi, Fujian, Hainan, Sicun, Yunnan, Guizhou, Zhejiang and Taiwan provinces, lying between 19 and 23° 50′N. It is mainly concentrated in the tract between 22 and 23° 40′N, from where most elite germplasm and varieties come. The existing elite mother tree germplasm-Edanli (Goose egg litchi), noted for its giant fruit and excellent flavour, is found around 19° 40′N within Hainan Province, and is known for more than 300 years. Germplasm of this type is tolerant to higher temperature. Systematically a sizable number of 150 accessions of litchi are conserved and maintained in the field gene bank of Tropical Fruit Tree Institute at Wushan, Guangzhou and data on 100 accessions are documented. It is observed that a systematic collection of litchi germplasm including both wild and cultivated forms in both *in situ* and *ex situ* conservation in Fujian, Guangdong and Guangxi regions of China (Chen and Huang, 2001; Liu and Ma, 2001). Numerous wild litchis can be found in the moist forests of Hainan Island (low land area) and in hilly areas of Leizhou Peninsula, western Guangdong and eastern Guangxi (Hsu *et al.*, 1964),There are large collections of litchi at Research Institutes *i.e.* The Chia-yi Agricultural Experimental Station in Taiwan (Yen *et al.*, 1984). The salient features of some of the native germplasms is given below.

Table 6.7: The Salient Features of some Native Germplasm of Litchi in China

Sl.No.	*Particulars*	*Germplasms/Cultivars/Varieties*
1.	Small seed	Nuomizhi (No Mai Tzi, Glutinous Rice Litchi), Lingshan Xiangli (Lingshan Fragrant Litchi), Hainan Xiaodingxiang (Hainan's Small Clove), Guangxi Zhangluoli (Guangxi Zhangluo Litchi).
2.	Crisp and sweet flavour with less tannin	Lingshan Xiangli.
3.	Early maturity	Shanyuehong (March) red, ripe in early May.
4.	Giant fruit	Edanli; 60-70 g fruit.
5.	Sustained higher yield	Heiye (Black Leaf), Baitangying (White Sugar Poppy)
6.	Ability for flower bud differentiation at higher temperature	Shanyuehong: better flower bud differentiation at 20°C as compared to 12°C for common varieties.
7.	Good canning quality	Heiye, Xuehuaizi.
8.	Drought and infertility tolerance, good adaptability	Tianyan (Sweet Stone), excellent performance in yield and fruit quality, high seed germination rate, fast growth, good for rootstock
9.	Good on-tree preservation	Huaizhi, fully ripe fruits that can be left on trees for fresh picking.
10.	Late maturity	Xuehuaizi, ripening during early/mid July, with high-yielding characteristics;
		Fijian Xiafanli (Fujian Xiafan Litchi), maturing during late July or early August enabling extended supply period in combination with early and intermediate maturity types.

India

Though, litchi was introduced to India from China during 18th century and its production has undergone substantial expansion here. There exists a narrow genetic base. India has collections of local selections of litchi in the states of Bihar, West Bengal, Uttarakhand and Punjab. In an effort for locating and collecting diversity in litchi, the study made on distribution of litchi genotypes in North Bihar in India has been outlined for diversity mapping. In India, 52 accessions are being maintained in the field gene bank at the National Research Centre for Litchi (NRCL), Muzaffarpur, Bihar and about 50 at Horticulture and Agro-forestry Research Programme (HARP), Ranchi, Jharkhand out of which 17 accessions were characterized for cluster, fruit and seed characters alongwith physico-chemical characteristics (Bhagmal *et al.*, 2005). The maintenance of germplasm is limited to field gene bank, little efforts are made on development of *in-vitro* condition. The Indian litchi industry is based mainly on two cultivars, Shahi and China. Due to short harvesting season and undesirable time of ripening of these cultivars, there is an urgent need for improved cultivars. All locally available, promising cultivars and selections are being evaluated in the main litchi producing areas. Almost all important genotypes of litchi available in the country are collected and are being maintained in field gene bank at different research centres/institutes. Maximum collections (52 accessions) are being maintained at National Research Centre for Litchi, Muzaffarpur, Bihar.

Table 6.8: Number of Accessions in Field Gene Bank at Different Centres

Sl.No.	*Centre/Institute/SAUs*	*No. of Accessions*
1.	Bidhan Chandra Krishi Vishwavidyalaya, Mohanpur, West Bengal.	16
2.	Rajendra Agricultural University, Pusa, Samsatippur, Bihar	20
3.	G.B. Pant University of Agriculture and Technology, Pantnagar, Uttarakhand	30
4.	ICAR Research Complex for Eastern Region (Horticulture and Agro-Forestry Research Programme, Plandu, Ranchi, Jharkhand)	51
5.	Central Institute of Sub-tropical Horticulture, Lucknow, Uttar-Pradesh	34
6.	National Research Centre for Litchi, Muzaffarpur, Bihar	52

Thailand

Litchi was first introduced into Thailand from China only 300 years ago through seedlings. These seedlings adapted to tropical conditions and named as low land litchi or tropical litchi do not require low temperature for flowering and fruiting, while another type *i.e.* Upland litchi or Subtropical type got introduced in the form of air-layers are truly subtropical types and require low temperature phase for proper flowering and fruiting. As reported by Yapwattanaphun and Subhadhrabandhu (2001) that Thailand has collections of litchi in Chiang Rai, Chiang Mai, Lamphun and Bangkok (in Kasetsart University, Bangkok).

Vietnam

Northern Vietnam is considered a part of the centre of origin of this crop. The distribution map of litchi was prepared based on collecting data using GIS tool. Litchi is growing wild as well as cultivated in Northern and Central Vietnam such as Ha Tay, Vinh Phu, Hai Phong, Hi Duong, Cao Bang and Lang Son (Rao *et al.*, 2005). The variability in terms of production is found in areas like Hanoi, Baçgiang, Haidong and Quangninh. In Vietnam, a total of 30 numbers of litchi accessions are systematically maintained and managed at the Research Institute of Fruits and Vegetables (RIFAV) in Hanoi (Papademetriou and Dent, 2002). The conventional approach to conservation of plant genetic resources in Vietnam is *ex situ* conservation (a total of 77 accessions : 32 in Vietnam Agricultural Science Institute, Hanoi + 17 in Phuho Fruit Research Centre, Vinhphu + 28 in Research Institute for Fruits and Vegetables, Hanoi) and the work on characterization, evaluation and utilization of litchi germplasms is in progress with limited number of cultivated types but the collection comprises of wild relatives, traditional cultivars, improved cultivars and improved/introduced materials. The conservation efforts in *Litchi chinensis* in Vietnam are concentrated on the highly endangered populations of this species, which were genetically distinct. The sustainable conservation involves *in situ* conservation methods which complement the *ex situ* conservation and *in vitro* conservation (Hoa *et al.*, 2005).

Australia

Promising cultivars from other litchi producing countries including Australia are included in the evaluation programme and it helps in broadening the genetic base of litchi cultivars (Oosthuizen,1994). The development of the litchi cultivars available to the Australian industry took place in last 20-25 years and in excess of about 40 cultivars are imported here from the significant growing regions of the world, including China, Taiwan, India, Thailand, South Africa and Mauritius, as well as the USA and these are held in arboreta in Queensland and New South Wales. In tropical America, there are adequate genetic resources available. (Campbell and Campbell, 2001). There are about 50 collections of litchi in Hilo, Hawaii and at various centres in Northern Australia (Aradhya *et al.*, 1995; Menzel, 2002) and from here new materials are still being taken to different locations, such as Central America for commercial evaluation/cultivation.

Israel

There are collections of litchi at various research centres and universities in Israel. The Volcani Centre and the Hebrew University of Jerusalem are evaluating about 80 accessions introduced from China, Thailand, Vietnam, India, South Africa and Australia (Goren *et al.*, 2001).

USA

In the western hemisphere, litchi is also significantly important in Florida and germplasm collections of litchi are housed and under continual evaluation at the USDA-ARS Subtropical Horticulture Research Station, Miami, Florida and USDA-ARS Tropical Plant Genetic Resources Management Unit, Hilo, Hawaii (Crane *et al.*, 2005). Florida has a collection of 27 litchi cultivars, being evaluated at the State University of Florida at Homestead and at the Fairchild Tropical Garden in Miami (Campbell and Campbell, 2001 a,b). Though recent but commercial clonal orchards with selected cultivars/varieties are established and within germplasm collections, the majority of the world's important cultivars can be found and in the genetic resources here are dominated by "Mauritius" and "Brewster", which show the most flowering success under tropical conditions. Apart from these two, Haak Yip, Emperor, Kawi Mai Pink and Kwai Mai received attention in the region, but they remain unproven for commercial production. University of California is currently assessing litchi cultivars suitablility in different locations throughout the state. The University of Puerto Rico Agricultural Experiment Station (UPR-AES) maintains a litchi germplasm collection and that too in the USDA Tropical Agriculture Research Station (USDA-TARS), Mayaguez, USA.

South Africa

South Africa has a collection of about 40 litchi cultivars at Nelspruit. There are two newly released varieties/cultivars Fay Zee Siu (early maturing) and No Mai Chee (late maturing) imported from Australia in the late 1980s. The cultivar Bengal is an important cultivar of commerce and found growing in Sao Paulo State, Brazil. CIRAD in Reunion has a collection of about 50 litchi cultivars at various locations on the Island (Normand and Bouffin, 1994).

To encourage the commercial development of litchi, many cultivars need to be introduced from around the world to evaluate their performance under local conditions. The reliability of fruiting, disease resistance, yield potential and commercial potential are priorities for the commercial suitability of a cultivar (Campbell and Campbell, 2001). The litchi germplasm evaluation programme is in advance stage at many places to identify the most suitable early, mid and late season cultivars for major growing areas, consisting of commercial cultivars, imported cultivars and clonal selections from the breeding programmes (Rheedar, 1999). IPGRI supported a project entitled "Development of cryopreservation techniques for long term conservation of recalcitrant seed and vegetatively propagated plant species in India" which is implemented at the National Bureau of Plant Genetic Resources (NBPGR), New Delhi, where the attempts for cryopreservation of litchi showed that the most responsive procedure as the desiccation freezing method of axes and an alternative method of *in vitro* slow growth was also successfully attempted in litchi (Rao *et al.*, 2005).

The developments of research on litchi germplasm resources are very slow inspite of considerable studies pertaining to origin and distribution, collection and conservation, classification and identification, selection and breeding of litchi germplasm resources in China. The classification of litchi can be divided into three stages: classification based on morphological traits, classification by isozyme analysis and DNA markers. The expression of morphological traits is often affected by environmental conditions; the number of loci that isozyme analysis can examine is small. Cultivars, which are closely related often have identical isozyme patterns. The limitations of classification based on morphological traits and isozyme analysis are then overcome by the utilization of molecular markers. At present, most new litchi cultivars are selected from superior seedlings obtained after open-pollination or sport; germplasm with desirable characteristics such as aborted seeds, rather big-fruit and rather late-bearing have not been fully exploited. In various efforts made to achieve the goal of improving fruit quality and agronomic characteristics, it is suggested that additional studies of genetic diversity on litchi and longan germplasm should be carried out with more powerful markers such as SSR and ISSR; marker assisted selection (MAS) should be employed to improve litchi through intra-inter species cross breeding efficiency. In India, genetic diversity was also assessed in very limited number of cultivars of litchi on the basis of six fruit characters and the study revealed enormous diversity in the material as indicated by the wide range of D^2 values (Tocher's method), it is expected to give maximum extent of heterosis whereas crosses between diverse clusters group should yield derivatives with better fruit characters (Dwedi and Mitra, 1996). Globally, the efforts are on to develop passport data and collected data on litchi has to be computerized. The information on existing and new accessions are being compiled and converted into electronic form for making that widely accessible. The spread of biodiversity can be traced back with the origin, history and spread of the crop.

Introduction and exchange of germplasm of litchi crop are faced with several difficulties. Although seed is the most ideal and convenient medium for import of germplasm but its recalcitrant nature poses great problem and it also does not breed true to the type. Import of germplasm through bud-sticks and cuttings is also not successful or has become viable in litchi. Apart from it, the procedural formalities, as first received by NBPGR, New Delhi for quarantine clearance and then sent to the Institute aids the complexities. Practically very little introduction/import or exchange of clones of evergreen sub-tropical fruit crops *viz.* litchi has taken place in past a few decades. As grafting protocol has not been standardised in litchi, the import of grafted plants is also not feasible, we may have to resort to introduction of some variability through seed as the second best alternative. But, fresh recalcitrant seeds lose viability slightly as they are dried, subsequently viability drops considerably at "critical moisture content." Further, drying results in complete death of the seeds. Critical moisture content levels vary among species, cultivars and seed lots (King and Roberts 1979; Chin 1988). Desiccation of excised embryos and embryonic axes has considerable potential for the *in vitro* conservation of recalcitrant embryos, since embryos are capable of surviving desiccation to lower moisture contents than the whole seeds (Chin 1988). The litchi in most of the producing countries outside China are based on introduced cultivars. The recalcitrant seeds of litchi are too short lived for germplasm collection, hence most of the germplasm collections are maintained as trees. The distribution of planting material can be made mainly as air-layers and rarely by cuttings and graft-wood.

Conclusion

The future of world litchi mainly depends on the ready availability of improved planting materials. A lot of improvement is warranted for adaptability to adverse weather conditions. To obtain any significant breakthrough in breeding, the litchi breeders should have a very strong assembly of diverse germplasm. One of the serious limitations of litchi breeding programmes in major producing countries is the narrow genetic base and that too maintained differently in varied places. Clones with unique characters are being collected through well organized explorations to the center of origin and neighbouring areas. Apart from this, many useful materials are being produced in different litchi research institutes. All these improved and diverse genotypes should be accessible to the breeder to make effective genetic manipulations and evolve more desirable genotypes. Wonderful results through breeding can be achieved only through systematic collection, conservation and utilization of biodiversity. In order to increase the commercial development of litchi, many cultivars need to be introduced from around the world to evaluate their performance under local conditions. The realibilityof fruiting, disease resistance, yield potential and commercial potential are considered priority for the commercial suitability of a cultivar.

References

Anupunt, P. and Sukhvibul, N., 2005. Lychee and longan production in Thailand. In: *Proceeding of 2nd International Symposium on Lychee, Longan, Rambutan and Other Sapindaceae Plants*, (Eds.) N. Chomchalow and N. Sukhvibul, p. 53–60.

Arora, R.K., Rao, V.R. and Rao, A.N., 1996. *Status Report on Genetic Resources of Litchi in China*. IBPGR Regional Office for Asia. The Pacific and Oceania, Singapore, p. 1–18.

Bailey, L.H., 1949. *Manual of Cultivated Plants*. McMillan Publishing Company, New York, p. 450.

Bailey, L.H., 1950. *The Standard Cyclopedia of Horticulture*. The McMillan Co., New York, 2: 1891–1893.

Banerjee, I. and Choudhary, K.L., 1944. A contribution to the life history of Litchi (*Litchi chinensis* Sonn). In: *Proc. Indian Acad. Sci.*, 19(B): 19–24.

Banta, E.S., 1952. Behold! The Lychee. *Amer. Fruit Grower*, 72(10): 10–11, 20–21.

Batten, D.J., 1986. Towards an understanding of reproductive failure in lychee (*Litchi chinensis* Sonn.). *Acta Hort.*, 175: 79–83.

Batten, D.J. and McChonchie, C.A., 1992. Pollination in lychee. In: *Proceedings of the Third National Lychee Seminar*. Australian Lychee Growers' Association, Bundaberg, Australia, p. 23–28.

Britton, N.L. and Wilson, P., 1926. Botany of Porto Rico and Virgin Isles. *Descriptive Flora–Spermatophyta, Scientific Survey of Porto Rico and Virgin Isles*. New York Academy of Science, 6(3).

Butcher, F.G., 1957a. Pollinating insects of lychee blossoms. In: *Proceeding of Florida State Horticulture Society*, 70: 326–328.

Chandler, W.H., 1958. Evergreen orchards. *Lea and Febriger*, Philadelphia, USA, p. 320–325.

Chen, W.H., 1949. The culture of the lychee. *Proc. Flo. State Hort. Soc.*, 61: 223–226.

Cobin, M., 1952. The lychee in Florida. *Fruit Variety and Horticulture Digest*, 6: 52–53.

Cobin, M., 1954. Lychee in Florida. *Bulletin Florida Agriculture Experiment Station*, 546: 1–35.

Chadha, K.L., 1968. Litchi cultivation in India. *Indian Horticulture*, 12: 13–16.

Chapman, K.R., 1983. Litchi (*Litchi chinensis* Sonn.) In: *Australian Tropical Tree Fruits*. Queensland Government Printer, Brisbane p. 179–191.

Chapman, K.R., 1984. Sapindaceae: Lychee (*Litchi chinensis* Sonn.). In: *Tropical Tree Fruits of Australia*, (Ed.) P.E. Page. Queensland Government Printer, Brisbane p. 179–191.

Chia, C.L., Hamilton, R.A. and Evans, D.O., 1997. Lychee Univ. Hawaii C/T/A/H/R *Hort. Commod. Fact Sheet* –1/. 4 pp.

Cull, B.W. and Lindsay, P., 1995. *Fruit Growing in Warm Climate.* Reed Books Australia, Chaiswood.

De Candolle, A., 1909. *Origin of Cultivated Plants,* 2nd edn. International Scientific Series. Kegan Paul, Trench, Triibner and Co. Ltd. London.

Degani, C., Beiles, A., El-Batsri, R., Goren, M. and Gazit, S., 1995. Identifying lychee cultivars by isozyme analysis. *Journal of the American Society of Horticultural Science*, 120: 307–312.

Deng, Y.C. and Ni, Y.Y., 1994. Photosynthesis of litchi tree and influence of potassium on accumulation of photosynthates. *Guangdong Agricultural Sciences*, 2: 20–22.

Dixton, T., Batten, D. and McChonchie, C., 2003. *Genetic Improvement of Lychee in Australia.*

Dutt, A.K., 1962. Response of Litchi to minor elements in laterite soils. *Fert. News*, 7: 24–26.

Dwedi, A.K. and Mitra, S.K., 1996. Divergence analysis of litchi (*Litchi chinensis* Sonn.) cultivars growing in West Bengal. *Indian J. Genet. Plant Breeding*, 56(4): 486–489.

Fivaz, J. and Robbertse, P.J., 1995. Possible pollination factors causing fruit drop in litchi (*Litchi chinensis* Sonn.). *Year Book of the South African Litchi Growers Association*, 7: 26–30.

Galan Sauco, V., 1989. *Litchi Cultivation*. Food and Agriculture Organization, Rome, p. 136.

Galan Sauco, V. and Menini, U.G., 1989. *FAO Plant Production and Protection Paper*, 83, Rome.

Ghosh, S.P. and Mitra, S.K., 2000. *Status Report on Genetic Resources of Litchi in India, Thailand and Mynmar*, (Eds.) S. Bhagmal, Ramanatha Rao and R.K. Arora. IPGRI South Asia Office, New Delhi, India.

Galan Sauco, V., Verdz Pirez, J.D. and Harnandez, P.M., 2000. Behaviour of different litchi cultivars in Canary Island. *Programme and Abstract, 1st International Symposium on Litchi and Longan, China, June* 19–23.

Goren, M. and Gazit, S., 2000. Litchi cultivars in Israel. In: *Programme and Abstract, 1st International Symposium on Litchi and Longan, China, June* 19–23.

Goren, M., Tomer, E. and Gazit, S., 2001. Litchi cultivars in Israel. *Acta Horticulturae*, 558: 129–133.

Goto, Y.B., 1960. Litchi and its processing. *Pacific Region Food Conference*, 1: 15–23.

Groff, G.W., 1943. Some ecological factors involved a successful lychee culture. *Proc. Flo. State Hort. Soc.*, 56: 34–55.

Hai, V.M and Dung, N.V., 2002. Lychee production in Vietnam. In: *Lychee Production in the Asia-Pacific Region*, (Eds.) M.K. Papadementriou and F.J. Dent. Food and Agricultural Organization of the United Nations, Bangkok, Thailand, p. 114–119.

Hamilton, R.A. and Yee, W., 1970. Lychee cultivars in Hawaii. In: *Proceedings of Tropical Region, American Society of Horticulture*, 14 : 7–12.

Hasan, M.A. and Chattopadhyay, P.K., 1990. Effect of different soil moisture regimes on growth and yield of litchi cv. Bombai. *Indian J. Hort.*, 47: 397–400.

Hooker, J.D., 1875. *The Flora of British India,* Vol. 1. L. Reeve and Company, London, p. 685–687.

Hong, L.T., Yuan, Y.P., Xi, O.L., Zhen, C.J. and Shu, L.J., 1996. *Status Report on Genetic Resources of Litchi in China*, (Eds.) R.K. Arora, R.V. Ramanatha and A.N. Rao. IPGRI South Asia Office, New Delhi, India.

Hsu, H.H., Peng, C., Hsuch, J.H. and Tsai, Y.F., 1964. The discovery of wild lychee trees. *Acta Horticultarae Sinica*, 3: 415.

Jawanda, J.S. and Singh, Sohan, 1977. Fruit yield, ripening time and fruit quality of litchi cultivars grown in Punjab. *Punjab Hort. J.*, 17(3/4): 122–125.

Joubert, A.J., 1986. The Litchi. In: *Handbook of Fruit Set and Development*, (Ed.) Shaul P. Monselise. CRC Press Inc., Florida, p. 233–246.

Katyal, S.L. and Chugh, D.V., 1961. Problems in extensive litchi cultivation. *Hort. Advances*, 5: 39–41.

Knight Jr., R.J., 2000. The lychee history and current status in Florida. In: *Programme and Abstract, 1st International Symposium on Litchi and Longan*, China, June 19–23.

Knight, R., 1980. Origin and World importance of tropical and sub-tropical fruit crops. In : *Tropical and Sub-tropical Fruits: Composition Properties and Uses*, (Eds.) S. Nagy and P.E. Shaw. Avi Publishing West Port (Connecticut), p. 1–20.

Knight, R.J., 1963. The lychee evaluation programme at the U.S. Plant Introduction Station, Miami, Florida. In: *Yearbook–Proc. Florida Litchi Growers' Assoc.*, 10: 47–51.

Lal, H. and Nirwan, R.S., 1980. Saharanpur selection: A new litchi cultivar. *Prog. Hort.*, 12(3): 75–77.

Leenhouts, P.W., 1971. A revision of *Dimocarpous* (Sapindaceae). *Blumea*, 19: 113–129.

Leenhouts, P.W., 1978. Systematic notes on the Sapindaceae–Nephelieae. *Blumea*, 24: 395–403.

Liang, J., 1981. The litchi: Origin, utilization and development of its culture. *J. Agric. Trinidad Botany Applied*, 28: 259–270.

Li, L.Y. and Chu, Y.C., 1948. Note on the Chen-Tzo Lychee of Henghwa, Fukian, China. *Proc. Florida State Hort. Soc.*, 60.

Liu, S.Y., 1954. Studies on *Litchi chinensis* Sonn. *Dissertation Abstract*, 14: 1018–1019.

Lu, L.X., Chen, J.L. and Chen, X.J., 1987. A study on chromosome number and meiosis in pollen mother cells in *Litchi chinensis* Sonn. *Journal of Fujian Agricultural College*, 16(3): 224–228.

McChonchie, C.A. and Batten, D.J., 1989. Floral biology and fruit set in lychee. In: *Proceedings of 2nd National Lychee Seminar*. Australian Lychee Growers Association, Cairns, Australia, p. 71–74.

McChonchie, C.A. and Batten, D.J., 1991. Fruit set in Lychee (*Litchi chinensis* Sonn.): Variation between flowers, panicles and trees. *Australian Journal of Agricultural Research*, 42(7): 1163–1172.

McGregor, S.E., 1976. *Insect Pollination of Cultivated Crop Plants*. US Department of Agriculture Handbook 496: 247–248.

Marloth, R.H., 1947. The litchi in South Africa. 1. Varieties and propagation 2. Cultural practices and marketing. *Farming South Africa*, 22: 824–870.

Mathura, Rai, Vishal Nath and Dey, Pradip, 2001. *Litchi*. CHES, Plandu, Ranchi, Jharkhand.

Menzel, C.M., 1991. *Litchi chinensis* Sonn. In: *Plant Resources of South-East Asia*, (Eds.) E.W.M. Verheij and R.E. Coronel. Wageningen, Pudoc, pp. 191–195.

Menzel, C.M. and Simpson, D.R., 1992. Flowering and fruit set in lychee (*Litchi chinensis* Sonn.) in subtropical Queensland. *Australian Journal of Experimental Agriculture*, 32: 105–111.

Menzel, C.M., Olesan, T. and McChonchie, C.A., 2000 Lychee, longan and rambutan. Optimizing canopy management. *Final Report to The Rural Research and Development Corporation*, Canberra, p. 92.

Menzel, C.M., Rasmussen, T.S. and Simpson, D.R. 1989. Effect of temperature and leaf water stress on growth and flowering of litchi (*Litchi chinensis* Sonn.). *J. Hort. Sci.* 64:739–752.

Mustard, M.J. 1954. Fundamentals of panicle differentiation. *Proceedings of the Florida Lychee Growers' Association*. 1:7–8.

Mustard, M.J., Nelson, R.O. and Goldweber, S. 1956. Exploratory study dealing with the effect of growth regulators and other factors on the fruit production of lychees. *Proc. Florida Lychee Growers Assoc.* 3:33–38.

Nakasone, H. Y. and Paull, R.E. 1998. Tropical fruits. In: *Crop Production Science in Horticulture Series, No. 7*. Wallingford, UK, CAB International, pp 173?207.

Nakata, R.C. and Watanable, Y. 1966. Effect of photoperiod and night temperature on the flowering of *Litchi chinensis*.Sonn. *Botanical Gazette* 127:146–152

Nakata, S., and R. Suehisa. 1969. Growth, and development of *Litchi chinensis* as affected by soil–moisture stress. *American J. Bot.* 56:1121–1126.

Nijjar, G.S. 1972. Litchi cultivation. Punjab Agricultural University, Ludhiana.

Ochse, J.J.; Soule, M.J. Jr.; Dijkman, M.J. and Wehlburg, C. 1961. Other fruit crops. In : *Tropical and sub–tropical Agriculture* Vol. 1. The Mac Millan Co., New York.

Toa, R. 1955. The superior lychees. *Proc. Florida. Lychee Grow. Assoc.*, 2 : 73–74.

Olesen, T., Menzel, C.M., Wiltshire, N. and McConchie, C.A. 2002. Flowering and shoot elongation of lychee in eastern Australia. *Australian Journal of Agricultural Research.* 53:977–983.

Pandey, R.M. and and Sharma, H.C. 1989. The litchi. Indian Council of Agricultural Research, New Delhi : 1–80.

Pandey, R.S. and Yadava, R.P.S. 1970. Pollination of litchi (*Litchi chinensis*) by insects with special reference to honey bees. *J. Agric. Res.*, 9 : 103–105.

Panhwar, F. 2000. Litchi cultivation in Sindh Pakistan. *Programme and Abstract, 1st International Symposium on Litchi and Longan,* China, June 19–23.

Penter, M., Kift, J. and Stassen, P.J.C. 2000. Chemical manipulation of litchi tree to improve fruit yield and quality. *Programme and Abstract, 1st International Symposium on Litchi and Longan,* China, June 19–23,2000.

Popenoe, W. 1920. Manual of Tropical and Sub–tropical Fruits, Mac Millian Co., New York.

Ray, P.K. and Sharma, S.B 1987. Growth, maturing, germination and storage of litchi seeds. *Scientia Horticulturae*. 33:213–221.

Palmer, G. 1956. Some aspects of the Lychee as a commercial crop. *Proceedings of the Florida State Horticultural Society*. 69:308.

Radlkofer, L. 1932. Sapindaceae. Pflanzenreich, Weinheim, Germany. P. 1539.

Roy, R. N., Rao, D.P., Mukherjee, S.K. and Chaterjee, B.K. 1987. Studies on the feeder root distribution of litchi (*Litchi chinensis* Sonn.). cv. Bombai. *Indian Agriculturist,* 31(1): 33–41.

Robbertse, P.J., Fivaz, J. and Menzel, C.M. 1995. The structure of the inflorescence in relation to the position and sex of the flower. *Year Book–South Africa Litchi Growers Association*, 7: 21–25.

Sengupta, B.N., Singh, Sanjay and Kumar, Rajesh, 2001. Cultivars. In. *Litchi: Botany, Production and Utilization,* (Ed.) K.S. Chauhan, p. 19–27.

Singh, L.B. and Singh, U.P., 1954. *The Litchi.* Superintendent Printings and Stationery, Allahabad, Uttar Pradesh.

Ray, P.K., Sharma, S.B. and Mishra, K.A., 1984. Important litchi cultivars of Bihar. *Indian Horticulture*, 30(1): 9–13.

Roy, R.S., 1952. *A Consolidated Report of the Fruit Research Scheme, Bihar (1933–1951).* Superintendent Government Printing, Bihar, Patna.

Simmonds, N.W., 1976. *Evolution of Crop Plants*. Longman, London, p. 321.

Singh, H.P., 1998. Genetic diversity, breeding and utilization of the genepool of Litchi. In: *Tropical Fruits Asia: Diversity, Maintenance, Conservation and Use*, (Eds.) Arora, R.K. and Rao, V.R. IBPGRI Office for South Asia, New Delhi, p. 185–195.

Singh, H.P. and Babita, S., 2002. Lychee production in India. In: *Lychee Production in the Asia-Pacific Region*, (Eds.) M.K. Papademetriou and F.J. Dent. Food and Agriculture Organization of the United Nations, Bangkok, Thailand, p. 55–67

Shukla, R.K and Bajpai, P.N., 1974. Blossom and bud differentiation and ontogeny in litchi (*Litchi chinensis* Sonn.). *Indian J. Hort.*, 31(3): 226–228.

Singh, K.K. and Jawanda, J.S., 1962. Litchi cultivation in the Punjab. *Punjab Horticulture Journal*, 2: 69–74.

Stephens, S.E., 1955. The Litchi. *Queensland Agriculture Journal*, 81: 19–20.

Stern, R.A., 1992. Elucidation of the factors affecting litchi productivity in Israel and the development of methods to improve its yield. *Ph.D. Thesis*, Faculty of Agriculture, Hebrew University, Rehovot, Israel.

Stern, R.A., Eisenstein, D., Voet, H. and Gazit, S., 1996b. Anatomical structure of two day old litchi ovules in relation to fruit set and yield. *Journal of Horticulture Science*, 71: 661–671.

Stern, R.A., Eisenstein, D., Voet, H. and Gazit, S., 1997a. Female Mauritius litchi flowers are not fully mature at anthesis. *Journal of Horticultural Science,* 72: 19–25.

Stern, R.A. and Gazit, S., 1996. Lychee pollination by honey bee. *J. American Soc. Hort. Sci.*, 121: 152–157.

Stern, R.A. and Gazit, S., 1998. Pollen viability in litchi. *J. American Soc. Hort. Sci.*, 123(1): 41–46.

Stern, R.A., Gazit, S., El-Batsari, R. and Degani, C., 1993b. Pollen parent effect on outcrossing rate, yield and fruit characteristics of 'Floridian', and 'Mauritius' lychee. *J. American Soc. Hort. Sci.*, 118: 109–114.

Stern, R.A., Gazit, S. and Goren, M., 1996a. Factors affecting litchi productivity (in Hebrew). Annual Report for 1996. Ministry of Agriculture, Israel.

Stern, R.A., Gazit, S. and Goren, M., 1997b. Factors affecting litchi productivity (in Hebrew). Annual Report for 1997. Ministry of Agriculture, Israel.

Storey, W.B., Hamilton, R.A. and Nakasone, H.Y., 1953. Groff: a new variety of lychee. *Circular of the Hawaii Agricultural Experimental Station,* 39: 1–8.

Tindal, H.D., 1994. Sapindaceous fruits; botany and horticulture. *Horticultural Reviews,* 16: 143–196.

Wang, B. and Qui, Y., 1997. Growth and fruiting In. *Litchi: Pictorial Narration of Cultivation,* (Eds.) Z.W. Zhang, P.Y. Yuan, B.Q. Wang and Y.P. Qui. Pomology Research Institute, Guangdong Academy of Agricultural Sciences, Guangzhou, China, p. 66–88.

Xuequin, Z., 1995. Genetic resources of tropical fruit species in China. In: *Proceedings of the Export Consultation on Tropical Fruit Species of Asia,* (Eds.) R.K. Arora and V.R. Rao. IPGRI Office for South Asia, New Delhi, p. 103–111.

Yamanishi, O.K. and Kavati, R., 2000. An overview of litchi production in Sao Paulo State, Brazil. In: *Programme and Abstract, 1st International Symposium on Litchi and Longan,* China, June 19–23.

Yang, Q.W., Huang, H.L., Wang, T.Y. and Ping, L., 1998. Exploration and research on crop germplasm resources in Sichuan, Shaanxi, Guizhou and Guangxi Province. *Crop Genetic Resources,* 1: 5–7.

Yokoyama, K.M., Wanitprapha, K., Nakamoto, S.T. and Chia, C.L., 1991. Lychee. University of Hawaii. *C/T/A/H/R Economics Fact Sheet,* 12: 4.

Yule, Sir Henry, Bunell, Arthur Coke and Hobsen-Jobson, 1903. Being a glossary of Anglo-Indian colloquial words and phrases, etymological, historical, geographical and discursive. New edition, William J., Murray, London.

Zhang, Z., Yuan, Wang B., Qui, Y. and Li, J., 1999. *Litchi: Pictorial Narration of Cultivation.* Pomology Research Institute, Guangdong Academy of Agril. Sciences, China.

2011, Biodiversity in Horticultural Crops Vol. 3 *Pages* **120–203**
Editor: **Professor K.V. Peter**
Published by: **DAYA PUBLISHING HOUSE, NEW DELHI**

Chapter 7

Mangoes of Western Ghats

S.P. Singh
ICAR Research Complex for Goa,
Ela Old Goa, Goa – 403 402,
E-mail: spsinghchandel@vsnl.com

Goa is the smallest state aligning the Western Ghats of India with a total area of 0.361 million ha of which the cropped area is 0.17 million ha, where as net area sown is 0.142 million ha. Out of total cultivable area of Goa, all horticultural crops occupy about 47.18 per cent. Goa has rich varietal diversity in fruits like mango. Several varieties of mango were introduced to the state. The total area under mango was 4,196 ha. in 2002-03. The productivity of mango in the state was 4,178 kg/ha. The total production has increased significantly from 10,000 tonnes in 1990-91 to 17,532 tonnes in 2002-03. Mankurad, Hilario, Musserat, Colaco Branca, Maxima, Fernandin, Goa Hapus and Udgo are a few of the important varieties available in the state. About 130 varieties are reported native of Goa. ICAR Research Complex for Goa, Ela, Old Goa maintains 172 total accessions in its germplasm collection of which 122 are local, 32 introduced varieties and 18 hybrids. About 40 local varieties are replanted and evaluated for their quality characteristics and yield performance. Growth and yield performance of 51 local mango varieties were studied in detail. Maximum height of 6.70 m. was in Maxima followed by Mankurad Keni (6.5m). Maximum number of fruits (247/plant) were in variety Toranja followed by Udgo (189 fruits/plant). Maximum fruit yield (kg/plant) was in variety Jesuit(49.70) followed by Udgo (45.1) and Maxima (39.34). Yield efficiency of the plant (kg/m^2) was maximum in variety Colaco Branca (2.34), followed by Costa (2.330) and Mankurad (1.632). Maximum individual fruit weight (520.67g) was in variety Jesuit followed by Bismi (520g). Mankurad Yellow recorded the average fruit weight of 264.00 g/fruit. Length of the fruit was maximum (20.90 cm), in Koita followed by Black Culas (12.23cm.) and Jesuit (11.89cm). Pulp weight of fruit was maximum (392 g) in variety Bismi. Pulp: stone ratio was maximum in variety Carrera Alphonso(6.7:1). Maximum pulp per cent (75.38) was in variety Bismi followed by Colaco Branca (73.24) and Mankurad Dodamarg (73.04).

T.S.S. of fruit was maximum in Hilario (4/15) (31° Bx) followed by Malgoa (23° Bx). Minimum T.S.S. (6°Bx) was in variety Mirand.

Gen bank of different mango varieties and landraces in Western Ghats is maintained in *ex situ* in States of Gujarat, Maharashtra, Goa, Karnataka, Kerala and Tamil Nadu. In Gujarat, gene bank is maintained at Balsar (177 accessions). Gujarat Agricultural University, Anand maintains 81 accessions/varieties. In Maharashtra, Regional Fruit Research Station, Vengurla maintains 379 varieties/accessions, Mahatma Phule Agricultural University, Rahuri 86 varieties/accessions, Marathwada Agricultural University, Parbhani 80 varieties/accessions and Fruit Research Station, Aurangabad, Maharashtra 130 accessions/varieties. The ICAR Research Complex for Goa maintains 172 accessions/varieties. In Karnataka, Indian Institute of Horticultural Research (IIHR), Bangalore maintains main genebank of 269 accessions/varieties. In Tamil Nadu, Horticultural College and Research Institute, Periyakulam, Theni (Tamil Nadu) maintains 75 accessions/varieties. TNAU Coimbatore also maintains a few accessions/varieties. In Kerala, Agricultural College, Padanakad, Kasargode (KAU) maintains 32 varieties including Venganapalli and Priyore. College of Horticulture, KAU maintains 32 of local and polyembryonic varieties of South West Coast of India including Alphonso.

Mankurad, the choicest variety of Goans has a lot of potential in the state in terms of variability. Mankurad accessions collected from Siolim (Bamanwada) are very promising with individual fruit weight ranging 234.00 g. to 406.00 g. Mankurad special has a very good size of 406 g, along with other appealing fruit qualities like less fibre, good keeping quality, better taste and flavour. Mankurad exhibited by growers at Mango Show at Panaji showed a lot of variability; fruit weight ranged from 194g to 340 g, where as T.S.S of the fruit ranged 14.6°Bx to 21.4°Bx in different accessions collected. Most of the types had better characters with unique taste. Because of unique flavour and taste, Mankurad and Hilario are the most favoured varieties of the state. Early Mankurad fruit gets maximum price of Rs. 100/fruit in the market. Fruit weight of Mankurad ranges 230.00 g. to 274.00 g. where as that of Hilario 216.00 to 220.00 g. In Mankurad, pulp (per cent) ranges 67.8 to 73.04 whereas in Hilario it ranges 62.04 to 70.64 per cent. T.S.S. of the fruit was maximum (20 per cent) in Mankurad yellow with better fruit quality. Pulp of the fruit was firm with less fiber in the pulp. A maximum T.S.S of 31 per cent was recorded in Hilario among all the accessions available in the Institute. Hilario variety bears much later than Mankurad. When Mankurad is in its fag end of harvest, Hilario starts appearing in the market. Among the varieties collected from the mango show of Goa, very attractive blush red colour on shoulder of Hilario was found. The weight of fruit was 220.00 g. with 18.2 per cent T.S.S with good taste and flavour. Over all variability in fruit characters and other details like organoleptic test and bio-chemical composition of the fruits in detail are recorded. Bardez Musserat and Salcette Musserat are other promising accessions with very attractive colour and fruit quality. Colaco Branca has very good fruit flavour and maximum yield efficiency. The Goan varieties produced in the state have a lot of biodiversity and potential with their unique characters for local markets and export.

Mango is one of the important fruit crops of tropical and subtropical climate of India. It is grown on 42.64 per cent of total area under fruit crops. Mango is known as king of fruits. A total of 41 species are found all over the world, and out of these *Mangifera indica* is the most common species widely adapted and about more than 1100 varieties are found all over the world. More than 1000 varieties are available in India. Most of the mango growing countries have their own varieties.In terms of average production, mango occupies fifth place after citrus, grapes, olive and apple and is followed by banana. More than 100 countries in tropics and subtropics grow mango. Egypt, Israel, Florida(USA), Australia,

Thailand and India are the main mango producing countries in the world. The total world production in 1994 was 18.450 x 10^3 metric tonnes. In 1995, it was 19.002 million tonnes (FAO, 1996). Production of mango increased substantially by 150.74 per cent in the last five years. Worldwide use of selected cultivars has not changed substantially but that of locally selected cultivars is increasing particularly in South East Asia, Australia and African countries. With the exception of Southeastern Asia, Australia and African countries cultivate locally selected varieties. Majority of the countries produce cultivars developed by Florida like Haden, Tommy Atkins and Keith. India produces 18 million tonnes of mango.The variety Alphonso accounts for 30-40 per cent of the Rs.750 million mango export market The biggest market lies in Japan, Australia, New Zealand and the Gulf.

In Goa, Mankurad, Hilario, Musserat, and a few other varieties are prominent. About 130 varieties are reported from Goa including above three varieties. The I.C.A.R. Research complex for Goa, Ela Old Goa maintains 172 accessions out of which 122 are local accessions, 32 introduced varieties and 18 hybrids. Important mango varieties grown in Uttar Pradesh are Dashehari, Langda, and Amrapali and in South India are Alampur Baneshan, Banglora, Neelam, NeelGoa, Kalapaddy, Suvarnarekha, Mallika and Arka Puneet. Mankurad possesses a special taste and flavour and is the most liked variety in Goa.

Western Ghat is one of the hot spots of Agro biodiversity in the world.Western ghats cover the states of Gujarat, Maharashtra, Goa, Karnataka and Kerala, with different agro climatic conditions. Annual rainfall in these areas is above 2000 mm. The soil is laterite and red and black soil in a fewer areas. Tropical evergreen, moist deciduous and dry deciduous vegetation are seen in Western Ghats. In Western Ghats, several horticultural crops - black pepper, cardamom, ginger- are originated. The Western Ghats, which sharply demarcate the region from the Deccan Plateau, have an elevation averaging 1,200 meters amsl with a number of peaks above 2,000 meters. The region enjoys tropical humid climate and receives heavy precipitation from 2,000 to 4,000 mm. While Kerala receives rainfall for more than six months through both Southwest and Northeast monsoons, the Konkan coast gets rainfall mostly through Southwest monsoons restricted from June to September. The annual temperature variation is between 17 and 35°C. The West Coast is traversed by a number of small and seasonal rivers, which are west flowing.

The Hindustan centre is one of the world's eight centers of crop plant origin and diversity, distinguished by Vavilov, the Russian Scientist. At least 166 food and crop species and 300 wild relatives of crops originated here (Tribhuvan Pant *et al.*, 1991). One species of mango (*Mangifera indica)* has over 1,000 varieties with fruit size ranging peanut to muskmelon size (Negi, 1993). Most of the Indian varieties are monoembryonic but about a dozen varieties from South India are polyembryonic namely, Olour, Bappakai, Muvandan, Chandrakaran, Mylepelian, Kitchner, Nekkare, Prior, Vellaikulumban, Peach, Starch and Kurukan which give true to type seedlings from nuceller embros.

The'Pane ka Aam' of Khamgaon-Maharashtra in India is a special variety. One of the oldest of these trees, well over 100 years, bears heavily five years out of 10 with two years of low yield. Average annual yield is 6,500 fruits, the highest record is 29,000.

Some time back, an investigation was made to find out the oldest existing mango tree in the world and it was found that the ancient one worshipped in the sanctum of the Sri Ekambranathar temple in Kanchipuram in South India, was 3500 years old. According to the local history- the words *Ek-amra-nathar* means Lord of the Mango Tree, and the divine couple Parameshwara and Parvati (of the Hindu Trinity of Gods) are said to have married below the tree. Scientists agree that this tree must be at least 2000 years old, as its history is recorded in the annals of the temple. Every year four varieties of mangoes ripen in the branches of this tree with different tastes and the local priests aver that this represents the four Vedas or Scriptures of the Hindus. The next oldest tree is the "crawling mango tree' at Sanjana in Gujarat-India, reputed to be 1300 years old. The annals of Parsis, who fled from Iran due to persecution in the seventh century A.D, have recorded, that this tree was extant, when they sought refuge in India. Unlike the usual mango tree, this tree has spread like a carpet on the ground (Swamy 2004).

The Western Ghats, extending along the west coast of India, covers an area of 180,000 square kilometers. The Western Ghats comprise the major portion of the Western Ghats and Sri Lanka Hotspot, one of 34 global biodiversity hotspots for conservation and one of the two on the Indian subcontinent. The area is extraordinarily rich in biodiversity. Although the total area is less than 6 per cent of the land area of India, the Western Ghats contains more than 30 per cent of all plant, fish, herpetofauna, bird, and mammal species found in India. Like other hotspots, the Western Ghats has a high proportion of endemic species. The region also has a spectacular assemblage of large mammals and is home to several nationally significant wildlife sanctuaries, tiger reserves, and national parks. The Western Ghats contain numerous medicinal plants and important genetic resources like the wild relatives of grains (rice, barley, *Eleucine coracana*), fruits (mango, garcenia, banana, jackfruit), and spices (black pepper, cinnamon, cardamom, and nutmeg).

The appemidi mango is native to the forest of Western Ghats where there are natural orchards of centuries old mango trees in the valley of Aghanashini, Kumudvati, Kali, Varada, Bedthi and Saraswati rivers in Uttar Kanada and Shimoga disricts. The trees are also found in places like Chittor and Khanapur in Belgaum and parts of Chikmagaloor district (Anita Pailoor 2007). Malanji Appe, a hug appemidi tree near Sirsi in Uttara Kannada, survives and has regular customers from three neighboring districts for over 50 years. Poornapranja Belur, who established the pickle industry in Sagar and Ripponpete, says that over 100 farm households are engaged in making pickle, producing around 100 tonnes a year. These homemade pickles have no artificial preservatives; the mango latex itsef gives the pickle a shelf life of around four years.

Table 7.1: Estimation of Area, Average Yield and Production of Mango in Goa

Year	*Area in ha*	*Area Increased in per cent*	*Av. Yield Kg/ha*	*Yield Increase/ Decrease in Per cent*	*Total Production in Tonnes*	*Total Production Increased/ Decreased in Per cent*	*Total Cropped Area in ha*
1998–99	4000	0	3000	0	12000	0	171455
1999–2000	4050	1.2	2469	–21.5	10000	–20	171455
2000–2001	4100	1.2	4269	42.16	18225	45.13	161907
2001–02	4140	0.96	4169	–2.39	17228	–5.78	161907
2002–03	4196	1.33	4179	0.21	17532	1.73	163846
2003–04	4179	–0.40	4259	1.90	17800	1.50	168634
2004–05	4226	1.11	4425	3.75	18700	4.81	169200
2005–06	4339	2.60	6242	29.08	27075	30.931	170327
2006–07	4414	1.69	4368	–42.85	19280	–40.43	172108
2007–08	4494	1.78	4204	–3.90	18894	–20.42	1,69,518

Source: Department of Agriculture, Government of Goa, Panaji–Goa.

The Western Ghats, known locally as Sahyadri Hills, are formed by the Malabar Plains and the chain of mountains running parallel to India's Western coast, about 30 to 50 km inland. They cover an area of 160,000 sq. km. and stretch for 1600 km from Indias southern tip to Gujarat in the North, interrupted only by the 30 km Palghat Gap, the report observed.

Table 7.2: Area, Production and Productivity of Mango in India

Year	Area		Production		Productivity Tonnes/ha
	Million ha	Per cent of Total Fruit Area	Million tonnes	Per cent of Total Fruits	
1991–92	1.078	37.5	8.752	30.4	8.1
1992–93	1.137	35.5	9.223	28.0	8.1
1993–94	1.127	38.2	10.113	27.1	8.3
1994–95	1.228	28.5	10.993	28.5	9.0
1995–96	1.283	38.2	10.811	26.0	8.4
1996–97	1.345	37.6	9.981	24.7	7.4
1997–98	1.385	37.5	10.234	23.7	7.4
1998–99	1.402	37.6	9.782	21.4	7.0
1999–00	1.487	37.3	10.504	23.0	7.1
2000–01	1.519	39.3	10.057	23.3	6.6
2001–02	1.576	39.3	10.020	23.3	6.4
2002–03	1.623	42.8	12.733	28.2	7.8

Table 7.3: State-wise Area, Production and Productivity of Mangoes

States	Area (Mill. ha)		Production (Mill. t)		Production (t/ha)	
	2000–01	2001–02	2000–01	2001–02	2000–01	2001–02
Andhra Pradesh	0.306	0.341	2.450	2.446	8.0	7.20
Uttar Pradesh	0.249	0.253	2.250	1.950	9.0	7.70
Bihar	0.139	0.139	1.113	1.253	8.0	9.00
Karnataka	0.134	0.116	1.291	1.131	9.6	9.80
West Bengal	0.063	0.065	0.380	0.585	6.1	8.90
Maharashtra	0.147	0.165	0.501	0.559	3.4	3.40
Gujarat	0.061	0.065	0.364	0.457	6.0	7.00
Tamil Nadu	0.108	0.111	0.556	0.439	6.6	4.00
Orissa	0.104	0.107	0.363	0.402	3.5	3.80
Others	0.208	0.214	0.789	0.798	–	–
Total	1.519	1.576	10.020	10.020	6.7	6.40

Source: Market News Service, UNCTAD/GATT, Geneva and APEDA, New Delhi.

The majority of Alliance for Zero Extinction (AZE) sites identified are in developing countries. There are 18 such sites in the United States and Australia. According to a report prepared by Ashoka Trust for Ecology and Environment, Banglore, though the total area of the Ghats is less than three per cent of the land area of India, it contains more than 3 per cent of the plant, bird and mammal species in India.(The Hindu 2006).

Table 7.4: Area, Production and Productivity of Mango in Major Mango Producing Countries of the World

Sl.No.	*Countries*	*Area*		*Production*		*Production Tonnes/ha*
		Mill. ha	*Prop. (Per cent)*	*(Mill. Tonnes)*	*Production (Per cent)*	
1.	India	1.600	43.36	10.800	40.64	6.75
2.	China	0.419	11.35	3.582	13.40	8.56
3.	Thailand	0.270	7.32	1.700	4.10	6.30
4.	Mexico	0.174	4.71	1.503	5.65	8.65
5.	Pakistan	0.105	2.85	1.089	4.10	10.37
6.	Indonesia	0.170	4.61	1.006	3.79	5.92
7.	Philippines	0.159	4.31	0.968	3.64	6.09
8.	Brazil	0.068	1.84	0.850	3.20	12.50
9.	Nilgeria	0.125	3.39	0.730	2.75	5.84
10.	Egypt	0.045	1.22	0.327	1.23	7.27
11.	Others	0.555	15.04	4.019	15.12	
12.	World	3.690	100.00	26.574	100.00	7.20

Source: Market News Service, UNCTAD/GATT, Geneva and APEDA, New Delhi.

Mango, being the most ancient fruit of India, is also very intimately connected with the Hindu religion. In Hindu mythology, it is highly valued as the fabled transformation of Prajapati (God of creation). Sanskrit poetry is full of praise for its bloom and fruit.The artistic shape of its cluster of leaves and the delicate fragrance of its bloom are immortalized in many a lyric and song. In one of the famous Sanskrit plays, *Shakuntala,* the mango flower is invoked as one of the five arrows of Kama Deva (Cupid). Even today, the Pancha Pallacva, or aggregate of five springs from mango trees, is used by the Hindus at various ceremonial functions. Mango blossoms are indispensable for the worship of Saraswati (Goddess of Wisdom and of the Fine Arts) in early spring, and of Shiva (one of the Hindu Trinity) which follows usually a month later. Festoons of mango leaves are used as decoration at almost all Hindu ceremonies and festivals. Dry twigs of the mango tree are used to light the sacred fire for Homa and other rituals. A mango grove was presented to Lord Buddha for repose beneath its graceful shade; all these show in what great esteem this tree was held in ancient India, and even now the tree is held in great respect.

Their use is particularly widespread in Hindu rituals of divine blessing, called pujas. A 'purnakumbha' or clay pot filled with water is topped with fresh mango leaves and a coconut. The pot signifies mother earth, water is the life-giver, mango leaves denote vibrant life, and represent divine consciousness. The whole object symbolizes Lakshmi, the goddess of fortune. There are famous Hindu temples in the Indian states of Orissa and Tamil Nadu where legend states that Shiva appeared as

Linga (the phallic form of Shiva) under a mango tree. Mango is also a rich part of Buddhist folklore. It features in the Jataka tales and frequently appears in Buddhist art. The Great Miracle of Sravasti, which is today on the border between India and Nepal, took place when Buddha converted people by miraculously reproducing himself in various forms in front of a mango tree. Buddha also caused a mango tree to sprout instantly from a seed to convince non-believers. A mango grove was his favorite place to rest and meditate and he was presented with one by one the courtesan Amrapali, who became a disciple (Mango spiritual).

Mango cultivation is nearly as old as Indian civilization. The lifelike representation of mango trees on the Buddhist stupas, including the famous Barhut Stupa of 110 B.C. testimonies to its antiquity. Mention is made of this tree in the *Charaka Samhita,* a great medico-scientific treatise. Alexander the Great (327 B.C.) spotted a pleasant mango garden in the Indus Valley, and probably this was the first occasion when a foreigner saw the tree. Later, Huien T'Sang (632-45) came to India from beyond its frontiers, mentioned this fruit in his notes under the name of *an-mo-lo* and *ambag,* respectively. Huien T'Sang is possibly the first writer who brought this fruit to the notice of people beyond India. These visits were followed by those of Ibn Batuta (1325-49), who described the mango as closely resembling an orange tree, except that it was larger and more leafy. Jordanus (1330), called the tree *amba* in his memorial and reached its climax when a Turkoman saint and poet, Amir Khusroo (*c.* 1330), composed an exquisite lyric in Persian, an imperfect translation of which would be:

> The choicest fruit of Hindusthan,
>
> For garden's pride the mango is sought;
>
> Era ripe, other fruits to cut we ban,
>
> But mango serves us, ripe or not.

Nicolo Conti (1419-44) also made a mention of this fruit, and added that it was thriving very well in Malabar (South India). Ludovici De Varthema (1503-8) noted its successful cultivation in Calicut (South India).

Round about the 16th century, when Mughal rule was established in India, the mango was honored with royal patronage. Babar (1526) described it as an excellent fruit. Akbar the Great (1556-1605) was so much in love with this fruit that he planted famous Lakh Bagh (orchard of 100,000 mango trees) near Dharbhanga, which also attested to the importance and esteem that this fruit had won at a time when large orchards of fruit trees were not even heard of. In the great work *Ain-a-Akbari,* it was said that a number of mango varieties were in cultivation in India during that period (Abul Fazal, 1590). Kairnah was the most famous variety of mango in Delhi in those days.

According to one European description of mangoes, "Apples of the Hesperides are but fables to them, for taste, the Nectarine, Peach and Apricot fall short"; and in another 'The Goa mango is reckoned the largest and the most delicious to the taste of any in the world, and I may add, the wholesomest and the best tasted of any fruit in the world' (Popenoe, 1920).

The mango originated in the Indo-Burma region (De Candolle, 1904; Popenoe, 1920; Vavilov, 1949-50; Mukherjee, 1951*a*). Based on their findings on the maximum number of allied species growing in Malaya, some workers believe that the Malaya region is the original home of the mango, for there, as many as twenty species grow. In my opinion, species growing at a particular place can only be a clue to the origin of a plant, and can in no case mean that the plant actually originated there. Moreover, the introduction of mango into certain islands of the Asiatic Archipelago was proved to have been made only in recent centuries (Rumphius, 1741), whereas it was undoubtedly under cultivation for more than four thousand years in eastern India and Burma (De Candolle, 1904). Besides, the number of the mango genera, occurrence of numerous wild and cultivated varieties, philological, archaeological and other literary evidence, number of ancient names, particularly Sanskrit ones, phytogeographical distribution, phytogenetic taxonomy of all the gardens of Bengal and Deccan since ancient times, all go to prove that the mango originated in the Indo-Burma region, and should leave no room for any doubt (Singh 1960).

Introduction and distribution of mango in other parts of the world are interesting. The first country to which mango was introduced happens to be Malaya. As stated before, the Malaya name for the mango, *mangga,* being the same as the Tamil word for mangoes, *mangga,* in South India, is evidence of the introduction of superior races of this fruit from India into Malaya. The size of its stone being quite large, it is inconceivable that the dispersal of mango stone from India in Malaya was brought about by birds or animals. Mango was almost certainly introduced to Malaya and other neighboring East Asia countries by Indians who used to voyage across the intervening seas for the sake of trade, commerce, and religion, during the Buddhist period, that is, in fourth or fifth century B.C.

The Sulu Archipelago and Mindanao were the first locations to be introduced to mango in the Philippines by Mohammedan missionaries and pirates between 1400 and 1450AD; but the island of Luzone, the Visayas, and the north coast of Mindanao received mangoes for the first time somewhere between 1600 and 1650. This introduction was brought about probably from Siam and the Dutch East Indies. Grafted varieties of mango were first imported from India and planted as late as in 1911. The regions of Cavite, Nueva, Ecija, Bulacan, Pampanga, Pangasinan, Cebu, Bohol, and Iloilo are well known for their mangoes, and Cavite, Cebu, Ilolio, Bohol, Bulacan and Pangapanga are famous for the quality of fruits (Wester, 1920).

Contact between people of India and those of the West, bought about by trade and commerce in the medieval period, gave an opportunity to the mangoes to travel West, as did many other tropical fruits and spices. The Portuguese were the first to establish trade contacts with India, particularly in spices and vegetable products, and they carried the plant and its seed from Goa (India) to East Africa, and then to West Africa. Gradually it found its way to the adjacent countries. Mangoes were reported growing in Somaliland in 1331 and in Yemen in the 19th century. The mango also reached England and was successfully grown under glass in 1690, and a few trees were found fruiting in Kew Gardens in 1808. It was found growing in the Moluccas in 1665, in Lisbon in the middle of 18th century, in Southern Italy in 1905, and in Queensland about 1870 (Burns and Prayag, 1916).

The Portuguese were also the first to introduce mango to Brazil by the beginning of the 18th century (Pope, 1929). Captain Cook found this fruit in plenty in Rio de Janeiro in 1768 (Popenoe, 1920). About 1742, it was introduced into Barbados in the West Indies where a few plants reached Santo Domingo. The introduction of mango in Jamaica is very interesting indeed. In 1782, captain Marshal of His Majesty's Ship *flora* captured off Jamaica a French ship that was sailing from Mauritius to Haiti. This ship contained, among other things, plants and seeds of various numbered species. No. 11, the mango was deposited by the gallant captain in the garden of Mr. Hinton East, situated near Gorda Town in the parish of St. Andrews, where it was grown. This was how mango was first introduced from Jamaica. Later, in 1869, for the first time, grafted varieties of mango were imported from India via Kew, and in all twenty-two varieties were introduced. The1884 grafted plants of 11 varieties of mango were introduced from the Botanical Gardens at Martinique. Since then mango was established itself so well in Jamaica that it looked almost indigenous to that island. In fact, it thrives so well there that it is impossible to think of Jamaica without its mango crop. The entire landscape is typically marked by the presence of mango trees right from sea level to about 1250 m a.m.s.l. Although mango thrives throughout Jamaica, it is not quite fruitful in humid districts and along the northern coast. Mango plantations in the districts of Castleton rarely fruit, but in the southern plains and in low, dry, lime and stony hills, it produces a heavy crop.

Mexico, where mango was quite successfully naturalized, owed its introduction to two sources, (1) the Philippines, where it was brought by Spanish galleons trading between Manila and Acapulco before 1779 (Popenoe, 1920), and (2) the West Indies, in the early 19th century. From Mexico, it spread to Hawaii (Higgins, 1906), although recent introductions in this century were direct from India, the Philippines, and Jamaica. In Costa Rica, it was introduced in 1796.

The first attempt to introduce mango in the United States was in 1833 by Henry Perrine in Florida, by procuring plants from Mexico. These plants, however, failed to survive due to neglect, war and fire. In 1861-1862, after a lapse of nearly 28 years, Dr. Fletcher made another attempt to introduce this fruit on the lower east coast of Florida by planting mango stones and nursuing the seedlings to bearing age. Of these, only one tree bore fruits in 1910 (Wester, 1920). In 1877, mango was successfully introduced to the west coast of Florida. The introduction of superior grafted varieties from India was first attempted in 1885 but the plants failed to survive. In 1889, however, another attempt was made by the Pomology Division of the United States Department of Agriculture by importing a shipment of 35 inarched trees from Calcutta which met with success. Miami, Florida, is credited with the introduction for the first time of the 'Manila' or 'Fillipino' mango in the United States in 1900 by getting a consignment of mango fruit of this type from Havana, Cuba (Wester, 1920).

Goa is a very important state lying in the Western Ghats of India and has lot of potential mango varieties. Four blocks of mango are being maintained at the ICAR Research Complex,Goa. First and second blocks were established in 1989 to 1991 and grafts were planted at a spacing of 5x5 m being a germplasm conservation and evaluation block. Growth and yield performance with detailed fruit characteristics of different varieties yielding in the block were recorded. Fertilizer doses of N 700 g,P 200 g and K 680 g/tree/year, in two split doses in June/July and October were given. Other than local varieties, 13 hybrids and 14 introduced varieties were also studied in detail.

Bordeaux paste and spray of insecticides and pesticides were used to control mango leaf hopper and some of the fungal disease respectively in the block. Total plant canopy and yield efficiency of the local varieties were computed and discussed. Sugar was estimated as per the AOAC method. Organoleptic test was conducted on scorecard methods. All the varieties were evaluated for their yield and growth parameter. Detailed parameters were described as per new descriptor of Mango published by IBPGR, Rome recently. Most of the varieties were described an fruit and seed (Stone) characters of Mango varieties available in the Institute. The yield efficiency of different varieties was estimated as per following methods:

1. **Yield Efficiency Using Total Cover Area**

   Yield efficiency/tree = yield (kg)/Total cover area (Πr^2)

   where, r = (NxS)+(ExW)/4

2. **Yield Efficiency Using Volume of the Tree**

   Yield efficiency = No. of fruits/Volume of Tree

   Volume of tree = $4/3A^2$ B or $4/3\Pi AB^2$

   A = (NxS)/2 and B = (ExW)/2, if height is more than (ExW) spreading, then

   B = height/2 and volume of tree = $4/3\ \Pi AB^2$

Mango is one of the important fruit crops of tropical and subtropical climate of India. It is grown on 42.64 per cent of total area under fruit crops. Mango is known as king of fruits. A total of 41 species is found all over the world. Out of these, *Mangifera indica* is the most common species widely adapted and about more than 1100 varieties are found all over the world. Among the varieties more than 1000 varieties are available within India. Most of the mango producing countries have their own variety, mostly in species *Mangifera indica*. In terms of average, mango occupies the fifth place after citrus, grapes, olives, and apple and is followed by banana. Egypt, Israel, Florida, Australia, Thailand including India are the main mango producing countries in the world. More than 100 countries produce mango in Tropics and Subtropics. The total world production in 1994 was 18.450×10^3 metric tonnes. In 1995 it was 19.002 million tonnes (FAO 1996). The production of mango increased substantially by 150.74 per cent in the last five years. The world wide use of selected cultivars from Florida has not changed substantially but locally selected cultivars are increasing particularly in Southeast Asia, Australia and several African countries. With the exception of southeastern Asia, Australia, and some African countries which cultivate mostly locally selected varieties, the majority of countries produce cultivars developed in Florida such as Haden, Tommy, Atkins and Keith. India produces 18 million tonnes of Mango. The Alphonso accounts for 30-40 per cent of the Rs.750 million Mango export market. The biggest market lies in Japan, Australia, New Zealand and the Gulf.

More than one thousand varieties are found in the country. Different varieties of mango are prominent in different parts of country. In Goa, Mankurad, Hillario, Mussarat and a few other varieties are prominent. About 77 varieties are reported from the Goa state including above three varieties. Presently, the Institute (ICAR, Goa) possesses a total 61 local accessions, 13 introduced varieties and 14 hybrids. Important Mango varieties grown in the country are in U.P-Dhashehari, Langda and Amrapali. A.P and South India have Alampur Benesan, Banglora, Neelam, Neelgoa, Kalipad, Suvernrekha, Mallika and Arka Puneet. In Goa, Mankurad possesses a special taste and flavour and is the most savoured fruit in Goa.

In Goa, which claims to produce 3700 tonnes of mangoes annually, three tonnes of fruit/ha. is considered a good harvest. But in Israel, the crop touches upto 75 tonnes/ha giving a potential to increase the production substantially by producing mangoes in higher scale.The variety Mankurad has fruit shape like Alphonso and has sugar acid blend with special flavour and taste. The variety comes in market even in first week of January and goes upto June. Variety Hilario starts a little late and Hilario goes into market upto July. Musserat has very good attractive color (reddish pink) and is mostly used for pulp purpose. Variety, Colaco Branca, possesses very good flavour and fruit shape is oblong. In Mankurad, a lot of variability is found with better fruit quality, color and flavour with potential for export. Hilario also shows a lot of variability with better quality and bears upto late season when Mankurad gets exhausted in the market. Karel mango, one of the pickling varieties, is highly acidic and is very good for pickling purpose and is available since beginning of the season from February onwards. Introduced varieties and hybrids also have the potential to increase mango production and productivity in the state. Among the introduced varieties, Kesar, Ratna, Neelgoa and Kalapaddy showed very good performance. Amrapali is performing very well under coconut garden and has given yield of 150 kg/tree with maximum fruitweight of more than 600 gm with T.S.S. of 26 per cent. Variability is found in a few of the clones of Amrapali with yellow varieties with diverse characteristics and biodiversity. Further it has the potential of growing introduced and Hybrid varieties (Arka Anmol, Amrapali) since most of them perform well in Goa and thus, may help to increase the overall production and productivity of the state.

Biological diversity of the Indian subcontinent is one of the richest in the world owing to its vast geographic area, varied topography and climate, and the juxtaposition of several biogeographical regions. Because of its richness in overall species diversity, India is recognized as one of the 12 'megadiversity regions' of the world. A very small number of countries, mainly in the tropics, possess a large fraction of world's species diversity and those countries, which possess the greatest species richness, are recognized as megadiversity countries, which attract special international attention. Of about 1.7 million species globally described and recorded in scientific literature, India has about 1,26,200 species (Khoshoo, 1995). It ranks tenth in the world both in respect of richness of flowering plants (17,000 spp.) and mammals (371 spp.).

Of the 12 Vavilovian centers of diversity and origin of major crop species, India is one. It is the Centre of diversity and origin of about 167 important cultivated plant species and domesticated animals. To name a few, the following cultivated plants arose in India and spread throughout the world: rice, sugarcane, jute, mango, citrus, banana, jack fruit, a few species of millets, Asiatic *Vigna*, egg plant, amaranths, edible *Dioscorea, Alocasia, Amorphophallus,* cardamom, black pepper, ginger, turmeric, several medicinal plants, aromatics and ornamentals. As far as agro-biodiversity is concerned, India ranks seventh among the centres of diversity and origin of crop plants (Khoshoo, 1995).

Of the 25 "Hot-pots" identified in the world, Eastern Himalayas and Western Ghats form two of them. Hot spots are recognized on the basis of presence of the greatest number of "endemic" or restricted range species. Since the endemic species are found in restricted areas, often requiring special niche for survival, many of them, particularly if their habitats are extinction prone and their habitats are distributed. At the global level, the Hot-spots are areas of high conservation priority, because if unique species are lost they can never be replaced.

Of the estimated 17,000 species of flowering plants reported from India, about 4500 species are found in the Western Ghats, of which 1720 (>one third) species are endemic. In other words, these endemic plants are found only in this region and nowhere else in the world. Nearly 33 per cent of these endemic plant species are rare or threatened and several are believed extinct or are on the verge to becoming so. Forests of Western Ghats, in view of their floristic diversity and numerous multipurpose species, are considered a varietal storehouse of economically important plants. In addition to the numerous timber and woodfuel plants, these forests provide a wealth of Non-Wood Forest Products (NWFPs) which are of particular significance to forest-dependent people and small-scale local industries. Non-Wood Forest Products provide fibre, food (edible fruits, nuts, spices, condiments etc.), extractive products (gums, resins and oleo-resins; tans and dyes; essential oils) and medicines. Western Ghats are also rich gene pools of relatives of many different crops like rice, sugarcane, black pepper, ginger, turmeric etc. These wild relatives play a vital role as genetic sources in plant breeding programmes. There are also in this region a large number of species of horticultural value. (Kaveriappa and Shetty 2001).

Midland of Kerala- Neriyamangalam and Adimali- have the common Mangoes, small Indian mangoes and striped neck mangoes. (Benny Kurian). Western Ghats is one of the 18 biodiversity hot spots recognized globally. The total number of species in the Western Ghats may be 10000 to 15000. RAPD and AFLP analyses can be useful tool for evaluating genetic diversity for breeding purposes, and for pursuing intellectual property rights of new varieties.

Kostermans and Bompard (1993) dealt with 69 *Mangifera* species of which 12 are of doubtful status, thus about 57 valid species occur in this genus. Maximum species diversity occurs in South-East Asia, particularly in Peninsular Malaysia, Sumatra and Borneo, and is well adapted to both monsoon and wet evergreen forests. Among these, 26-30 species bear edible fruits. Section/subgenus *Mangifera* (stamens with free filaments) has species, such as *M. indica, M. altissima, M. griffithii, M. laurina, M. minor, M. monandra, M. pentandra, M. quadrifida* and *M. similes* with maximum diversity in cultivated gene pool occurring in India and other parts of south Asia *i.e.* Bangladesh, Sri Lanka, Pakistan, Nepal; and in South-East Asia - Myanmar, Indo-China, Thailand, Indonesia and the Philippines-, less so in Malaysia. In East Asia, a limited diversity occurs in Southern China and in Taiwan. Section/subgenus (stamens often with joint/connate ailments) includes species like *M. caesia, M. foetida, M. kemanga, M. odorata* and *M. pajang.* These wild species and less known cultigens are distributed in Malaysia, Kalimanthan, Sumatra and Borneo. Natural hybrids occur in East Kalimanthan.

The cultivated diversity is more where monoembryonic forms are grown as in India/South Asia, as seedling populations present enormous variability. About 1000 types occur in India (Chadha, 1995). Most varieties originated as chance seedlings. Commercially grown varieties number around 30. A few famous monoembryonic varieties are Alphonso, Dushahari, Langra and Totapuri. A polyembryonic variety occurs in coastal regions of India. Overall, maximum diversity of Indian origin occurs in monoembryonic varieties well distributed in South Asia, lesser so in South-East and East Asia. Polyembryonic forms occur as in the Philippines and diversity is limited in monoembryonic varieties. Hybrids of *M. indica* and *M. foetida* occur in Borneo and Java; within species variation occurs more in *M. foetida* (its fruits possess good flavour ad firmness). *M. gedebe* is adapted to wet areas/water logged conditions (Arora 1995).

The Southern humid tropical areas of Andaman and Nicobar Islands, South Madhya Pradesh and Western Ghats are the secondary centres of origin of Mango in the world. Two of the smallest and largest varieties - Motidara (literaly pearl sized weighting 2 g.) and Jam e Jam (4 kg) are grown by fruit farmer Javed Farukh Faridi in his farm at Raul, Delhi. Alphonso from Maharashtra is often considered the Royal among all the Mangoes. Payri from the Konkan coast is more expensive than the Alphanso. Kesar from Gujarat is fleshy and large. Dalimbi is a colourful mango from Valsaal region. (Kusumlata, 2002).

Goa is associated with the choicest and sweetest of quality mangoes. The Kadambas (1000-1350 A.D.) and later the Governors of Vijayanagara empire (1380-1472 A.D.) promoted mango orchards close to temple complexes and in their capitals. The local self -governing village associations of Goa - the gaunkaris also brought huge areas under mango cultivation (Kamat,1998).

Goa has over 100 cultivars of mango grown over an area of 3700 ha and yielding about 35-40,000 tonnes. The diversity of mango cultivars reflects the years of efforts in grafting. Although crude methods of grafting were already known in India, Portuguese helped perfect the art of mango grafting in Goa.

Today, only the dominant and popular varieties enter the market. These include *Mankurad, Hilario, Salcete mussarat, Malgesh, Nicolau afonsa, Xavier, Udgo, Culas, Fernandin, Goa alphanso, Furtad, Costa, Sakri* and *Rosa, Bishop.* Mankurad is very popular as table fruit. Hilario is the sweetest mango in India. Karel is preferred for pickles. Musserat is good for jams and jellies.

Garcia de Orta in his *Colloquios* devoted an entire chapter to the mango. He reported the medicinal uses of mango: the use of baked mango seeds for congestion and the bitter mango kernel as an intestinal deworming agent. A British traveler, Dr. John Fryer (1673), East India company surgeon, praised the mango varieties in Goa. Captain Alexander Hamilton (1727) wrote "The Goa mango is reckoned the largest and most delicious to the taste of any in the world and the wholesomest and the best tested of any fruit in the world. The French Doctor Bernier (1765) after tasting the mango jam/jelly wrote: "There is no more delicious jam or jelly in the world."

The Mango in diplomacy: Mango diplomacy is something unheard of in India. But the documents in Goan archives tell a different story. Renowned Indo-Portuguese historian Pandurang Pissurlenkar reported that during 16-17th century Alfonso mangoes were sent to Delhi for obtaining the favours of Mughal emperors and their influential Nawabs. The Bhonsle's of Sawantwadi also engaged such mango diplomacy. The Portuguese Governors of Goa used to send baskets of Goan Alfonsa and Fernandin mangoes to the Peshwas of Pune. Portuguese diplomat at Pune, Vithalrao Valaulikar wrote in 1792 to the Governor of Goa to ban all private trade in mangoes from Goan mango varieties (Kamat 1998). The Portuguese introduced a system of special permits for private traders to export of prized varieties of mangoes from Goa.

Goan Mango graft: It is not known whether the permit system was applicable to Goan Mango grafts. The work done by the Portuguese missionaries and the Goan cultivators popularized Goan mango grafts. The first reference to the grafting of mango tree is in a 1710 publication by Jesuit Francisco de Souza. Father Clemente da Ressureicao in his *Tratado de Agricultura* (1872) described grafting techniques. Bernardo Francisco da Costa in his manual *Practico do Agricultor Indiano (1872)* wrote on mango cultivation methods. He founded the first canning factory in 1882 and made a case for exporting Goan mangoes in the shape of slices in syrup as well as jelly forms. Bernado da Costa could be considered the pioneer of modern mango processing in India. Another Portuguese mango-lover, General Joao de Sampayo (1902) in his booklet *A mangueira* listed 33 mango varieties in Goa.

A recent publication by ICAR Research Complex,Goa listed more than 100 varieties. Mangoes from Goa were exported to Brazil in 1811. The British took the Goa Priori variety to Bombay. It came to be called "Bombay Priori". From Bombay,the mango reached the islands of the West Indies under British control. Famous voyager Captain Cook found mangoes in Jamaica in 1788. The Jamaicans still call their mangoes "Bombay mangoes".

Mango cultivators are currently identified on the basis of morphology of fruits like leaf and fruit characters. Recently, reliable genetic markers are developed and introduced for mango cultivar identification. These include isozymes, random amplified polymorphic DNA's (RDAs) and Variable Number Tandam Repeats (UNTRs). Genetic markers are not only to be useful for cultivar identification but also for the purpose of mango breeding (Jafer *et al.*, 2007).

During the sixteenth century, when Mughal rule was established in India, the Mango was honored with royal patronage. Babur (1526) described it as an excellent fruit. Akbar the Great (1556-1605) was so much in love with this fruit that he planted the famous Lakhbags.

Mango is the most popular fruit among millions of people in the Orient, particularly in India, where it is the choicest of all indigenous fruits. It occupies relatively the same position in the tropics as is enjoyed by apple in temperate America or in Europe. In fact, it will be no exaggeration to say that because of its excellent flavour, attractive fragrance, beautiful shades of colour, delicious taste and healthful value, mango is now recognized as one of the best fruits in the world. Although, at present, it is being cultivated commercially in a number of countries of the world, nowhere does it achieve the same premier position as in the subcontinent of India, where it thrives throughout the length and breadth of the country. Hardy nature of the tree, low cost of maintenance and profuse yield, mango is known as the poor man's fruit, and thus posseses a mass appeal.

It can safely be admitted that no fruit is more beautiful or delicious than some of the class-one varieties of mango. They are gradually becoming known by the people of the West, and in their mother country, India, they are so popular that, on the whole, they may justly merit being called the national fruit. At present, it is a common practice in rural India for the owner of a mango orchard to invite his friends and relatives for a picnic in his orchard during the season. It is almost everywhere a common sight to see crowds of men, women, and children going round mango orchards helping themselves to the pickings, as gleaners do during the harvest. Until the beginning of the twentieth century, collection of superior grafted varieties of mango was the hobby of kings and nobles only, but now it is a source of honorable living for anyone. These days mango growing is a profitable enterprise and well-planned and well-kept mango orchards are a source of both pleasure and profit.

Not many people in Goa would be able to tell where on earth is the Isle of Mangoes. The Mango groves, on account of which, it was so called, have died and the term Isle of Mangoes is forgotten in the corridors of time. This small island is right in the center of Goa, within the district of Tiswadi, between Zuem, Divar and the mainland.

Capt. Pedro Barreto de Rezenda circa in 1646 wrote about Isle of Mangoes in a manuscript found in the National Library of Paris. During 1635, Capt. Rezende was the Secretary of Portuguese Viceroy, Conde de Linhares. He left a detailed description among other things of the forts, settlements, churches and budgets incurred from Muscat to China and Africa including Goa.

This is what Rezende wrote about the Isle of Mangoes:

"Its length is as far as a gunfire shot can reach. Its width is much less. There is nothing more than palm trees and mango groves on this island. The owner of this island lives there and has a few houses with towers, which serve as a fortress of defence against thieves and marauders venturing from the mainland. There are about 80 native Christians who tend the groves during the day and take to arms during the night."

The description given by Capt. Rezende about the adjoining islands is also very illuminating. This is what he wrote about the island of Divar. "It is located in front of the City of Goa. Its length is about a league and the width about a fourth part of a league. The fort of Narvem is partly located on this island, which has around 4,000 residents. About 2,000 of these are cultivators during the day and armed fighting men during the night. The rest are fishermen. All the residents are Christians having their churches in charge of Vicars duly paid."

About the Island of Charao, he says, "It is located across the river NE of Divar with a league and a half for its length and about two leagues around. It is much more inhabited than Divar. It has nearly 14,000 Christian residents out of which 6,000 are meant for defence purposes. There are two churches, a great many palm and mango groves apart from rice fields where rice and vegetables are grown".

Rezende praises the people of Zuem, "Because whenever its residents are called to take up arms, they do not sulk or pretend illness. They respond promptly." He describes it as "a little less than a league for length and its width is as far as a falcon gunshot goes". On the side of the City of Goa, it has many beautiful houses belonging to Portuguese, built of morter, brick and lime.

Its population is about 1,200 residents, all of them Christians having a church called St. Stephen in charge of a Vicar. With regard to the larger island of the group, called the island of Goa, his description is extensive. It is divided into passes. Each pass or entry to the island has its fortification and check post. The important among these entry points are Narvem, Benastari, Daugim, Ribandar, Aggaxem, Panjim, Cabo Gaspar Dias, Dry Passage and Korbolli.

In a recent issue of a Hong-Kong-based publication, a new variety of luscious Goa mangoes is at Ela gardens, near Old Goa, by crossing Neelam with Alphonso variety. The fruit itself was imported by the Portuguese from its African possessions where it grows wild, large and not so tasty. Perhaps, the Isle of Mangoes could be revived again, where the best mangoes were grown in by-gone days (Fernandes 1985).

Mango (*Mangifera indica)* is native of India. Based on the detailed study of the genus *Mangifera*, Mukherjee (1949, 1885) recorded that although this genus might have originated in the region covering Burma, Siam, Indo-China and Malaya peninsula, the mango itself had its birth in Assam-Burma region. Clonal selection, selection from chance seedlings and breeding efforts resulted in identification of elite improved varieties of mango for commercial cultivation. In addition to large genepool of mango (*M.indica*) available in the country, three other related species *viz. M. andamanica*, *M. sylvatica* and *M. camptosperma* were reported from the country (Mukherjee 1985).

(*a*) Seedling Races, both Wild and Cultivated Types

This group consists both of Monoembryonic and Polyembryonic types. Most of the present day commercial varieties are selections from seedlings. A number of Polyembryonic varieties are reported from the coastal areas.

(b) Horticulture Races, both Wild and Cultivated Types

Distinct clonal differences were observed in the same variety when grown under different agro climatic regions and propagated vegetatively from different types of parent materials and on different rootstocks. In a famous North Indian variety 'Dashehari', more than 50 types were under study at the Central Institute of Sub-tropical Horticulture (CISH), located in Lucknow, U.P. Improved clones of commercial varieties like Dashehari, Langra and Himsagar are already made. (Ghosh 1997).

The genepools of mango available in the country have been evaluated and from systematic studies, the horticulturally important characters like dwarfness, regularity of bearing, season of cropping, fruit characters (size, colour, quality), and resistance to pests and diseases are employed in breeding programme, which resulted in release of a number of mango hybrids in the country (Chadha and Yadav 1997).

Although evaluation reports are available for reaction of cultivars to insects and diseases in fields, no donor sources of resistance have yet been identified. Through clonal selections, high TSS containing varieties (Sahib Pasand, Laxshman Bhog), regular bearing Dashehari (clone No.53), off-season selection (Niranjan), heavy bearing Langra (about 780 kg/tree), dwarf selections of South Indian varieties Rumani and Totapuri and dwarf polyembryonic types (Manipur I and Manipur II) were identified as promising ones.

In India, about 30 commercial varieties are grown on large scale and there is a distinct consumer preference of varieties in different regions. Isozyme studies revealed that North Indian varieties differ genetically from South Indian varieties. Some of the prominent commercial varieties in different regions of India are:

1. Northern India–Dashehari, Langra, Bombay Green, Chausa, Fazil.
2. Central India–Langra, Dashehari, Chausa, Safeda and Sunderja.
3. Western India–Alphonso, Kesar, Mankurad, Vanraj.
4. Southern India–Banganapalli, Totapuri, Neelum, Pairi, Mulgoa, Kalepad, Rumani.
5. Eastern India–Langra, Kishan Bhog, Zardalu, Himsagar, Bombai, Malda.

Table 7.5: Mango Varieties Under Cultivation in Different States of Western Ghats

States	*Variety Grown*
Goa	Fernadin, Mankurad, Hillario, Salcete Musserat, Bardez Musserat, Bemcorad, Cota, Cuals, Maxima, Sakri, Xavier, Jesuit, Aphonso.
Gujarat	Alphonso, Kesar, Rajapuri, Vanraj.
Karnataka	Alponso, Banglora, Mulgoa, Neelum, Pairi.
Kerala	Mundappa, Olour, Pairi, Alphonso, Bannet Alphonso, Bangalora, Benganapalli, Neelum, Kalapadyx, Guddadat and Prior.
Maharashtra	Alphonso, Mankurad, Mulgoa, Pairi, Kesar.
Tamil Nadu	Banganapalli, Benglora, Neelum, Rumani, Mulgoa

Source A: I Kisan Mango Varieties, Important commercial Varieties

<http://www.kisan.com/links/ap mango varieties.shtml>,

Source B: Annual report of ICAR Research Complex for Goa. 1998-2008., Mangoes in India Varieties.

http://www.horticultureworld.net/india-mango.htm.

Source C: Status of Mango Cultivation in Kerala. Radha,T and Nair,S.R. ISHS Acta Horticulturae 509: VI International Symposium on Mango.

Table 7.6: Major *ex situ* Collections of *M. indica* Germplasm in India

Sl.No.	*Centres*	*No. of Accessions*
1.	Agricultural Experiment Station, Paria, Gujarat, Gujarat Agricultural University (GAU), Shahi Bagh,	177
2.	Ahmedabad, Gujarat	109
3.	Fruit Research Station, Aurangabad, Maharashtra	130
4.	Gujarat Agricultural University, Anand Campus	81
5.	Gujarat Agricultural University, Navsari Campus	109
6.	ICAR Research Complex Goa, ela, Old Goa.	172
7.	Indian Institute of Horticultural Research, Bangalore	269
8.	Mahatma Phule Agricultural University, Rahuri	86
9.	Marathwada Agricultural University, Parbhani	80
10.	PK- Horticultural College and Research Institute, Periyakulam, District-Theni (Tamil Nadu)	75
11.	Regional Fruit Research Station, Vengurla	379
12.	University of Agricultural Sciences, Dharwad	54

Source: Part III-PGR Conservation and Use- Germplasm introduction, exchange, evaluation and Mango resource information system http://www.mangifera.org

Mango cultivars are classified into two groups: Monoembryonic type or Indian type and Polyembryonic type or Indo-Chinese type. In India, majority of the cultivated types are Monoembryonic.

Surprisingly, Polyembryonic types were grown only in Southern India, especially in coastal parts of Kerala, Karnataka and Goa. '

Emergence of multiple seedlings from a single seed is referred to as Polyembryonic. This was observed in 59 plant families, 158 genera and 239 species. In mango, nucellar embryos were observed in 19 mango varieties. Using crosses of Polyembryonic and Monoembryonic cultivars and their segregating hybrids, it was demonstrated that Polyembryony in mango is controlled by a single dominant gene. Polyembryonic and monoembryonic types are intercrossable, and cultivation of Polyembryonic varieties is confined to the West coast adjacent to the Western Ghats, which is one of the hot spots of biodiversity (Ravishankar *et al.*, 2004).

Vasco da gama's discovery of the sea route to India in 1498 and Alfonso de Albuquerque's invasion and conquest of Goa in 1510 turned this city into the centre of all Portuguese colonial activities in the Orient. During the first two hundred years of the Portuguese presence in that area, one of the most positive aspects was the wealth of information gathered on India, its people, customs and habits and its flora and fauna. These were genuine spiritual, cultural and scientific values which were little known in the west and would have a profound impact on modern history (Rego 2002).

One of India's greatest natural assets is its flora. The country stretches from the Himalayas to Cape Camorin (Kanyakumari), covering areas which have temperate, sub-tropical and tropical climates. This leads to a broad range of plant species, including many which have, over time, been of worldwide interest.

Ever since antiquity, many people showed interest in certain aspects of Indian flora. The mango tree, one of its most important components, has been the subject of special study and enjoys a reputation throughout the Orient. It is grown in India for about four thousand years, with studies of its origin leading to the conclusion that it first appeared in the area embracing Assam, Burma (now Myanmar) and Thailand, as some wild plants of both *Mangifero indica* and *Mangifera sylvatica* are found in this vast region. Given this ancient history, it is no surprise that references to mango trees and mangoes themselves appear in Indian Vedic books and in descriptions of orchards in epic works such as the *Ramayana* and the *Mahabharata.*

Plants were already studied in ancient India. The exact science called *Vrikyshiuveda* was one of the subjects studied at several learning centres, and the criteria used to classify the plants were generally according to their medicinal uses. This method was adopted by Van Rheede (1680-1703) in *Hortus Indicus Malabaricus* and would later be the inspiration for Carl Linnaeus (1753) when he extended the use of Latin names to Indian plants in his *Species Plantarum.*

The earliest literary references to mangoes were apparently made by the Chinese authors Huen Tsang (624-645 AD) and FA Hien (409-411 AD), while Persian literature called it " *Ambag*", as recorded by Ibn Haukal (902-968 AD).

The first western reference to the plant came in 1428, when Friar Jordanus Catalani or Jourdain de Severac mentioned in his *Rec. de Voyage* book number 42, but it was Ludovico de Varthema who first used the word mango in 1510, when relating his travels. The first Portuguese reference to the fruit probably appeared in the early sixteenth century, in Gabriel Rebelo's description of the Moluccas. Garcia de Orta repeated the name in 153 in *Coloquio Trigesimo Quarto sobre as Mangas.*

Cristovao da Costa then produced a description and image of the mango tree. Linschoten adopted the same words, and in 1653, Niccolao Manucci mentioned the various types of mangoes and named a few of them. Only in 1912, were the plant and its precious fruit described by a Goan author, Caetano Xavier Gracias. "Originally from India, where it is abundant and widely grown". Better mangoes are obtained when the tree is grown using grafting or layering. The best mangoes in India came from Goa, where they were grown and the use of grafting was widely spread by the Jesuits, and it was from Goa that some of the finest types were taken to Brazil (sic), during the reign of King Joao Vi at the start of the last century. This is why it is grown and found almost throughout equatorial and Southern America. It has a tall trunk, with a rough bark, large and numerous branches, alternating petiolate and lanceolate leaves, thick and smooth, measuring 3 to 6 cm long, thick at the base. It does not have stipules, but ends in large or upright or ascending panicles which rigid, curve slightly and are reddish in colour. It has small, solid pedicles and concave oval bracts. Its flowers are small and yellow with red lines at the base of petal. A large number of perfect male flowers are mixed in among the hermaphrodite ones all over the panicles. The calyx has 5 sepals; concave oblong follies, 5 lanceolate petals, three times longer than the calyx; 5 yellow nectarines, filaments which are weakly located between the lower lobe of the nectary and the germs; a purple oval anther. The germs are obliquely oval, 1 cell contains just one egg. The style is alongside the germ and has the length of the stamen, with a small, inward-curving stigma. The oblong seed is kidney-shaped and varies in size. The endocarp is like an almond, very long and woody, having one seed and two valves, covered in fleshy pulp and the skin. It blossoms from January-March, and produces fruit from April to June (Rego, 2002).

It remains to be seen when, why and how mangoes were "baptized" with Portuguese names. It seems very probable that the mangoes–like the new converts to Christianity–got their names from the Portuguese missionaries who (apart from their evangelical activities) devoted baptizing the results with surnames and other names in Portuguese. This tradition was later developed by the landlords of local palm tree groves, where high quality mango trees which merited cultivation were grown alongside the palm-trees. Thus, the different varieties with Portuguese names appear in orchards and yards belonging to Catholic, Hindu and Muslim families.

Perhaps the first record comes from Sebastiao Gonsalves (1553-1619), who reported that the name of the *Malgessa* mango was derived from the Malijaz Temple in Kenheri, a village near Bombay. Many of the varieties listed either as a result of personal observation during my career in the Goa Agriculture Services or collected from existing literature have vanished over the centuries, probably due to their poor quality, which led to the end of their production. Currently, mango trees cover around one million ha or some 60 per cent of the entire fruit-growing area within the Indian Union.

In 1962, the Botanical Survey of India set up a team under Rolla Seshagiri Rao to draw up an exhaustive study on Goan flora. The team detected 115 angiosperms belonging to 657 genera and divided into 146 families. The families include the Anacardiaceae, which in turn includes the most important species for Goa's economy, the *Anacardium occidentalis,* L. (cashew tree) and the *Mangifera indica,* L. (mango tree).

Mango trees form an additional source of income for the *batcar* (landlord). During the mango season, which lasts from early March to mid June, the trees are rented out to the traditional renters *(renters),* who pay the landlord according to a contract that varies according to each year's production. One part is paid in cash (rend), while the other is in mangoes *(pensaum)* of each variety found in the orchard. The *rendkar* picks the mangoes, takes them to the local markets or sends them to the Bombay market, where the *Malcorada* (locally called *mankurad)* is very popular.(Rego 2002) Local preserve factories also use the most appropriate varieties.

The following is a description of some of the most important varieties of the Goan mango, *Malcorada*, which is the first to bear fruit (in mid-March), and the most popular variety in Goa. In contrast to what the name seems to indicate (poorly coloured) and to the *Bemcorada* variety, it actually has a very attractive colour. There are several subtypes, some of which have a very fibrous pulp. This characteristic may vary from year to year and seems to be linked to climate conditions, soil types and age of plant. For example, many mango trees which initially produce fibrous fruits, later give excellent fruit as the tree ages. The *Hilario* variety is relatively recent, having discovered fifty years ago in the property of Hilario Fernandes in the village of Siolim. This variety, the sweetest in India, has since spread throughout Goa and bears fruit at the end of May. Like *Hilario, Fernandina* is also a late variety that produces fruits of excellent consistency, and therefore widely used for home made preserves, using slices of the fruit mixed with honey. The taste is similar to that of peaches, although the pulp's yellow colour makes it less popular with factories, which prefer a more orange colour. *Monserafe vamelho* are *Moserrate de Salcette* are very popular and productive. They have a strong scent of arsine and excellent consistency, which explain the high demand from Goan preserve factories. *Alfonsa* or *Alfonso* is undoubtedly the dominant variety in Goa. The Production expanded to the village of Ratnagiri (Maharashtra), where it is now better known at home and abroad as *Ratnagiri Alphonso* (Rego, 2002).

Alfonso–considered by many the Royal among all mangoes–is grown in Maharashtra. It rates number one on the popularity chart of 'favourite mangoes'. British actor Terence Stamp was once asked by an interviewer what his favourite food was. "Alfonso mango" came the prompt reply. He was asked 'Why?' His answer was "Because unless you've had an Alfonso mango you've never had a mango." Known for its delicate and subtle flavour, Alfonso is the most expensive mango in India (Salman Saeed 2002).

It was probably the Portuguese who introduced grafting into Goa to reproduce the finest varieties of mango. The *Mangifera indica,* L. has two strains: the mono-embryonic and the poly-embryonic. In the former, the seed comes from an allogamic fertilization: crossing two plants of the same or of two different varieties. Thus, the seed and the resulting future plants offer no guarantee of having the quality of a specific variety. In the latter case, the seed produces small plants identical to the mother-plant and can be separated from it and planted. This is the result of the embryo-even if it comes from different varieties–being generally suppressed, while the plants derive from the vegetative development of the ovule's tissues and therefore retain qualities of the mother-plant.

All the varieties in India are mono-embryonic, which means that the material obtained from the seed is very variable, imposing the need for grafting to maintain the original characteristics. The polyembryonic varieties only exist on the Malabar Coast. Only ten varieties are known, the important ones being the *Olour* and *Chandrakaran.* Therefore, to maintain the quality of any specific mono-embryonic variety, for many years, it was essential to use vegetative methods of reproduction.

It may be that the Indian "gurus" did not know about these methods of grafting prior to the Portuguese arrival, although Hayes (1945) claims that grafting was already used. Thus, it fell to the Portuguese to bring the science of grafting–already widely used within Portugal–to India. Francisco de Souza (1750) was the authority to credit the Portuguese with the improvement in mangoes due to grafting. However, Braganca Pereira (1886) disagreed, claiming that the Portuguese merely improved the existing method used in Goa at the time. This argument was based on a *Portaria das ilhas* (Regulation of the Islands) from 1553, but there appear to be a problem with the document's date, since there were neither regulations nor a "Council of the Islands" at that time (Rego 2002).

Figure 7.1: Local Mango Varieties

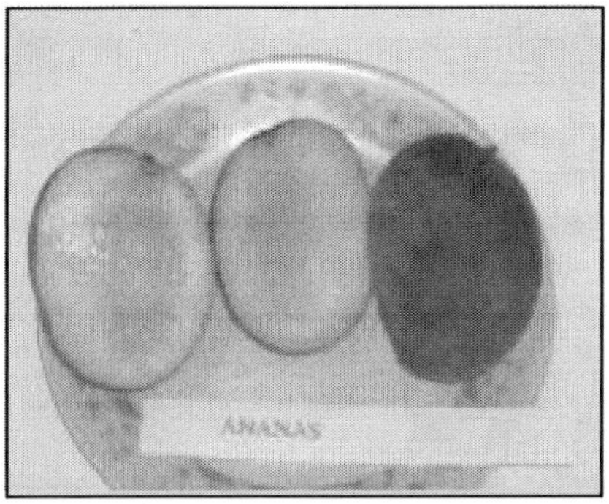

Ananas

Carrera Branca

Ball

Bemcurado

Contd...

Figure 7.1–Contd...

Colaco Branca

Cota

Contd...

Figure 7.1–Contd...

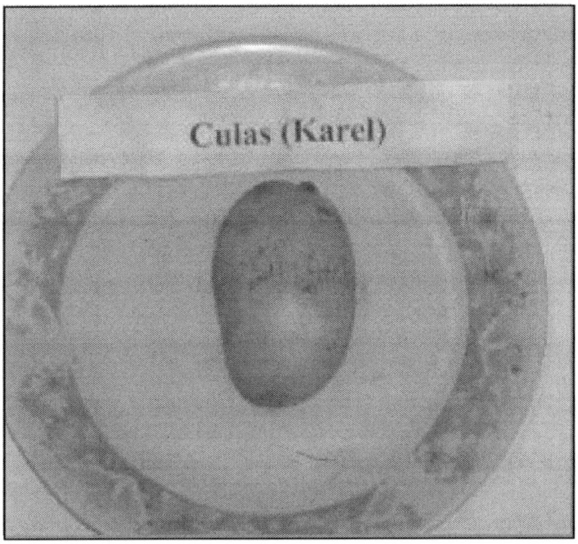

Culas (Karel)

Fernandin Porvorim

Furtad

Koita

Contd...

Figure 7.1–Contd...

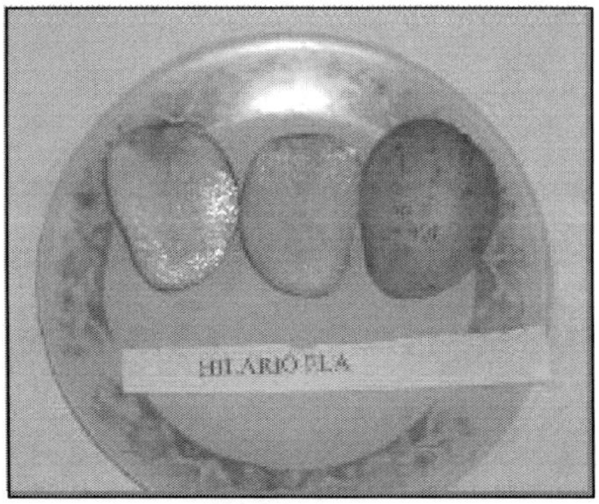

Hilario Ela

Hilario Mpsa

Elayvelangan

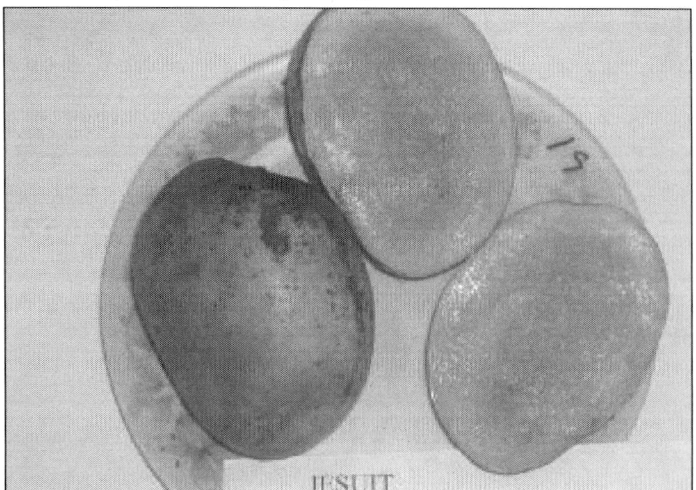

Jesuit

Contd...

Figure 7.1–Contd...

Italario Volform

Malgesh

Malgoa

Contd...

Figure 7.1–Contd...

Manga Rosa

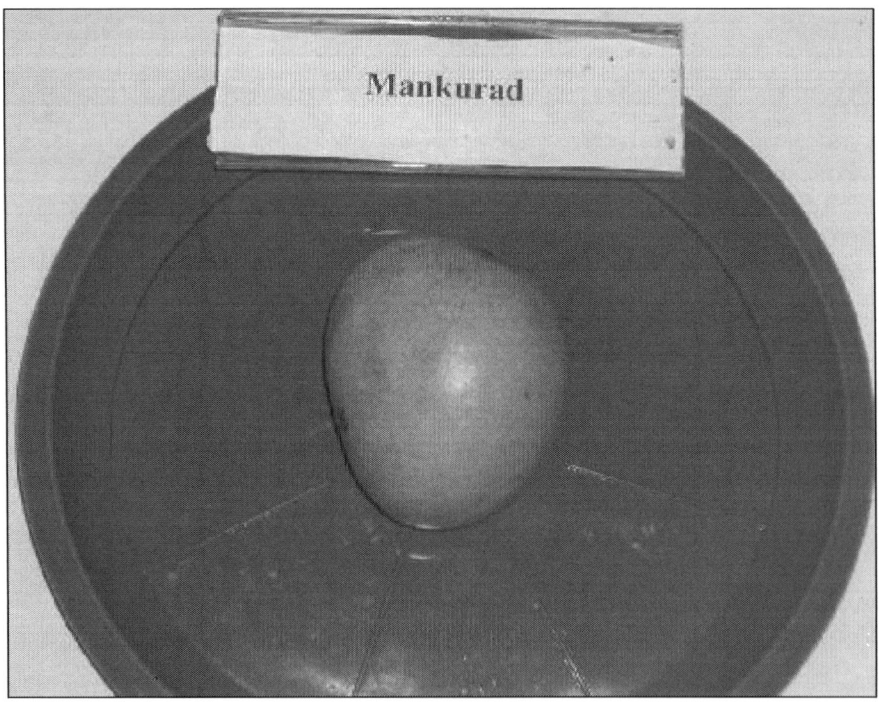

Mankurad

Contd...

Figure 7.1–Contd...

Mankurad Dodamarg

Mankurad Keni

Contd...

Figure 7.1–Contd...

Mankurad PWD

Mankurad Sports

Contd...

Figure 7.1–Contd...

Mankurad I/C

Mirand

Maxima

Contd...

Figure 7.1–Contd...

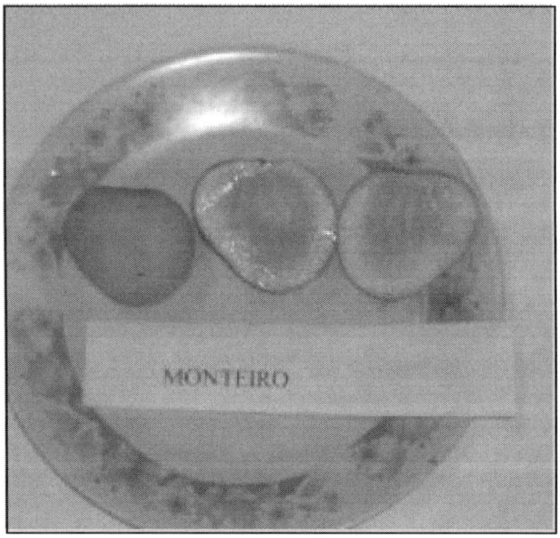

Monteiro

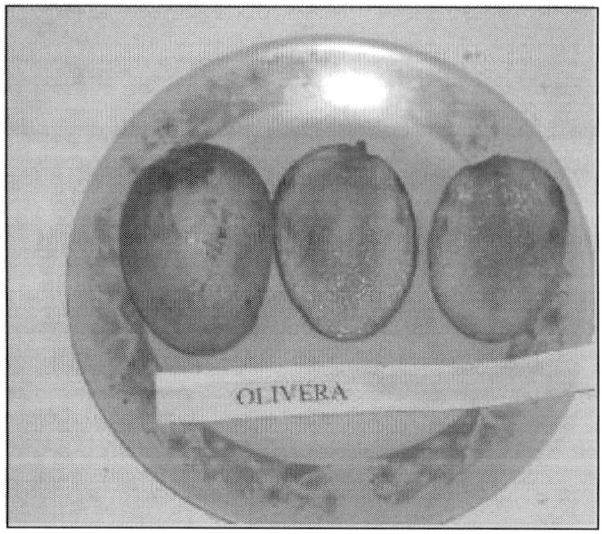

Olivera

Papel Branco

Contd...

Figure 7.1–Contd...

Rebbelo

Toranja

Contd...

Figure 7.1–Contd...

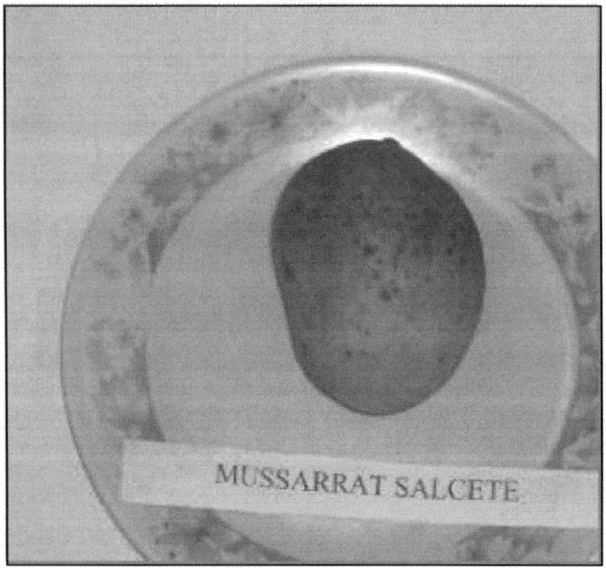

Mussarrat Salcete

Salgadin

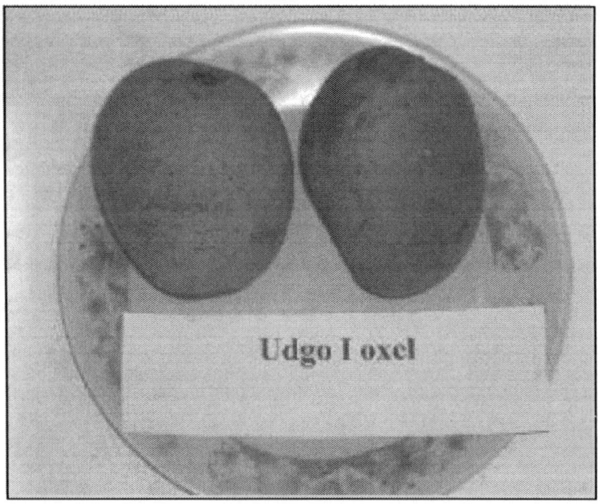

Udgo Ioxel

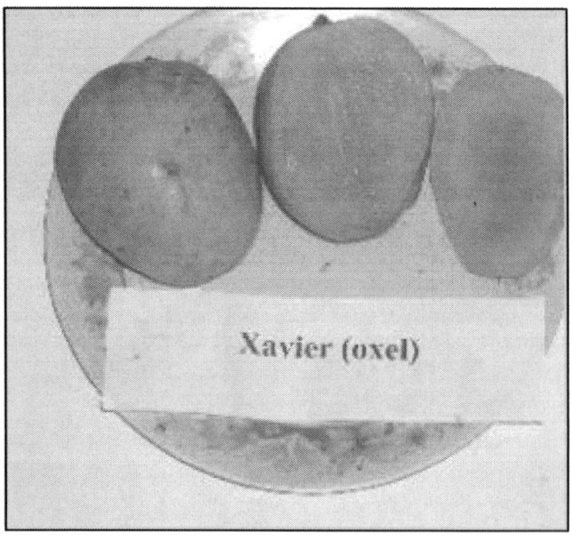

Xavier (Oxel)

The grafting techniques used on mango trees were perfectly established in 1872, as shown by two publications from that year. Friar Clemente da Ressurreicao (1773), whose book was published in 1872, noted that grafting could be carried out both during the Sravana months (from the Hindu calendar, corresponding to July and August) and the south-west monsoon months (June to September).

In his *magnum opus* " Practical Manual for the Indian Farmer", Bernado Francisco da Coasta, the man who pioneered India's canned fruit through his *Costa e Companhila* factory (the first in Asia) provides detailed information on methods of vegetative reproduction, including grafting of mango trees. This method unquestionably revolutionized mango production, first in Goa and later throughout India. The means of reproducing mangoes through hybridization was first introduced into India in 1942 at the Sabour (Bihar) Agricultural Centre, and excellent varieties have since been produced.

There is no reliable information on the spread of mango trees outside Asia. Some claim that the Persians took them to the eastern coast of Africa in the tenth century, but Ferrao believes that it may have been the Portuguese, since the fruit was not mentioned in Vasco da Gama's report, whereas several other fruits from the coast of Vasco da Gama arrived, the authors of the report would certainly have seen it. The spread of the Goan mango to the American continent was one consequence of the discoveries. According to Ferrao, the earliest information on the introduction of mango tree to Brazil mentions "seven seeds with endocarps of mango trees and each one of them has four to five sprouts". They were send in January 1683 by D. Francisco Tavera, the Viceroy of India and Count of Alvor on board the *Sao Francisco Xavier*. The *Afonsa* variety is one of Goa's most famous, and spread to several other states of Indian Union. Owing to the rigorous selection procedures done in the Ratnagiri (Maharashtra) district, the Goan *Afosna* is now widely known in world markets as *Ratnagiri Alphonsa* (Rego 2002). The *Pires* variety is now called *Pari* in India, and was introduced in the subtropical area of the United States in 1902 by the state department of agriculture. The mango tree is also closely connected with the Hindu religion. As a patheist religion, this includes plants and animals which contribute to human life as part of its belief system. Legend has it that it was Hanuman–the close friend of Rama who helped him to rescue Sita after she had been kidnapped by the king of Lanka–who brought the sacred tree to India. The *Chaturmas Mahatma* considers mango an incarnation of Prajapati, father of all creatures. As it is so deeply rooted in Goan culture and a part of Goan identity, the mango tree was also a muse for Goan poets. There are many examples but I have chosen an extract from a poem by Manohar Sardesai, originally written in Konkani:

> Like a drop of honey
>
> Or soft as a bride's lips
>
> Our pride, our wealth,
>
> Is the Goan, Mango.

Growth and yield performance of 54 local mango varieties were evaluated during 2003-04. Maximum height of 6.70m was found in Maxima followed by Mankurad Keni (6.50 m). Maximum periphery of the plant, 97.00 cm was found in the variety Toranja followed by Bencorado (95.00cm). Maximum spreading of the plant was found 10x7 m as NS x EW in variety Sakari (Nevra) followed by Maxima 8.4x8 m as NS x EW. Total cover area was also found maximum 56.72 sq.m. in variety,Sakari followed by Maxima (52.78 sq.m.). Maximum number of fruits, 247 was in variety Toranja followed by Udgo (189). Fruit yield (kg/plant) was maximum(49.70 kg)in variety Jesuit followed by Udgo (45.10 kg/plant) and Maxima (39.34 Kg/plant). Yield efficiency of the plant was maximum 2.340 Kg/m^2 in variety Colaco Branca, followed by Costa 2.330 Kg/m^2 and Mankurad 1.632 Kg/m^2.

Table 7.1: Growth Performance with Yield and Yield Efficiency of Local Mango Varieties during 2003–2004

Sl.No.	Name of Varieties	Height of Plant (m)	Girth Periphery of Main Stem (cm)	Spread of Plants NxS (m)	Spread of Plants ExW (m)	Total Cover Area (πr^2) m^2	Volume of Tree (m^3)	Fruits/ Plant	Fruit Wt. (kg)/ Plant	Yield Efficiency (Kg/m²)	Yield Efficiency (Kg/m³)
1.	Annanas	4.80	50	3.90	5.70	18.09	66.310	10.0	2.600	0.143	0.15
2.	Aranjo	4.60	40	3.80	3.10	9.340	42.080	5.00	1.400	0.149	0.11
3.	Ball	4.40	52	5.20	6.00	24.62	97.970	75.0	20.70	0.840	0.77
4.	Bencorado	4.85	95	6.00	4.50	21.64	73.860	18.00	4.846	0.97	0.24
5.	Bindao	5.10	49	3.50	3.40	9.340	47.640	5.00	1.925	0.206	0.10
6.	Bismi (Soreus)	6.00	68	6.80	4.50	25.06	128.11	2.00	1.000	0.039	0.02
7.	Black Culas	5.24	98	5.8	4.3	20.02	83.385	10.0	2.916	0.145	0.11
8.	Caria Alphonso	5.60	58	7.20	5.50	31.65	118.16	20.0	6.485	0.204	0.16
9.	Caria Branca	3.80	34	2.10	3.40	5.940	15.870	15	5.00	0.55	0.57
10.	Colaco Branco	4.50	55	3.60	4.40	12.56	38.150	125	29.40	2.340	3.28
11.	Costa	4.55	50	3.10	3.50	8.550	33.590	58.0	19.90	2.330	1.73
12.	Cota	4.50	59	4.80	5.00	18.85	62.800	85.0	18.30	0.970	1.37
13.	Culas	3.30	37	1.80	2.80	4.150	10.260	5.00	1.500	0.360	0.49
14.	Elavaygan	5.50	72	6.3	5.1	25.50	99.730	15.00	10.00	0.51	0.21
15.	Fernandin	5.10	57	5.10	4.20	16.97	69.420	18	4.700	0.216	0.259
16.	Furtad	6.00	82	7.50	7.70	45.34	232.71	242	34.31	0.524	2.99
17.	Goa Hapus	5.00	40	2.70	2.20	4.710	35.330	–	–	–	–
18.	Hilario (Ela)	3.10	40	4.60	2.40	9.621	23.15	35.0	7.000	0.72	1.51
19.	Hilario (MPSA)	4.75	54	4.50	4.60	16.25	137.09	54.0	10.50	0.646	0.39
20.	Hilario (Volfran)	5.90	92	5.70	8.00	36.83	190.91	40	12.5	0.339	0.29
21.	Hilario Mariana	5.60	52	5.20	4.30	17.72	74.000	24.0	3.000	0.169	0.32
22.	Hilario Volfarm	6.50	83	6.50	6.60	33.69	12.900	40.0	12.50	0.371	3.10
23.	Imp. Mankurad	4.60	62	3.80	3.90	11.64	42.080	35	7.87	0.67	0.83
24.	Jesuit	6.40	88	5.60	6.90	30.66	139.53	125	49.70	1.621	0.90
25.	Koita	4.00	52	4.80	3.90	14.85	40.190	12.00	5.366	0.361	0.29
26.	Malgesh	5.15	77	4.70	6.00	22.47	88.550	13.0	3.300	0.146	0.15
27.	Malgoa	3.80	57	5.10	5.50	22.05	80.740	14.0	2.333	0.105	0.17
28.	Manga Rosa	6.20	85	6.00	5.20	24.63	120.70	18.0	4.850	0.196	0.14
29.	Mankurad	5.20	45	2.70	2.30	4.910	38.210	35.0	8.000	1.632	0.92
30.	Mankurad (IC)	5.20	63	5.20	5.50	22.47	82.320	49	15.800	0.70	0.96

Contd...

Table 7.1–Contd...

Sl.No.	Name of Varieties	Height of Plant (m)	Girth Periphery of Main Stem (cm)	Spread of Plants NxS (m)	Spread of Plants ExW (m)	Total Cover Area (πr^2) m^2	Volume of Tree (m^3)	Fruits/ Plant	Fruit Wt. (kg)/ Plant	Yield Effi-ciency-cy (Kg/m^2)	Yield Effi-ciency-cy (Kg/m^3)
31.	Mankurad Assolna	3.65	32	1.70	1.60	2.140	11.850	2.00	0.50	0.07	0.06
32.	Mankurad Benaulim	5.20	51	3.00	2.00	4.910	42.450	–	–	–	–
33.	Mankurad Diwar	5.10	44	2.40	2.50	4.710	32.670	–	–	–	–
34.	Mankurad Do-damarg	4.90	62	4.20	5.00	16.61	54.950	81.0	18.63	1.12	1.754
35.	Mankurad improved	6.3	89	7.0	6.5	35.45					
36.	Mankurad Keni	6.50	90	7.00	5.80	32.15	154.78	2.00	0.50	0.03	
37.	Mankurad PWD	3.20	28	6.00	1.40	10.75	32.150	63.00	14.216	0.18	
38.	Mankurad Sports	4.00	39	3.00	3.80	9.070	25.120	5.00	1.10	0.12	0.198
39.	Mankurad yellow	4.35	83	4.70	4.10	15.20	46.540	6	1.584	0.104	0.128
40.	Maxima	6.70	84	8.40	8.00	52.78	281.34	182	41.39	0.784	0.14
41.	Mirand	3.98	52	3.30	3.20	8.290	27.360	31.00	5.960	0.718	1.13
42.	Monteiro	5.10	52	3.20	2.70	6.830	43.560	121	15.492	2.283	2.77
43.	Olivera	4.60	50	3.50	4.40	12.25	38.760	46.0	11.20	0.915	1.19
44.	Papel Branco	3.95	68	3.80	4.00	11.94	31.820	2.00	0.250	0.021	0.06
45.	Rebello	4.50	60	5.10	5.00	20.02	66.730	64.0	13.70	0.684	0.95
46.	Reonald	3.70	43	2.60	2.50	5.100	18.630	2.00	0.790	0.154	0.10
47.	Sakar Ambo	3.68	28	3.05	1.75	3.27	–	–	–	–	–
48.	Sakare (Nevra)	5.90	80	10.00	7.00	56.72	256.43	76.0	21.50	0.379	0.30
49.	Sakri	5.30	65.00	6.70	7.00	36.834	163.67	328.00	59.942	1.627	2.00
50.	Salcette Muss.(D)	5.20	55	4.20	4.40	14.51	59.430	103	13.50	0.837	1.73
51.	Salgadin	5.45	53	3.50	5.20	14.85	49.530	40.00	10.00	0.79	0.84
52.	Toranja	4.60	97	6.30	7.70	38.47	195.48	247	32.20	0.840	1.26
53.	Udgo	5.30	70	7.60	8.00	47.76	254.55	189	45.10	0.944	0.74
54.	Xavier	3.20	33	2.60	2.10	4.340	13.930	2	0.510	0.117	0.143

1. Yield efficiency using total cover area.

   Yield efficiency/tree = kg yield/total cover area (πr^2) where r = (N x S) + (E x W)/4

2. Yield efficiency using volume of tree.

   Yield efficiency= Fruits/volume of tree

   Volume of tree=$4/3\pi A^2B$ or $4/3\pi AB^2$

   A = (N x S)/2 and B = (E x W)/2, if height is more than (ExW) spreading, then B = height/2 and volume of tree = $4/3\ \pi AB^2$

Characterization and preliminary evaluation of 56 different local mango varieties were done in 2003-04. Maximum fruit weight of 520.67g. was in variety Jesuit followed by Bismi (520.00 g.) and Koita (470.00 g.) Mankurad Yellow gave an average fruit weight of 264.00 g. Fruit length was maximum (20.90cm) in Koita followed by Black Culas (12.23 cm) and Jesuit (11.89 cm). Fruit width was maximum

Table 7.2: Fruit Characters of Different Local Mango Varieties

Sl.No.	Name of Varieties	Weight of Fruit (g)	Length of Fruit (cm)	Width of Fruit (cm)	Girth of Fruit (cm)	Stone Weight (g)	Length of Stone (cm)	Width of Stone (cm)	Girth of Stone (cm)	Skin Weight (g)	Pulp Weight (g)	Pulp Stone Ratio	Pulp (%)	T.S.S (%)
1.	Annanas	465	11.88	8.78	7.71	51.0	10.04	3.67	2.22	65.0	349	6.84:1	75.05	22.00
2.	Araujo	248.00	8.12	6.07	5.37	22.00	6.43	3.14	1.84	30.00	196.00	8.9:1	79.03	16.00
3.	Ball	440.00	10.18	7.89	8.08	53.30	7.69	4.17	2.40	85.23	301.47	5.66:1	68.52	20.06
4.	Bencorado	762.00	11.7	11.1	10.40	52.00	8.1	4.3	2.10	88.00	622.00	11.9:1	81.62	17.00
5.	Bindao	168	8.59	6.85	5.63	50	7.00	4.23	2.50	48	70	1.4:1	41.67	17.8
6.	Bismi	520.00	10.55	9.74	8.24	58.00	8.50	4.54	1.54	70.00	392.00	6.76:1	75.38	12.00
7.	Black Culas	300.00	12.23	7.30	5.33	60.00	10.97	3.96	2.22	70.00	170.00	2.83:1	56.67	10.00
8.	Carica Branca	444.0	11.21	7.20	3.31	42.00	8.57	2.09	4.11	117.3	284.70	6.77:1	64.12	15.1
9.	Cariera Alphonso	389.30	10.29	7.61	8.34	42.00	7.89	1.98	4.06	63.30	284.00	6.76:1	72.95	18.50
10.	Colaco Branco	221.20	10.50	5.91	6.39	26.40	8.18	1.79	3.28	32.80	162.00	6.14:1	73.24	15.08
11.	Costa	210.0	8.46	6.38	6.53	36.00	6.52	1.83	4.04	50.00	124.00	3.44:1	59.04	18.0
12.	Cota	251.00	10.82	6.31	6.69	27.00	8.65	1.59	3.52	43.00	181.00	6.70:1	72.11	18.00
13.	Culas	256.00	10.6	6.2	5.8	36.00	9.4	3.7	1.80	42.00	178.00	4.94:1	69.53	12.00
14.	Elavangan	225	9.64	7.08	6.46	48				45	132	2.75:1	58.67	9
15.	Elavangan	684.00	10.22	10.17	10.32	76	8.65	5.32	3.06	128.00	480.00	6.31:1	70.17	23.1
16.	Fernandin	396	11.15	8.01	7.32	60	9.37	3.37	2.55	128				17.9
17.	Furtad	251.33	9.13	6.92	6.30	45.33	6.92	4.03	2.26	42.66	163.34	3.60:31	64.99	17
18.	Goa Hapus	212	8.11	7.06	5.7	40.0	6.46	2.26	3.86	40.0	132.00	3.3:1	62.26	12.00
19.	Hilario 10/10	218.00	8.360	6.49	7.07	26.00	6.14	3.19	2.05	38.00	154.00	5.92:1	70.64	23.00
20.	Hilario 4/15	216.00	8.740	6.82	6.24	34.00	6.58	3.24	1.91	48.00	134.00	3.94:1	62.04	31.00
21.	Hilario Ela	156	8.09	6.56	5.61	20	6.24	3.22	2.32	43	93	4.65:1	59.62	19
22.	Hilario MPSA	220.00	8.920	6.22	6.85	32.00	6.99	2.08	3.40	48.00	140.00	4.38:1	63.64	23.00
23.	Hilario Volfarm	246.00	9.36	7.23	6.63	40.00	5.9	3.5	2.10	50.00	156.00	3.9:1	63.41	26.00
24.	Imp. Mankurad	230	9.64	6.46	6.04	27	8.02	4.30	3.41	51	152	5.6:1	66.09	23
25.	Jesuit	520.67	11.88	8.34	9.22	64.00	9.35	2.30	4.67	93.33	363.34	5.68:1	69.78	18.00
26.	Koita	470.00	20.90	6.71	6.15	83.00	19.25	3.59	2.26	95.00	292.00	3.52:1	62.13	12.00

Contd...

Table 7.2–Contd...

Sl.No.	Name of Varieties	Weight of Fruit (g)	Length of Fruit (cm)	Width of Fruit (cm)	Girth of Fruit (cm)	Stone Weight (g)	Length of Stone (cm)	Width of Stone (cm)	Girth of Stone (cm)	Skin Weight (g)	Pulp Weight (g)	Pulp Stone Ratio	Pulp (%)	T.S.S (%)
27.	Malgesh	200.00	9.200	6.34	7.31	38.00	6.65	2.21	4.00	54.00	108.00	2.84:1	54.00	16.50
28.	Malgoa	148.00	7.700	4.97	6.17	20.00	6.17	1.48	3.57	34.00	94.00	4.70:1	63.51	23.00
29.	Manga rosa	234.00	9.950	6.29	9.11	42.00	7.58	2.28	4.08	50.00	142.00	3.38:1	60.68	14.00
30.	Mankurad	194.00	10.2	8.62	7.42	42.00	9.12	5.08	2.36	46.00	106.00	2.52:1	54.63	19.80
31.	Mankurad Diwar	227	8.89	6.93	6.13	48	6.96	3.96	3.10	61	118	2.45:1	51.98	23
32.	Mankurad Dodamarg	230.00	8.810	6.55	6.12	26.00	7.57	3.62	1.50	36.00	168.00	6.46:1	73.04	19.00
33.	Mankurad IC	358	10.45	7.96	6.95	38	8.46	4.24	3.54	60	260	6.8:1	72.63	19
34.	Mankurad Keni	255	8.96	7.23	6.57	42	7.53	4.12	3.67	45	168	6.8:1	65.88	21
35.	Mankurad Koregaon	274.00	9.460	7.10	7.54	40.00	7.34	1.93	3.97	46.00	188.00	4.70:1	68.61	18.00
36.	Mankurad PWD	206.00	8.34	6.80	6.38	28.00	6.72	3.72	1.79	42.00	136.00	4.85:1	66.01	14.40
37.	Mankurad yellow	264.00	8.210	7.27	7.01	37.00	6.98	3.98	1.82	48.00	179.00	4.84:1	67.80	20.00
38.	Maxima	214.00	8.660	6.19	7.17	34.00	6.80	2.59	3.16	44.67	135.33	3.98:1	63.24	19.67
39.	Mirand	200.00	8.740	5.87	5.90	50.00	7.65	4.50	2.20	42.00	108.00	2.16:1	54.00	6.00
40.	Monteiro	138	7.15	6.75	5.33	19.4	5.10	4.06	1.28	32.8				24.75
41.	Olivera	180.00	9.210	5.69	6.29	32.00	7.43	2.18	2.94	48.00	100.00	3.13:1	55.56	22.00
42.	Papel	300	10.07	8.23	7.81	51	7.64	3.84	3.01	97	152	2.98:1	50.67	15
43.	Papel Branco	70	6.34	4.72	3.97	22	5.93	3.21	2.74	20	28	1.4:1	40.00	13
44.	Rebello	208.0	9.68	5.85	6.64	40.00	7.93	2.01	2.49	60.00	108.00	2.7:1	51.92	15.00
45.	Reonald	398	11.19	7.66	8.28	36	9.18	3.14	2.12	86.00	276.00	7.66:1	69.34	14.7
46.	Sakare (Nevra)	232.00	8.5	7.7	6.2	44	6.8	4.1	2.2	40.00	148.00	3.36:1	63.79	9.00
47.	Sakari	282.00	8.910	6.81	5.56	40.67	6.87	2.02	3.56	62.67	178.66	4.39:1	63.35	18.33
48.	Sakari Nevra	280	9.46	7.78	6.94	57.0	6.85	3.79	2.33	82.8	137.2	2.40:1	49.00	21.00
49.	Salcette Musserat	380.00	11.89	7.81	9.32	60.00	9.13	2.33	4.57	88.00	232.00	3.87:1	61.05	12.00
50.	Salgadin	400.67	11.47	7.12	8.10	45.00	8.72	2.33	3.55	103.3	252.37	5.6:1	62.98	22.00
51.	Toranjo	172.00	8.100	6.19	6.54	42.00	6.55	2.76	3.69	70.00	60.00	1.43:1	34.88	15.00
52.	Udgo	220.00	7.700	7.33	6.51	30.00	5.59	3.91	1.94	40.00	150.00	5.00:1	68.18	18.00

(9.74cm) in Bismi followed by Jesuit (8.34 cm). Fruit girth was maximum(9.32 cm)in Salcette Musserat followed by Jesuit (9.22 cm). Skin weight of fruit was maximum (95.00 g) in variety Koita followed by Jesuit (93.33 g). Stone weight was maximum (83.00g) in Koita followed by Jesuit 64.00g. Length of the stone was also maximum (19.25 cm) in Koita followed by Black Culas (10.97 cm). Pulp weight of fruit was maximum(392 g) in variety Bismi followed by Jesuit 393.34 g. Pulp Stone ratio was maximum(6.76)in variety Cariera Alphonso and Bismi followed by Cota (6.70) and Mankurad Dodamarg (6.46). Maximum pulp percentage (75.38) was in variety Bismi followed by Colaco Branca(73.24) and Mankurad Dodamarg (73.04). T.S.S. of the fruit was maximum(31 per cent)in Hilario (4/15) followed by Malgoa (23 per cent), Hilario 10/10 (23 per cent) and Hilario MPSA (23 per cent). The minimum T.S.S. was 6 per cent in variety Mirand.

Maximum yield was recorded (12.5 kg/plant) in Hilario Volfarm (10/10) followed by Hilario MPSA (8.00 kg/plant) whereas maximum yield efficiency was 1.51 kg/m^3 in variety Hilario Ela (7/6) with an yield of 7 kg/plant in 2003-04.

Table 7.3: Variability in Hilario at ICAR Research Complex, Goa

Sl.No.	*Name of Varieties*	*Height of Plant (m)*	*Growth Periphery of Main Stem (cm)*	*Spreading of Plants*		*Fruit no.*	*Fruit Wt/ Yield (kg)*	*Yield Efficiency (kg/m³)*
				NxS (m)	*ExN (m)*			
1.	Hilario Mariana-4/11	6.10	68	3.80	4.00	24	3	0.32
2.	Hilario MPSA-4/15	4.75	54	4.50	4.60	43	8	0.28
3.	Hilario (Ela)-7/6	3.10	40	4.60	2.40	35	7	1.51
4.	Hilario 10/10 Volfran	5.90	92	5.70	8.00	40	12.5	0.29

Genetic Variability in Mankurad

During 2003-04, four varieties yielded and maximum yield recorded in Mankurad Dodamarg (17.3 kg/plant) with 81.00 fruits. Maximum (T.S.S. 20 per cent) was found in Mankurad Yellow. Maximum yield efficiency was recorded from Mankurad Dodamarg (1.754 kg/m^3)during this period.

Table 7.4: Variability in Mankurad at ICAR Research Complex, Goa

Sl.No.	*Name of Varieties*	*Height of Plant (m)*	*Growth Periphery of Main Stem (cm)*	*Spreading of Plants*		*Fruit No.*	*Fruit Weight/ Yield (g)*	*Yield Efficiency (kg/ cubic meter)*
				NxS (m²)	*ExN (m²)*			
1.	Imp. mankurad	4.60	62	3.80	3.90	35	228.57	0.831
2.	Mankurad Assolna	3.65	32	1.70	1.60	56	11.30	1.096
3.	Mankurad Benaulim	5.20	51	3.00	2.00	NA	NA	NA
4.	Mankurad Diwar	5.10	44	2.40	2.50	NA	NA	NA
5.	Mankurad Dodamarg	4.90	62	4.20	5.00	81.00	230.00	1.754
6.	Mankurad IC	5.20	63	5.20	5.50	3	0.62	0.96
7.	Mankurad Keni	6.50	90	7.00	5.80	30	7.25	0.2432
8.	Mankurad PWD	3.20	28	6.00	1.40	11	2.50	0.2833
9.	Mankurad Sports	4.00	39	3.00	3.80	5.000	220.00	0.198
10.	Mankurad Yellow	4.35	83	4.70	4.10	6	264.00	0.128

NA: Not available.

Figure 7.2: Genetic Variabilities in Varieties of Mango

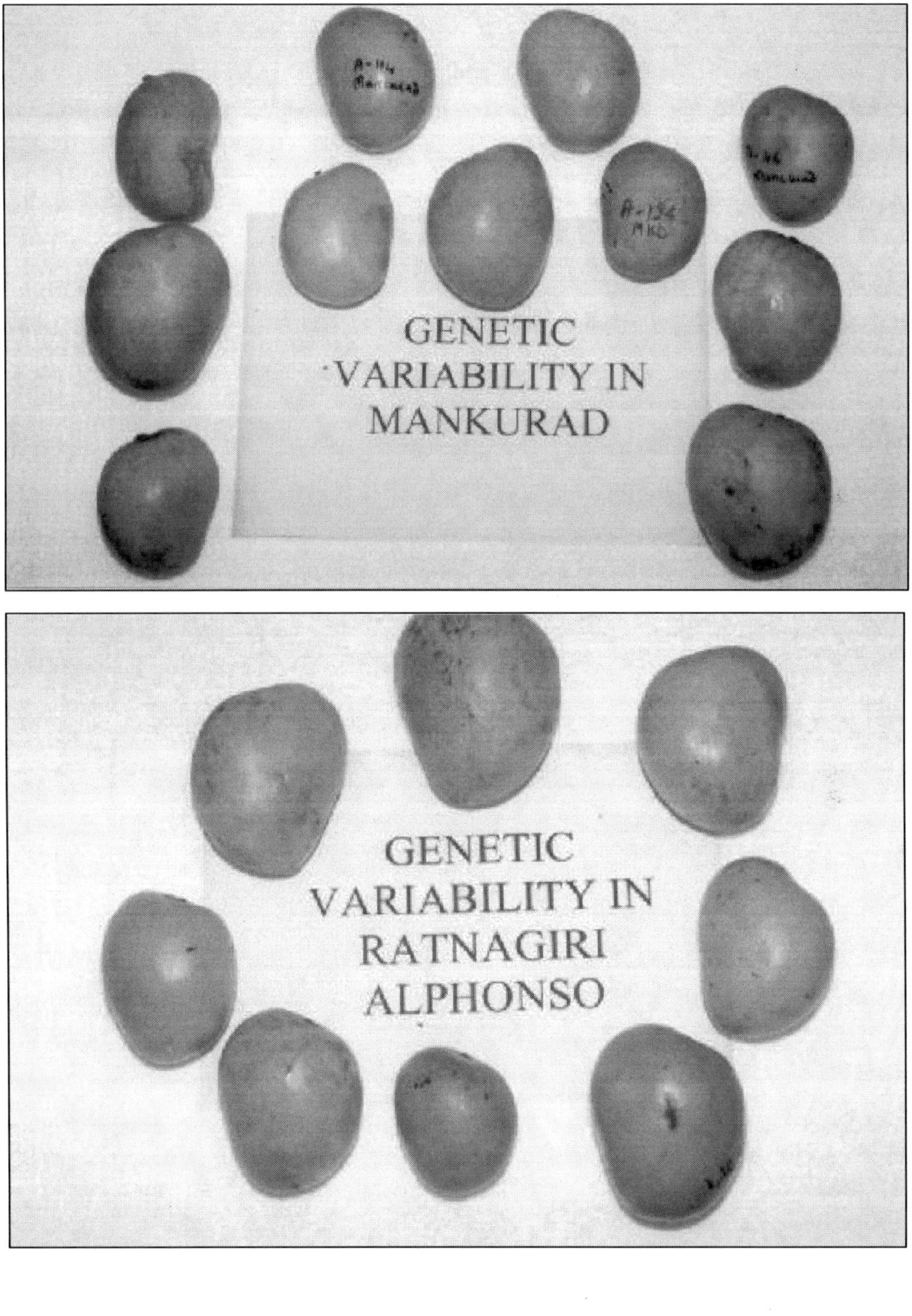

Contd...

Figure 7.2–Contd...

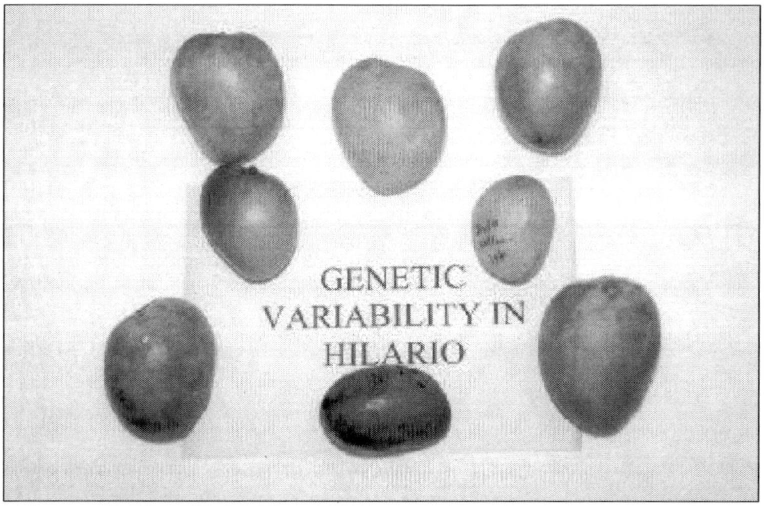

Table 7.5: Variability in Mankurad Collected from Mango Show

Sl.No.	Name of Varieties	Fruit Characteristics				Stone Characteristics				Peel Weight (g)	T.S.S. (^{o}Brix)	Pulp Weight (g)	Pulp (%)	Pulp : Stone Ratio	Remarks
		Fruit Weight (g)	Fruit Length (cm)	Fruit Width (cm)	Fruit Girth (cm)	Stone Weight (cm)	Stone Length (cm)	Stone Width (cm)	Stone Girth (cm)						
							Mankurad								
1.	Mankurad (P)	300.00	9.49	7.62	7.11	32.00	6.67	3.73	1.91	42.00	18.2	226.00	75.33	7.06:1	Good flovour slight spongy tissue
2.	Mankurad A-114	280.00	9.49	7.17	6.85	30.00	7.15	3.75	1.77	40.00	15.0	210.00	75.0	7:1	Fibreless, good quality pulp, dark yellow
3.	Mankurad A-115 (H) Mankurad	234.00	8.63	7.11	6.46	32.00	6.01	3.85	2.07	44.00	21.4	158.00	67.52	4.94:1	Small fruit, skin dark yellow, fibre present
4.	Mankurad A-124 (K)	194.00	8.54	6.68	6.17	40.00	6.93	4.11	2.18	34.00	14.8	120.00	61.85	3:1	Good quality with fibre
5.	Mankurad A-125 (G)	250.00	9.11	7.44	6.69	40.00	6.93	4.11	2.18	34.00	14.8	176.00	70.4	4.4:1	Good quality with fibre
6.	Mankurad A-46	216.00	7.93	7.04	6.50	40.00	6.83	4.33	2.14	62.00	20.2	114.00	52.78	283:1	Fibrous, spoiled
7.	Mankurad B-79 (O)	340.00	10.36	7.78	7.33	20.4	7.38	3.55	1.69	50.00	14.6	269.6	79.29	13.2:1	Oval shape skin dark yellow, good quality
8.	Mankurad D-21	220.00	9.08	7.40	6.30	24.00	6.55	3.53	1.91	52.00	18.2	144.00	65.45	6.00:1	Dark skin sindoori at shoulder good taste and flavour

Contd...

Table 7.5–Contd...

Sl.No.	Name of Varieties	Fruit Characteristics				Stone Characteristics				Peel Weight (g)	T.S.S. (^oBrix)	Pulp Weight (g)	Pulp (%)	Pulp : Stone Ratio	Remarks
		Fruit Weight (g)	Fruit Length (cm)	Fruit Width (cm)	Fruit Girth (cm)	Stone Weight (cm)	Stone Length (cm)	Stone Width (cm)	Stone Girth (cm)						
							Hilario								
1.	Hilario D-10	158.00	7.59	6.32	5.77	28.00	5.82	3.45	1.94	42.00	24.4	88.00	55.69	3.14:1	Dark yellow skin on shoulder good flavoured less fibre, pulp yellow
2.	Hilario D-14	220.00	8.58	7.26	6.43	26.00	6.22	3.55	2.16	44.00	23.4	150.00	68.18	5.76:1	Good flavour, less fibre, apical skin orange
3.	Hilario D-15	220.00	8.38	6.86	6.39	28.00	6.37	3.32	2.05	48.00	24.00	144.00	65.45	5.14:1	Pulp yellow and less fibre
4.	Hilario Maye C-53	178.00	8.57	6.97	5.54	30.00	6.71	3.54	2.17	38.00	18.00	110.00	61.79	3.66:1	Fruit spoiled
5.	Hilario Shyam Dhargal	480.00	15.67	9.53	7.71	84.00	8.37	4.57	2.58	98.00	17	298	62.08	3.54:1	Fibre present

Table 7.6: Variability in Mankurad Collected from Different Places

Sl.No	Name of Varieties	Fruit Characteristics				Stone Characteristics				Skin Weight (gm)	Pulp Weight (g)	Pulp Weight (g)	Pulp : Stone Ratio
		Weight (gm)	Length (cm)	Width (cm)	Girth (cm)	Weight (gm)	Length (cm)	Width (cm)	Girth (cm)				
1.	Mankurad Assolna	292.5	2.92	7.59	1.98	51.5	8.18	4.62	7.59	53.76	187.24	64.07	3.63:1
2.	Mankurad Bamanwada	320	9.44	7.41	4.10	36.0	7.14	1.87	8.44	52.00	234.00	72.5	6.44:1
3.	Mankurad Special	406	10.49	7.73	1.90	42.0	8.00	4.31	7.73	80.0	284.00	69.95	6.76:1
4.	Mankurad Yellow	234	7.86	6.78	2.20	30.0	6.15	3.68	6.78	40.0	164.00	70.08	5.47:1

Eight variants of Mankurad were evaluated for their fruit characters. Maximum fruit weight was recorded(340.00 g) in Mankurad B -79.The fruit had oval shape and skin of the fruit dark yellow. Pulp of the fruit was without fibre and of good quality. T.S.S of fruit ranged 14.6 to 21.45 per cent. Pulp weight was maximum in variety Mankurad Keni (227.2g) followed by Mankurad Assolna (187.5g). Maximum pulp stone ratio was in Mankurad Keni (6.31:1) followed by Mankurad Assolna 5/7 (4.47:) Maximum pulp per cent was in variety Mankurad Keni (72.08) followed by Mankurad Assolna (64.43).

Five variants of Hilario were evaluated for fruit characters. Maximum fruit weight(220.00 g) was found in D-21, D-15, D-14, D-10, and D-21. Hilario is found very unique, dark red skin was found on the shoulder with good taste and flavour. T.S.S. of fruit ranges 18.2 to 24.4 per cent and was maximum in D-10.The skin of the fruit was yellow on the shoulder. Fruits had very good flavour with less fibre and pulp was yellow (Table 7.5).

Mankurad collected from different places also showed a lot of variability in fruit size, weight, appearance, and quality. The fruit weight ranged 234.00 to 406.00 g in Siolim area. The T.S.S of the fruit ranged 14 to 19 per cent. Mankurad collected from Banaswada, Siolim, was excellent in colour and appearance with red colour on shoulder. The fibre content was less and fruit weight was 320.00 g, the ideal size for mango (Table 7.6).

Table 7.7: Growth Performance with Yield and Yield Efficiency of Hybrid Mango Varieties Introduced at ICAR Research Complex (2003-2004)

Sl.No.	Name of Varieties	Ht of Plant (m)	Girth Periphery of Main Stem (cm)	Spreading of Plants		Total Cover Area (Πr²)	Volume of Tree (m³)	Fruit No/ Plant	Fruit Wt (kg)/ Plant	Yield Effi- ciency (kg/m²)	Yield Efficiency/ cubic meter (m³)
				NxS (m²)	ExW (m²)						
1.	Amrapali	6.15	103.0	7.20	7.00	39.57	184.63	425	60.50	2.302	2.302
2.	Arka Anmol	6.75	73.00	6.70	4.80	25.95	159.76	19	6.687	0.088	0.118
3.	Arka Aruna	4.63	74.00	5.40	5.10	21.63	73.500	17	13.91	0.095	0.231
4.	Arka Puneet	4.80	75.00	7.00	5.80	32.15	123.23	31	10.22	0.211	0.251
5.	HB-117	3.60	31.00	2.40	2.30	4.33	16.280	73	11.10	4.485	4.485
6.	HB-151	4.40	81.00	4.50	5.40	19.23	68.670	205	23.00	2.985	2.985
7.	HB-45	5.20	82.00	5.10	6.10	24.61	99.310	54	22.50	0.544	0.544
8.	HB-56	6.00	133.0	6.75	6.80	36.03	163.34	54	16.00	0.331	0.331
9.	HB-87	5.35	85.00	7.20	7.40	41.83	206.34	51	13.30	0.247	0.247
10.	KXB	3.60	52.00	4.6	5.2	18.84	65.090	260	31.80	3.994	3.994
11.	Mallika	4.10	48.00	3.10	2.40	5.93	27.270	5.0	1.000	0.168	0.183
12.	Neelgoa	5.25	95.00	6.70	8.50	45.34	253.46	20	11.78	0.259	0.078
13.	Ratna	4.60	60.00	4.00	5.00	15.89	52.330	40	9.000	0.764	0.764

Growth and yield performance of 13 hybrid varieties were evaluated at ICAR Research Complex, Goa in 2003-04. Maximum plant height was in hybrid Arka Anmol(6.75m) followed by Amrapali(6.15 m.) Maximum girth on the periphery of main stem was in variety HB-56(133cm) followed by Amrapali(103.00 cm). Maximum spread of 7.20 m² of the plant was found towards NxS in variety Amrapali and HB-87. It was found maximum 7.40 m² towards ExW in variety HB-87. Maximum cover

Figure 7.3: Hybrid Variety

Amrapali

Arka Anmol

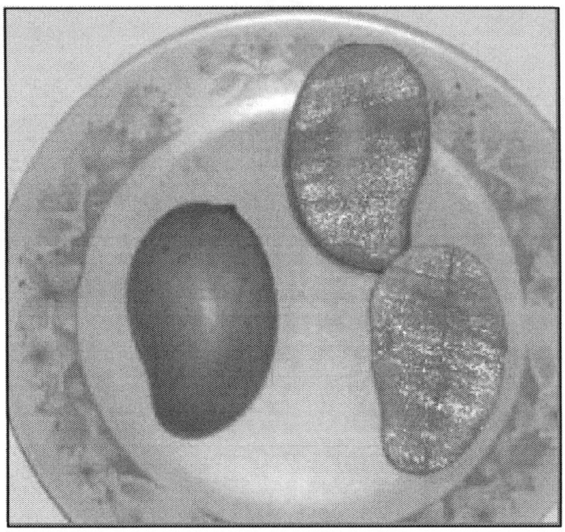

Amrapali

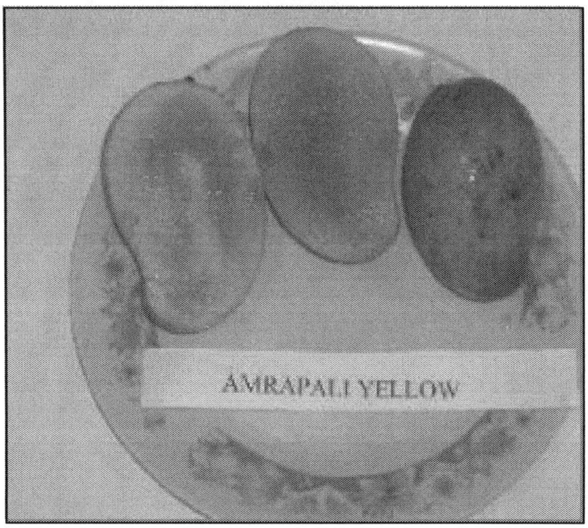

Amrapali Yellow

Contd...

Figure 7.3–Contd...

Ratna

Arka Aruna

Contd...

Figure 7.3–Contd...

HB-117

HB-151

Contd...

Figure 7.3–Contd...

HB-45

HB-87

HB-56

Contd...

Figure 7.3–Contd...

K x B

Neelgoa

area was found 41.83 sq. m. in hybrid HB-87 followed by Amrapali 39.57 sq.m. Maximum number of fruits(425) was in Amrapali followed by KXB (260). Fruit weight/plant was maximum (60.50 kg/plant)in Amrapali followed by KXB (31.80 Kg/plant). Yield efficiency was maximum(2.563 kg/m^2) in hybrid HB-117 followed by HB-151 (1.916kg/m^2) and KXB (1.687 K kg/m^2) (Table 7.7).

Thirteen varieties of Hybrid Mango were evaluated in 2003-04. Maximum fruit weight (562.00 g.) was found in Arka Aruna followed by HB-56 (444.00 g). Length of fruit was maximum in Arka Aruna(16.17 cm) followed by Neelgoa (11.26 cm) and Arka Puneet (10.85 cm). Width of fruit was maximum in hybrid Arka Aruna((9.69cm) followed by HB-56 (8.63cm) and Ratna (8.27cm). Girth of the fruit was maximum (10.98 cm) in variety Arka Aruna followed by HB-56 (10.05cm). The smallest fruit was in 138.25 g in variety HB-151(138.25g). Maximum pulp weight 334.00g. was found in HB-56 followed by Arka Aruna(330.03g). Pulp Stone ratio was maximum in HB-56 (11.13) followed by Neelgoa (8.33) and Arka Aruna (6.78). Maximum pulp percentage (75.23) was in HB-56 followed by Amrapali (71.06) and Arka Anmol (70.33). T.S.S. of the fruit was maximum(28 per cent)in HB117 followed by Ratna (25 per cent) (Table 7.8).

Growth and yield performance of 8 introduced varieties were studied in 2003-04. Maximum height of the plant was 7.15m in Vallaicolamban followed by Ratnagiri Alphonso (6.8m). Girth on the periphery of the main stem was maximum 95cm in Ratnagiri Alphonso followed by Vallaicolamban 93.00cm. Spread of the plant was maximum 6.3m^2 towards NxS in variety Elavangan, followed by Vallaicolamban (5.4 m^2 as NxS). Maximum spread towards ExW was 6.9m^2 in Vallaicolamban followed by Ratnagiri Alphonso (6.8m^2). Maximum cover area was 37.37sq.m.in Ratnagiri Alphonso followed by Vallaicolamban(22.69 sq.m.). Maximum number of fruits(176) with an yield of 33.75 kg/plant was in Ratnagiri Alphonso whereas a total fruit number of 159 with yield of 28.3kg/plant was found in Vallaicolamban. Yield efficiency of the variety was maximum in Neelam(1.707 kg/m^2) followed by Vallaicolamban (0.953Kg/m^2) (Table 7.9).

Fruit and other characteristics of eight introduced varieties of Mango were studied in 2003-04. Maximum fruit weight of 660 g/fruit was in variety Himayuddin followed by Totapuri (600 g/fruit). Minimum fruit weight (167.20 g) was recorded in variety Vallaicolamban. Maximum length of fruit (23.00cm) was recorded in Totapuri followed by Himayuddin (12.72 cm). Width of fruit was also found 9.60cm in Totapuri followed by Himayuddin (9.38cm). Girth of fruit was maximum (8.75cm) in Himayuddin followed by Kalapaddy (7.86cm). Maximum skin weight(110.00g) was found in Totapuri followed by Himayuddin (85.00g). Stone weight was maximum (52.00g) in variety Totapuri followed by Himayuddin and Neelam (44.00g). Maximum length of stone was 13.62cm in Totapuri followed by Himayuddin (10.80g). Pulp weight of the fruit was 531.00g in Himayuddin followed by Totapuri (438.00g). Pulp stone ratio was maximum 12.07 in Himayuddin followed by Totapuri (8.42). Pulp (per cent) is maximum (80.45) in Himayuddin followed by Ratnagiri Alphonso (74.07). T.S.S. of the fruit was maximum (26 per cent) in Mallika followed by Himayuddin and Neelam (22 per cent) (Table 7.10).

Seven varieties collected from Siolim were evaluated for their fruit and other characters in 2003-04. Maximum 1000.00g fruit weight was in variety Bismi or Bismi followed by Karel (900.00g). Length of fruit was maximum(15.50cm) in Karel followed by Bismi (14.60cm). Width of fruit was maximum (11.79cm) in variety Bismi followed by Karel (9.30cm). Similarly girth of fruit was also maximum (10.50cm) in Karel followed by Bismi (9.32cm). Skin weight of fruit was maximum (160g) in Karel followed by Bismi (160g). Stone weight was maximum(138g) in Bismi followed by Karel (80 g). Maximum T.S.S. of the fruit was in Hilario Maye (22 per cent) followed by Pairi (18 per cent). Minimum T.S.S. of the fruit (11 per cent) was in variety Karel used for pickling purpose (Table 7.11).

Table 7.8: Fruit Characters of Different Mango Hybrids

Sl.No.	Name of Variety	Weight of Fruit (g)	Length of Fruit (cm)	Width of Fruit (cm)	Girth of Fruit (cm)	Stone Weight (g)	Length of Stone (cm)	Width of Stone (cm)	Girth of Stone (cm)	Skin Weight (g)	Pulp Weight (g)	Pulp : Stone Ratio	Pulp (%)	T.S.S (%)
1.	Amrapali	364.50	9.350	7.73	6.550	47.00	11.22	3.81	1.73	58.50	259.00	5.51:1	71.06	20.63
2.	Arka Anmol	300.00	10.24	6.80	6.66	49.00	8.25	2.54	2.84	40.00	211.00	4.31:1	70.33	16.00
3.	Arka Aruna	562.00	16.17	9.69	10.98	48.67	14.90	4.53	2.05	183.30	330.03	6.78:1	58.72	15.00
4.	Arka Puneet	350.00	10.85	7.78	7.480	49.33	8.940	2.23	4.18	65.33	235.34	4.77:1	67.24	19.33
5.	HB-117	352.00	10.54	7.26	8.680	42.00	8.540	1.98	4.93	90.00	220.00	5.24:1	62.50	28.00
6.	HB-151	138.25	8.290	5.06	5.830	26.00	6.710	2.01	3.51	29.00	83.250	3.20:1	60.22	25.25
7.	HB–45	314.00	9.820	7.10	8.140	40.00	7.580	2.14	4.04	66.67	207.33	5.18:1	66.03	16.33
8.	HB-56	444.00	10.68	8.63	10.05	30.00	8.300	1.83	4.81	80.00	334.00	11.13:1	75.23	20.00
9.	HB-87	324.00	10.29	7.16	8.420	36.00	8.100	2.24	4.52	66.60	221.40	6.15:1	68.33	20.00
10.	KXB	166.00	10.14	4.68	5.830	26.00	8.840	1.38	3.26	50.00	90.000	3.46:1	54.22	15.00
11.	Neelgoa	400.00	11.26	7.63	9.320	33.00	9.020	1.07	5.32	92.00	275.00	8.33:1	68.75	23.00
12.	Ratna	370.00	10.06	8.27	6.68	54.00	8.25	5.96	125.0	60.00	256.00	4.74:1	69.19	25.00

Table 7.9: Growth Performances with Yield and Yield Efficiency of Introduced Mango Varieties in 2003–04

Sl.No	Name of Variety	Ht. of Plant (m)	Girth Periphery of Main Stem (cm)	Spreading of Plants		Total Cover Area (πr^2) (m^2)	Volume of Tree (m^3)	Fruits/ Plant	Fruit Wt. (kg/plant)	Yield Efficiency (kg/m^2)	Yield Efficiency (kg/cubic meter) kg/m^3
				NxS (m^2)	ExW (m^2)						
1.	Himayuddin	3.90	52	5.1	4.0	16.25	40.600	21.0	10.03	0.617	0.39
2.	Kesar	4.30	70	4.7	4.5	16.61	45.480	54.0	19.32	1.163	1.07
3.	Neelam	4.80	50	4.1	5.0	16.25	49.440	156	27.75	1.708	3.16
4.	Ratnagiri Alphonso	6.80	95	7.0	6.8	37.37	169.39	176	33.75	0.903	1.04
5.	Totapuri	2.90	33	3.1	2.2	5.512	13.640				2.47
6.	Totapuri (A.B.)	4.70	53	5.1	3.7	15.19	58.960	10.0	4.850	0.319	3.88
7.	Vallaicolamban	7.15	93	5.4	6.9	22.69	144.47	159	28.30	0.953	1.10

Figure 7.4: Introduced Varieties

Alphanso Joseph (Parvorim)

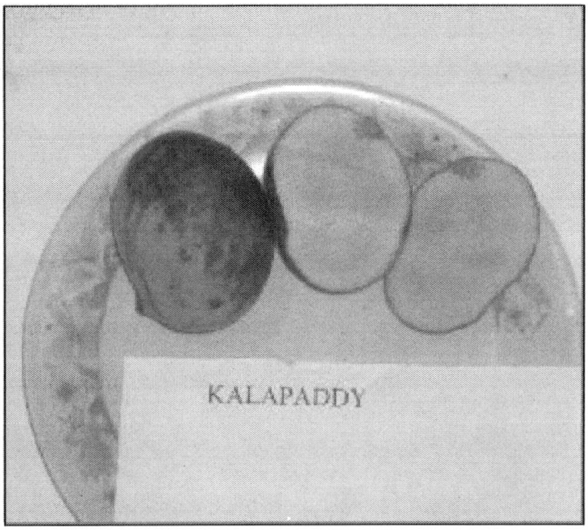

Kalapaddy

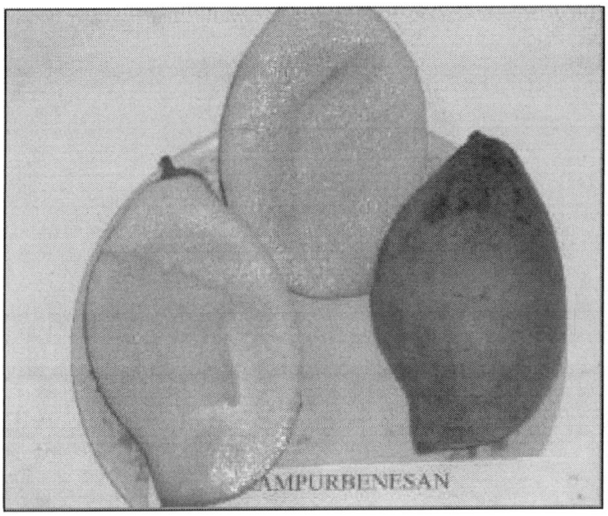

Alampurbenesan

Neelam

Contd...

Figure 7.4–Contd...

Madhulika

T.C.C.

Contd...

Figure 7.4–Contd...

Totapuri

Himayuddin

Contd...

Figure 7.4–Contd...

Sensation

MDCH-1

Contd...

Figure 7.4–Contd...

STH-1

Vallaicolamban

Contd...

Table 7.10: Fruit Characters of Different Introduced Mango Varieties

Sl.No.	Name of Variety	Weight of Fruit (g)	Length of Fruit (cm)	Width of Fruit (cm)	Girth of Fruit (cm)	Stone Weight (g)	Length of Stone (cm)	Width of Stone (cm)	Girth of Stone (cm)	Skin Weight (g)	Pulp : Weight (g)	Pulp Stone Ratio	Pulp (%)	T.S.S (%)
1.	Himayuddin	660.00	12.72	9.38	8.75	44.00	10.80	4.350	2.45	85.00	531.00	12.0:1	80.45	22.00
2.	Kalapaddy	373.00	12.19	6.79	7.86	42.00	10.26	3.680	1.98	75.00	256.00	6.10:1	68.63	16.00
3.	Kesar	326.67	11.62	7.63	6.71	41.00	9.130	3.720	1.84	45.00	240.67	5.87:1	73.67	19.00
4.	Mallika	170.00	10.45	5.55	4.57	20.00	8.920	2.780	1.22	36.00	114.00	5.70:1	67.06	26.00
5.	Neelam	300.00	10.36	7.81	6.75	44.00	8.300	4.600	1.95	70.00	186.00	4.22:1	62.00	22.00
6.	Ratnagiri Alphonso	277.30	9.420	6.90	7.59	33.30	7.190	2.130	3.76	38.60	205.40	6.17:1	74.07	18.00
7.	Totapuri	600.00	23.00	9.60	7.50	52.00	13.62	5.500	1.92	110.0	438.00	8.42:1	73.00	14.00
8.	Vallaicolamban	167.20	10.37	4.94	6.19	29.20	10.37	2.124	3.14	27.60	110.40	3.78:1	66.03	15.55

Table 7.11: Fruit and Other Characteristics of Local Varieties Collected from Siolim

Sl.No.	Varieties	Fruit Weight (g)	Length (cm)	Width (cm)	Girth (cm)	Skin Weight (g)	Stone Weight (g)	Pulp Weight (g)	Pulp Stone Ratio	Pulp (%)	T.S.S (%)
1.	Bismi	1000.00	14.6	11.79	9.32	160	138	702.00	5.09:1	702.00	12.00
2.	Fernandin	220.00	8.65	6.77	6.38	38	41	141.00	3.44:1	141.00	13.00
3.	Goa Hapus	186.00	8.99	6.81	5.63	30	32	124.00	3.88:1	124.00	16.00
4.	Hilario Mange	210.00	8.36	6.30	6.40	30	28	152.00	5.43:1	152.00	22.00
5.	Karel	900.00	15.5	9.30	10.5	180	80	640.00	8.00:1	640.00	11.00
6.	Pairi	243.30	8.48	6.83	7.01	42	36	165.30	4.59:1	165.30	18.00
7.	Xavier Maye	280.00	9.51	7.84	5.37	52	48	180.00	3.75:1	180.00	17.00

Fruits from eight varieties from IIHR Bangalore were evaluated during 2002-03. Maximum fruit weight was 358.00g in Tommy Atkins followed by Arka Aruna (320 g/fruit). Length of fruit was maximum(11.82 cm) in Arka Aruna followed by Tommy Alkins (7.94cm). Fruit girth was maximum (7.82cm) in Mundappa Black followed by Rumani (7.00cm). Skin weight was maximum (60.00g) in Tomy Alkins and Arka Aruna followed by Arka Nilkiran (58.00g). Stone weight of fruit was maximum (52g) in Tommy Alkins followed by Arka Aruna (48.00g). T.S.S. of the fruit was maximum (15 per cent) in Mundappa Black followed by Sensation (14 per cent). Minimum T.S.S. of the fruit was 7 per cent in Arka Aruna and Arka Neelkiran and Janardan Pasand recorded 8 per cent.

Table 7.12: Fruit and Other Characteristics of Varieties Collected from IIHR

Sl.No.	Varieties	Fruit Weight (g)	Length (cm)	Width (cm)	Girth (cm)	Skin Weight (g)	Stone Weight (g)	Pulp Weight (g)	Pulp Stone Ratio	Pulp (%)	T.S.S (%)
1.	Arka Aruna	320.00	11.82	7.25	6.05	60	48	212.00	4.42:1	66.25	7.00
2.	Arka Nilkiran	244.00	8.93	7.48	6.22	58	38	148.00	3.89:1	60.66	8.00
3.	Janardan Pasand	192.00	8.82	6.35	5.52	46	42	104.00	2.48:1	54.17	8.00
4.	Mundappa Black	282.00	8.65	7.65	7.82	50	44	188.00	4.27:1	66.67	15.0
5.	Raspari	193.00	7.89	7.08	5.62	38	24	131.00	5.46:1	67.88	10.0
6.	Rumani	280.00	8.45	8.89	7.00	32	34	214.00	6.29:1	76.43	9.00
7.	Sensation	244.00	9.19	6.57	6.62	42	40	162.00	4.05:1	66.39	14.0
8.	Tomy Alkins	358.00	10.05	7.94	6.97	60	52	246.00	4.73:1	68.72	11.0

Important Varieties Performing Better at Fruit Research Station–Vengurla, Maharashtra

1. *Austeen*: Oblong Red blush on shoulder
2. *Badami model*: Roundish oblong, yellowish red on shoulder.
3. *Badarakalsha*: Roundish fruit, Dark yellow on shoulder.
4. *Baganpalli*: Roundish fruit yellow on shoulder

Figure 7.5: Mangoes from Vengurla

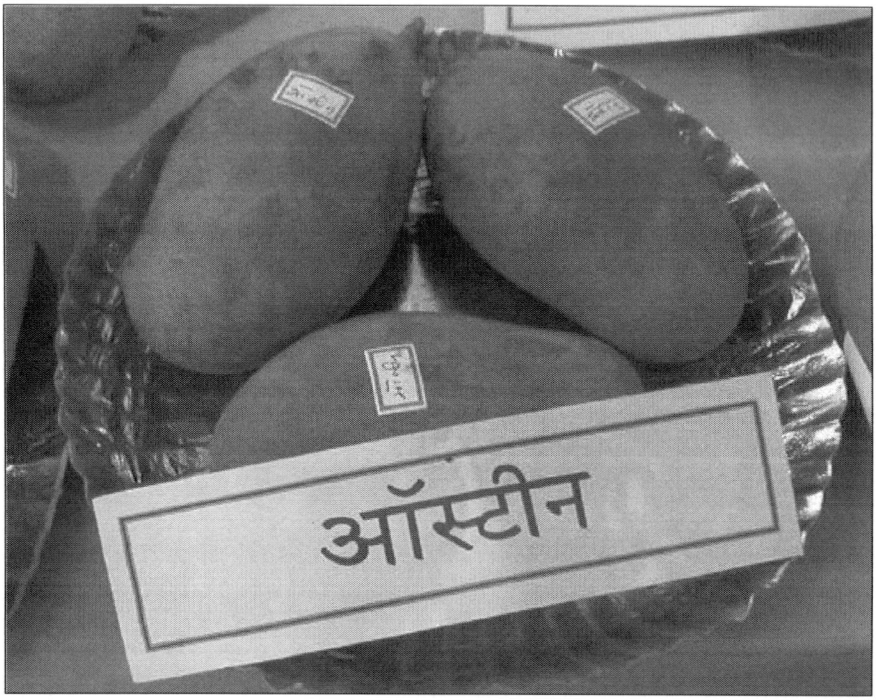

Austeen

Badami Model

Contd...

Figure 7.5–Contd...

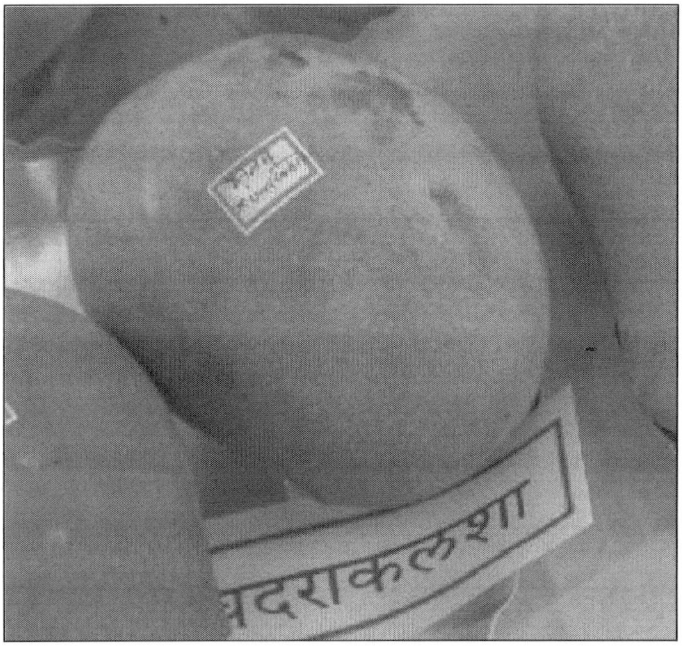

Badarakalsha

Baganpalli

Contd...

Figure 7.5–Contd...

Black Columban

Hamlet

Contd...

Figure 7.5–Contd...

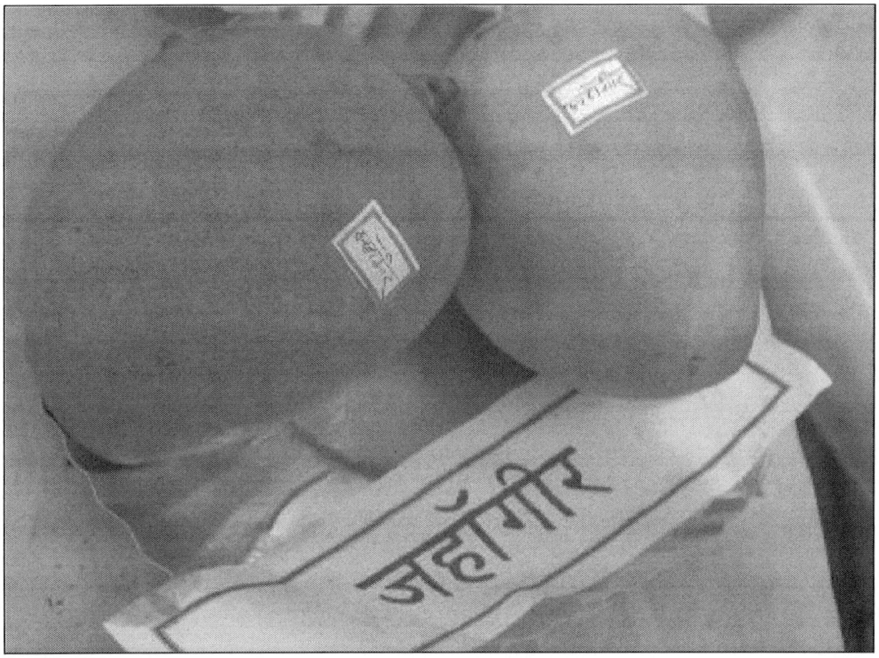

Jahangir

Kingfon

Contd...

Figure 7.5–Contd...

Kondur Goa

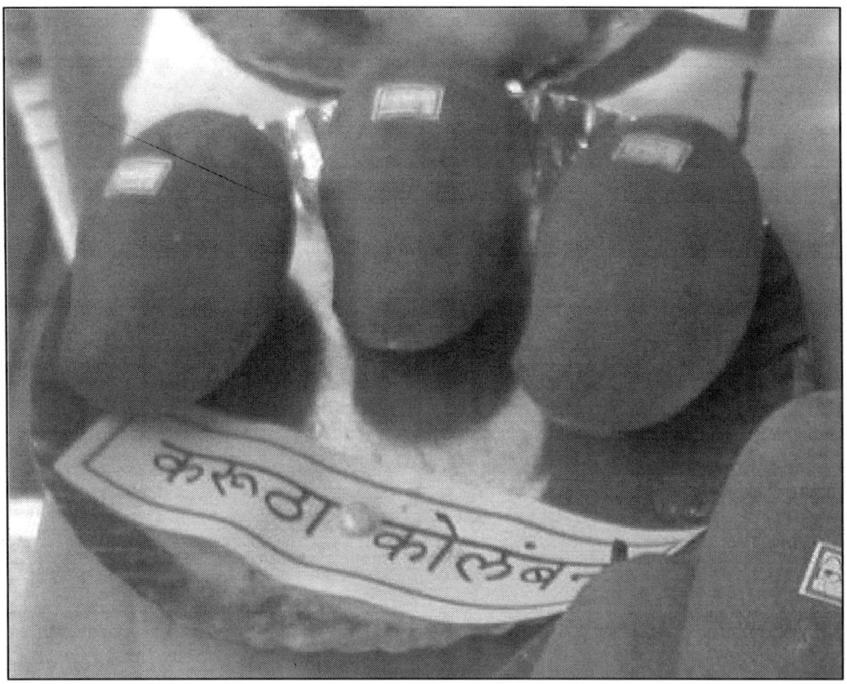

Kurtha Columban

Contd...

Figure 7.5–Contd...

Ladaio

Mulgoa

Contd...

Figure 7.5–Contd...

Mussarad Sasti

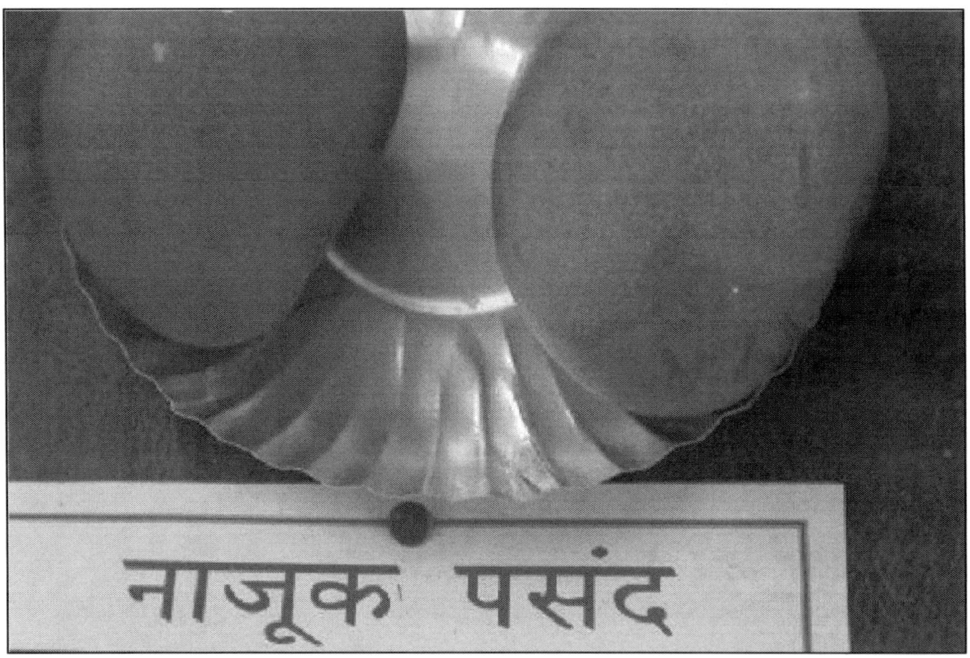

Najuk Pasand

Contd...

Figure 7.5–Contd...

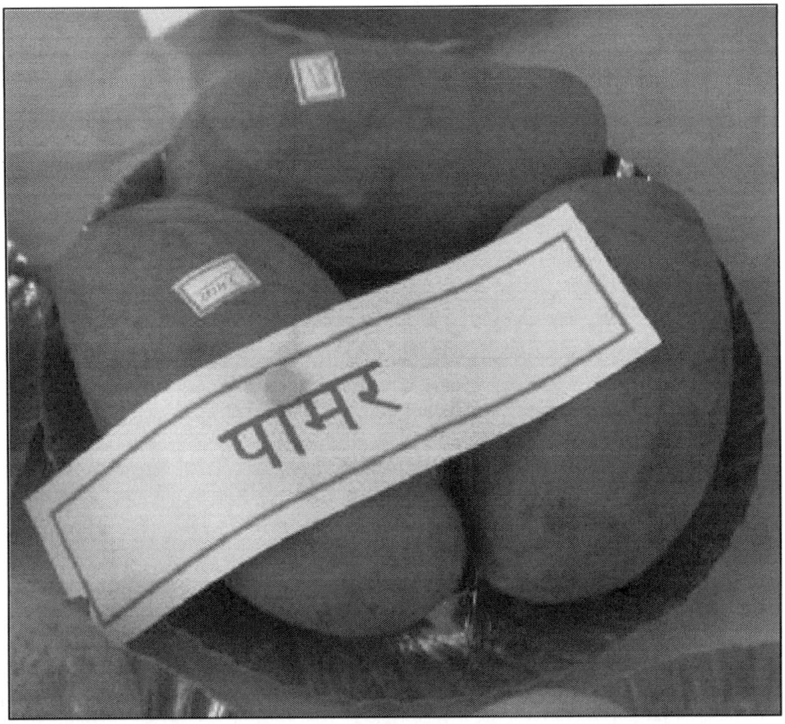

Pamer

Parmalgoa

Contd...

Figure 7.5–Contd...

Pedherbalm

Red Maral Totapuri

Contd...

Figure 7.5–Contd...

Ruse

Sannkulu

Contd...

Figure 7.5–Contd...

Vatganga

5. *Black columban*: Oblong fruit, Reddish yellow on shoulder
6. *Hamlet*: Oblong roundish, yellowish on shoulder.
7. *Jahangir*: Oblong ovate yellowish on shoulder.
8. *Kingfon*: Oblong cylindrical Red blush on shoulder.
9. *Kondur Goa*: Round fruit, yellow on shoulder.
10. *Kurtha columban*: Oblong yellowish orange on shoulder
11. *Ladaio*: Oval fruit yellowish on shoulder.
12. *Mulgoa*: Roundish fruit reddish on shoulder.
13. *Mussarad sasti*: Oblong roundish fruit-Dark Red on shoulder.
14. *Najuk Pasand*: Oblong fruit yellowish on shoulder.
15. *Pamer*: Oblong fruit bluish red on shoulder
16. *Parmalgoa*: Roundish fruit reddish yellow on shoulder.
17. *Pedherbalm*: Oblong beak pointed yellow on shoulder
18. *Red Maral Totapuri*: Oblong, Red blush on shoulder
19. *Ruse*: Roundish fruit, yellowish Red on shoulder.
20. *Sannkulu*: Oblong roundish yellow on shoulder.
21. *Vatganga*: Roundish fruit yellowish green on shoulder.

Table 7.13: Characters of Elite Varieties of Mango Identified in Goa

Sl.No.	Name of Varieties	Fruit Size (g)	Fruit Shape	Fruit Attractive-ness	Skin Colour of Ripe Fruit	Colour of a Pulp	Pulp Texture of Ripe Fruit	Quantity of Fibre in Pulp	Pulp Juiciness	Pulp Aroma	Presence of Turpentine Flavour	T.S.S (%)
						Local						
1.	Bardez Musserat	388	Roundish	Excellent	Yellow red	Golden yellow	Firm	Absent	Slightly juicy	Intermediate	Absent	12
2.	Bishop	742	Ovate Roundish	Excellent	Yellow green	Orange yellow	Intermediate	Low	Juicy	Mild	Absent	11
3.	Black Culas	560	Oblong	Good	green	Yellow	Firm	Low	Slightly juicy	Strong	Mild	16
4.	Culas	290	Oblong Oblique	Good	Yellow	Whitish yellow	Intermediate	Low	Juicy	Intermediate	Absent	13
5.	Fernandin Porvorim	156	Oblong	Good	Yellow	Light yellow	Intermediate	Low	Slightly juicy	Intermediate	Absent	15
6.	Goa Alphanso	416	Roundish oblique	Excellent	Yellow	Yellow	Medium	Low	Juicy	Intermediate	Absent	12
7.	Hilario Ela	204	Ovate Oblong/ Ovovoid	Excellent	Yellow	Light yellow	Firm	Low	Slightly juicy	Strong	Mild	20
8.	Hilario Mapusa	186	Ovate Roundish	Excellent	Yellow	Light yellow	Firm	Low	Slightly juicy	Strong	Absent	18
9.	Hilario Volfarm	214	Ovate Oblong	Good	Yellow	Yellow	Firm	Low	Slightly juicy	Strong	Absent	19
10.	Hilario Lobo	212	Ovate Oblique	Excellent	Dark Yellow	Yellow	Intermediate	Low	Slightly juicy	Strong	Mild	19
11.	Hilario Mark Saligao	216	Obovoid	Excellent	Yellow	Whitish yellow	Intermediate	Low	Slightly juicy	Strong	Absent	16
12.	Hilario P. Naik	82	Obovoid	Good (Small size)	Yellow	Light yellow	Firm	Absent	Juicy	Strong	Mild	21
13.	Hilario Shiolim	316	Ovate Oblique	Excellent	Yellow	Yellow	Firm	Low	Slightly juicy	Strong	Mild	13

Contd...

Table 7.13–Contd...

Sl.No.	Name of Varieties	Fruit Size (g)	Fruit Shape	Fruit Attractive-ness	Skin Colour of Ripe Fruit	Colour of a Pulp	Pulp Texture of Ripe Fruit	Quantity of Fibre in Pulp	Pulp Juiciness	Pulp Aroma	Presence of Turpentine Flavour	T.S.S (%)
14.	Jesuit	660	Ovate oblong	Excellent	Light yellow	Yellow	Intermediate	Low	Slightly juicy	Intermediate	Absent	14
15.	Mankurad (G.S. Potekar)	238	Roundish	Excellent	Yellow	Yellow	Firm	Absent	Slightly juicy	Strong	Absent	18
16.	Mankurad Giant	480	Roundish Oblique	Excellent	Yellow	Golden yellow	Firm	Absent	Slightly juicy	Strong	Absent	19
17.	Mankurad I/C	328	Oblong oval	Excellent	Yellow	Yellow	Firm	Absent	Slightly juicy	Strong	Absent	18
18.	Mankurad K	262	Roundish	Good	Yellow	Yellow	Firm	Low	Juicy	Intermediate	Absent	22
19.	Mankurad Lobo	252	Ovate Roundish	Excellent	Yellow orange	Yellow	Intermediate	High	Juicy	Intermediate	Absent	23
20.	Mankurad SLG	302	Roundish	Excellent	Golden yellow	Golden yellow	Intermediate to firm	Low	Slightly juicy	Intermediate	Absent	17
21.	Musserat	234	Obovoid/ Ovate Oblong	Excellent	Pink yellow	Whitish yellow	Firm	Low	Slightly juicy	Strong	Different unique flavor	17
22.	Musserat Salcette	408	Ovate raniform	Excellent	Yellow	Bright yellow	Firm	Absent	Slightly juicy	Strong	Mild	17
23.	Porvorim Color Fruit	212	Roundish	Excellent	Red yellow	Orange yellow	Firm	Low	Slightly juicy	Strong	Mild	14
24.	Sakri	406	Ovate	Good	Dark Yellow	Dark Yellow	Intermediate	High	Juicy	Intermediate	Absent	19
25.	Salgadin	374	Ovate Oblong	Excellent	Yellow	Yellow	Firm	Low	Slightly juicy	Intermediate	Absent	20
26.	Udgo	586	Roundish oblique	Excellent	Yellow	Golden yellow/ Bright yellow	Intermediate	Absent	Slightly juicy	Strong	Mild	19

Contd...

Table 7.13–Contd...

Sl.No.	Name of Varieties	Fruit Size (g)	Fruit Shape	Fruit Attractive-ness	Skin Colour of Ripe Fruit	Colour of a Pulp	Pulp Texture of Ripe Fruit	Quantity of Fibre in Pulp	Pulp Juiciness	Pulp Aroma	Presence of Turpentine Flavour	T.S.S (%)
						New mango variety						
27.	Chonchoro	192	Roundish	Good	Yellow	Orange yellow	Intermediate	Low	Juicy	Intermediate	Mild	14
28.	Manjoram	318	Oblong oblique	Good	Purple green	Light yellow	Soft	Intermediate	Juicy	Intermediate	Mild	12
29.	Rare variety from Shiolim	276	Ovate	Rare Excellent	Yellowish green	Whitish yellow	Intermediate	Low	Slightly juicy	Intermediate	Mild	10
30.	Sckraddin	160	Cordate	Good	Yellow	Golden yellow	Inter-mediate	Intermediate	Juicy	Intermediate	Absent	18
31.	Shantiago	262	Ovate oblique	Excellent	Greenish yellow	Light yellow	Forum	Low	Slightly juicy	Strong	Absent unique flavour is there	15
						Hybrid						
32.	Amrapali	214	Oblong	Excellent	Green yellow	Orange	Intermediate	Low	Juicy	Strong	Absent	19
33.	Amrapali Yellow	224	Oblong Raniform	Excellent	Yellow bright	Orange/Bright orange yellow	Firm	Low	Slightly juicy	Strong	Absent	19
34.	Arka Anmol	364	Oblong Oval	Excellent	Yellow	Bright yellow	Firm	Low	Slightly juicy	Intermediate	Absent	13
35.	HB-117	250	Ovate Oblique	Excellent	Yellow	Orange yellow	Firm	Low	Slightly juicy	Strong	Absent	23
36.	HB-151	345	Ovate Raniform	Good	Yellow green	yellow	Firm	Low	Slightly juicy	Strong	Absent	21
37.	HB-87	210	Ovate roundish	Good	Green yellow	Light yellow	Intermediate	Absent	Slightly juicy and crispy pulp	Strong	Absent	17
38.	KXB	250	Oblong Oblique	Good	Yellow	Yellow	Firm	Low	Slightly juicy	Intermediate	Mild and unique flavour	13

Contd...

Table 7.13–Contd...

Sl.No.	Name of Varieties	Fruit Size (g)	Fruit Shape	Fruit Attractive-ness	Skin Colour of Ripe Fruit	Colour of a Pulp	Pulp Texture of Ripe Fruit	Quantity of Fibre in Pulp	Pulp Juiciness	Pulp Aroma	Presence of Turpentine Flavour	T.S.S (%)
						Introduced variety						
39.	Alampur Benesan	818	Oblong Elliptic	Excellent	Redish yellow	Yellow	Firm	Low	Slightly juicy	Intermediate	Mild	13
40.	Kalapaddy	204	Oblong Oval	Good	Yellow	Yellow	Firm	Absent	Slightly juicy	Intermediate	Very mild other flavour	17
41.	Madhulika (STH-1)	250	Oblong	Excellent	Yellow	Golden yellow	Firm	Absent	Slightly juicy	Mild	Absent	17.5
42.	Madhulika (MDCH-1)	245	Oblong	Excellent	Yellowish red	Orange red	Firm	Absent	Slightly juicy	Intermediate	Absent	18.0
43.	Ratnagiri Alphanso Joseph Porvorim	206	Ovate Oblique	Excellent	Yellow	Dark orange	Intermediate	Low	Juicy	Intermediate	Absent	14

Figure 7.5: Local Mango Varieties

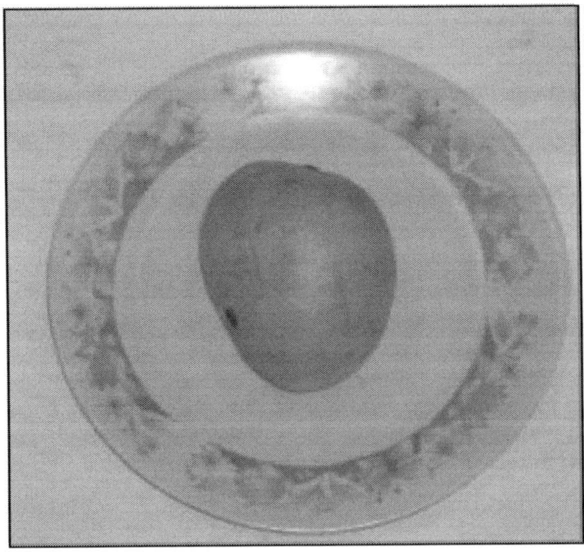

Mankurad Giant Siolim

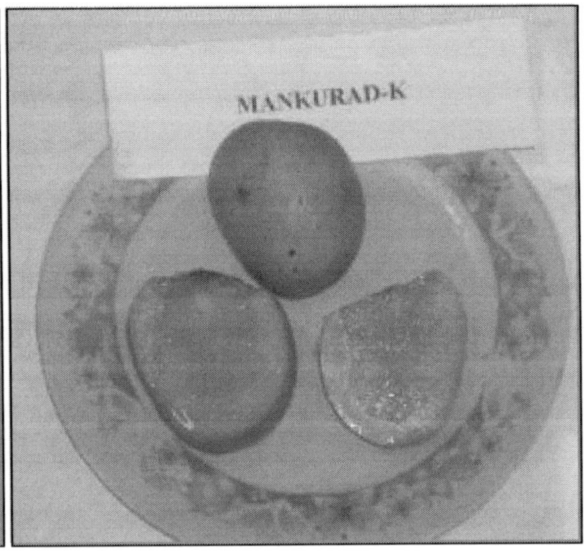

Mankurad-K

Mankurad Lobo

Bondez Mussreat (Parvorim)

Contd...

Figure 7.5–Contd...

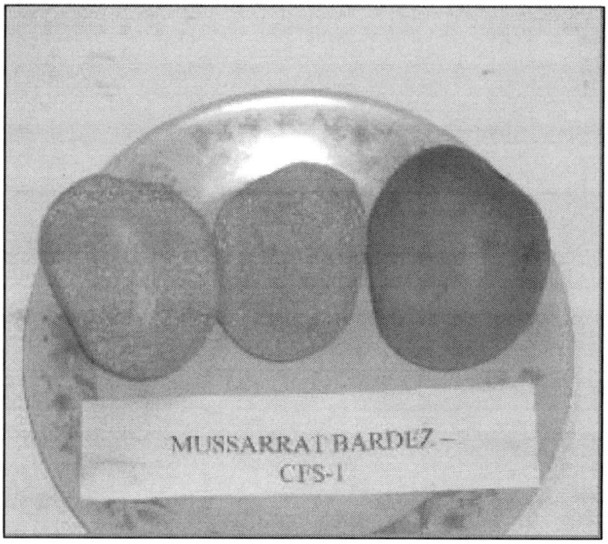

**Mussarrat Bardez-
CFS-1**

**Mussarrat Bardez-
CFS-2**

**Mussarrat Bardez-
CFS-3**

Bardez Musserat Parvorim

Contd...

Figure 7.5–Contd...

Bardez Musserat Siolim

Rotto

Gadgo

Contd...

Figure 7.5–Contd...

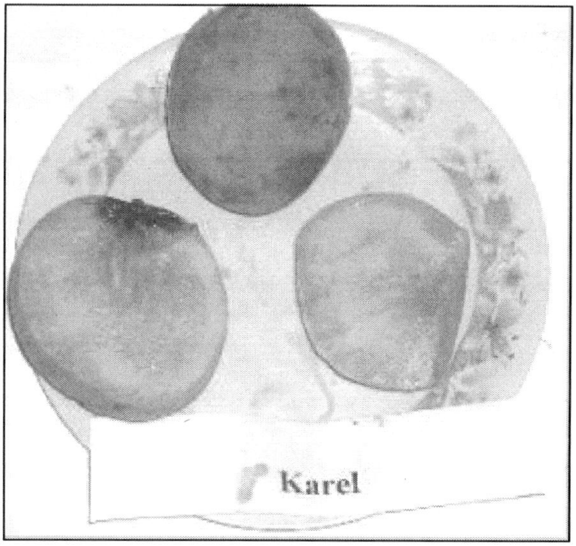

Karel

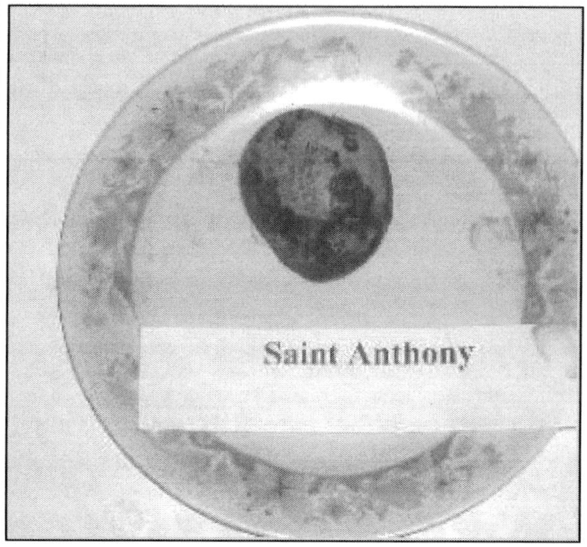

Saint Anthony

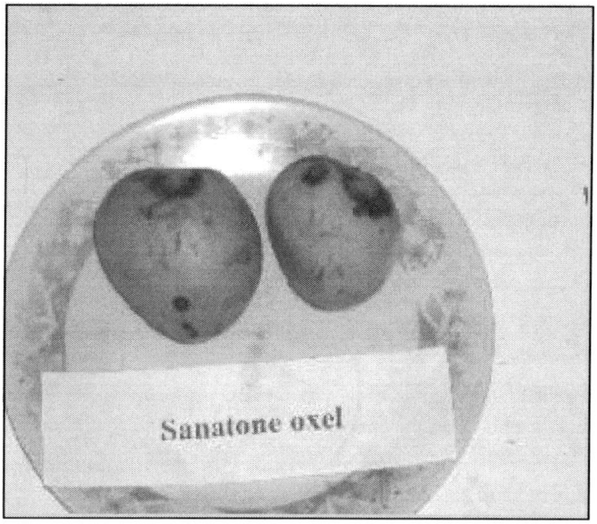

Sanatone Oxel

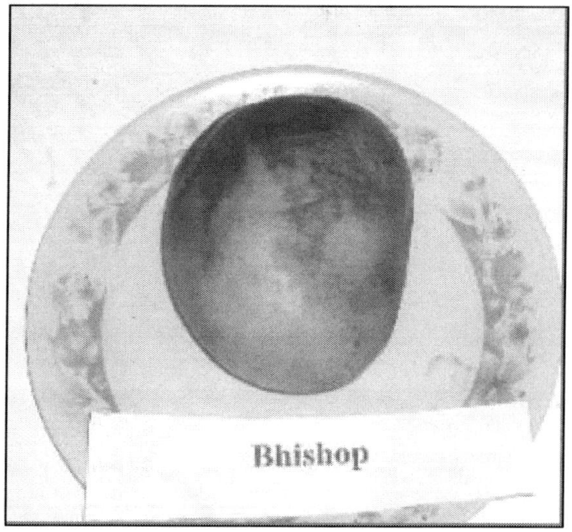

Bhishop

Contd...

Figure 7.5–Contd...

Cruz Fernandes

Hilario Siolim

Figure 7.6: New Mango Varieties

Goa Alphanso

Hilario Lobo

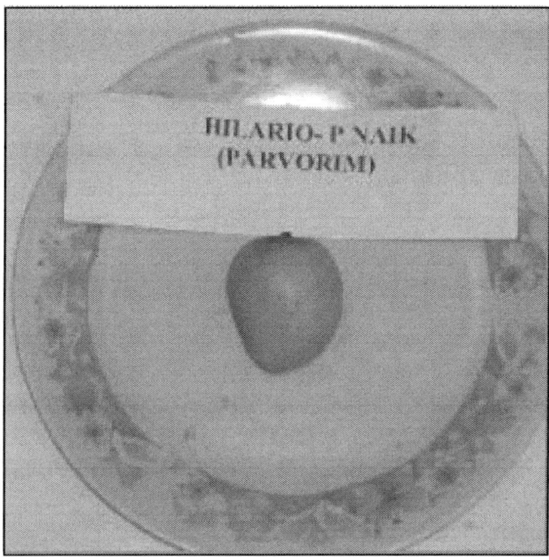

Hilario-P Naik (Parvorim)

Chonchoro

Contd...

Figure 7.6–Contd...

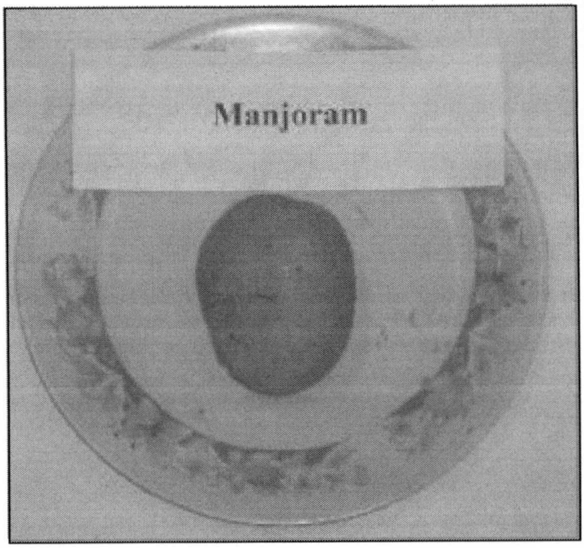

Manjoram

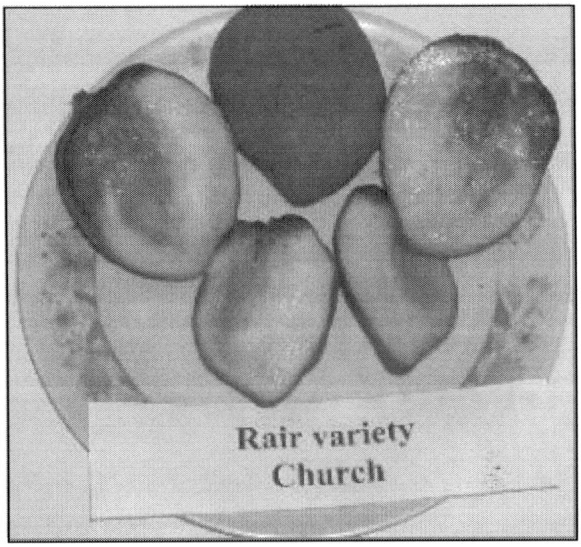

Rair Variety Church

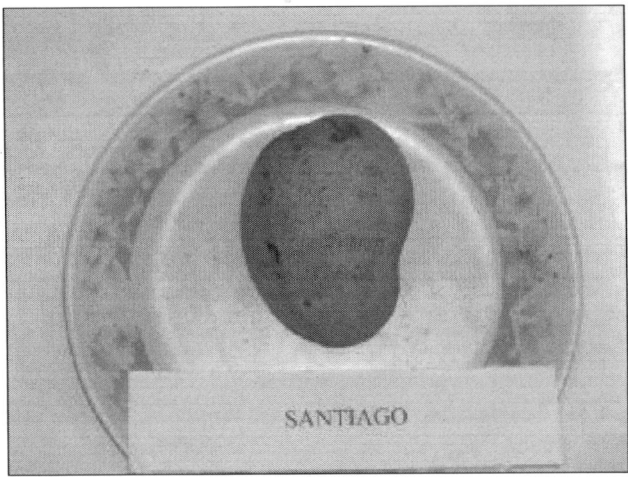

Santiago

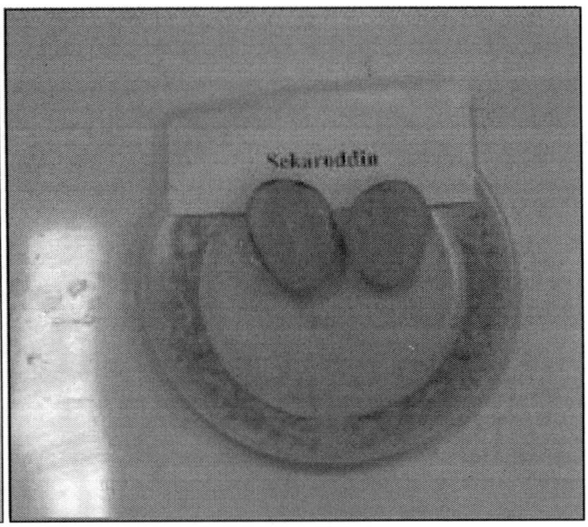

Sekaroddia

Hybrid variety introduced at the ICAR Research Complex-Amrapali, performs very well. One of the clonal selections Amrapali yellow is also giving better fruit quality and attractive colour. Other Hybrids introduced from IIHR like Arka Anmol, Arka Aruna, Arka Puneet, Arka Neel Kiran are also performing well in Goa condition. Other Hybrids introduced to the Institute are HB 117, HB151, HB 45, HB 56, HB 87, KXB and Mallika. Ratna was also introduced and characterized in detail. Introduced varieties Alampur Benesan, Benesan, Himayuddin, Kalapaddy, Kesar, MDCH I (Madhulika), Neelam, Pedda Rasam, Ratnagiri Alphanso, Royal Special, STH I (Madhurima), Totapuri (A.B.), Vallaicolamban are also described and characterized in detail. Twelve accessions from Andamans were described. *Mangifera andamanica*, which bears fruit in bunches in small size is collected and characterized. Another species collected from Andamans *Mangifera camptosperma* is characterized by very thin and flat square seeds/stones which can be used in breeding programme. Another rare species *Mangifera griffithii* is Monoembryonic, with red seed colour is collected and described along with other Polyembrionic varieties which come in bunches. Two exotic collections Sensation I and Sensation II were also described and characterized in detail. The Goa has a lot of genetic resources in Mango with different size, shape, colour, quality and other genetic variability of the fruit. Effort is going on to find out more varieties of mango which are rare and extinct to the region.

References

Alliance for Zero Extinction. The Hindu *"Call to Conserve the Western Ghats"* p. 1–2. http://www.hindudonet.com/2006/02/09/stories/20060209012402.htm

Annual Report of ICAR Research Complex for Goa, 1998–2008.

Anonymous, Goa foundation/biodiversity– Current (Known) Range and status of Biodiversity. p. 14–16. http://www.goacom.com/goafoundation/biodiversity/3rdbasp.1.htm

Anonymous. *Production and Productivity of Different Horticulture crops.*

Arora, R.K., 1995. Genetic resources of native tropical fruits in Asia: Diversity, distribution and IPGRI's emphases on their conservation and use: Genetic Resources– Importance of study, genetic resources and priority species. p. 43–47. http://www.biodiversityinternational.org/publications/Web_version/510/ch04.htm.

Burns, W.E. and Prayag, S.H., 1920. *The Book of Mango*. In: Bulletin of the Department of Agriculture. Bombay, p. 103.

Chadha, K.L. and Yadav,L.S., 1997. *Mango in 50 Years of Crop Science Research in India*. ICAR, New Delhi, p. 419–432.

Chadha, K.L., 1995. Status report on fruit species germplasm conservation and utilization in South-East Asia. In: *Proceedings of the Expert Consultation on Tropical Fruit Species of Asia,* (Eds.) R.K. Arora and V. Ramanatha Rao. IPGRI Office for South Asia, Pusa Campus, New Delhi, India, p. 85–102.

De Canddle, 1904. *Origin of Cultivated Plants*. Kegan Paul, Research, New Delhi. 3–4. Department of Agriculture Government of Goa. 2001–2002: 1

Ecosystem Profile, *Western Ghats and Sri Lanka.* http://www.cepf.net/xp/cepf/where_we_work/western_ghats/eco_profile/profilesection.xml

Ferrao, 1993. In Costa, 1872, pp.281–375: Tratado de Agrcultura escripto por Fr.Clemente da Russurreicao., p.188. See Laufer, 1929.

Ferrao, Jose E, Mendes, 1993. A Aventura das Plantas e os Descorbriumtos Portugueses. Lisbao; Instituto de Investigacao Gientfica Tropical/Comissao National para as Comemoracoes dos Descobrimentos Portugueses/Fundacao Berardo. http://midimaavu.info/objective.htm. http://www.ipgri.cgiar.org/publications/HTMLPublications/510/ch06.htm <http://www.kisan.com/links/ap_mango varieties.html>

Fernandes, J.M., 1985. *The Isle of Mangoes*. Navhind Times January 15th.

Ghosh, S.P., *Fruit Wealth of India, Part I: Genetic Resources–Importance of Study, Genetic Resources and Priority Species*. p. 1–4. http://www.biodiversityinternational.org/publications/Web_version/510/ch04.htm

Jafar Jaskni and Haider Abbas, Yousaf, 2007. Advisory biotechnology, Mango biotechnology, Home Biotechnology, Pakisthan.com. Dr. Jafar Jaskni Incharge Plant tissue culture laboratory Dept. of Horticulture, university of Agriculture (Faisalabad) Pakistan 2007. http://www.pakistan.com/english/advisory/biotechnology/mangoes.biotechnology.shtml

Lakshinarayana Hegde, K.V. and Kalave, Shivanand, 2007. Appemedi Fest. Tender Mango fest in Karnataka "midi maavu 2007" April 23rd–27th 2007. http://farmedia.org/release/appemidi.html

Kamat, Nand Kumar, 1998. From the preface to the Book *Nilima's Mango's Recipes*. Page 14–16 of 24 Current known.

Kaveriappa and Shetty, B.V., 2001. Mangalore University describe biodiversity of the Western Ghats and the work on Conservation of Plants around Kaiga nuclear power plant. "Biodiversity of the Western Ghats with special reference to conservation of plant Diversity at Kaiga". *An International Journal of Nuclear Power*, 15(1 to 4): 40–42.

Khoshoo, T.N., 1996. Concern mounts over rapid erosion of Himalayan biodiversity. *Diversity*, 12(3): 24–25.

Kisan, I., *Mango: Varieties, Important Commercial Varieties*, p. 1–11.

Kostermans, A.J.H. and Bompard, S.M., 1993. *The Mangoes: Their Botany, Nomenclature, Horticulture and Utilization*. IBPGR and Linnean Society of London. Academic Press, London.

Kuriaan, Benny. *Exploring Biodiversity Hot Spots and Village Culture in the Western Ghats*. p. 1–4. <http://www.earthfortong/P2 in 014.htm>.

Kusumlata, 2002. *The Kaas Asm: Ode to the Mango, India's Regal Fruits.*

Leenhouts, P.W., 1971. A revision of *Dimocarpus* (Sapindaceae). *Blumea*, 19: 113–131.

Mango Resource Information System. http://www.mangifera.org.

Mangoes in India Varieties. http://www.horticultureworld.net/india–mango.htm.

Mangoes in India Waste Utilization. http://www.horticultureworld.net/mango–india.htm.

MANUCCI, 1907–1908, Vol. III, p.170–171. GRACIAS, 1912, pp.32–33.

Mathew, P.A. and Dhandar, D.G., 1995. *Mangos of Goan Origin with propogation and Culture: Description of Goan Varieties*. Indian Council for Agricultural Research– Research Complex for Goa, Old Goa, Technical Bulletin, 1: 5–48.

Mukherjee, S.K., 1951b. Origin of mango. *Indian Journal of Genetics and Plant Breeding*, 11, 56.

Mukherjee, S.K., 1985. Systematic and eco-geographic studies on crop geenpool (*Mangifera* L.). *International Board for Plant Genetic Resources,* Rome Italy, 86 p.

National Biodiversity Strategy and Action Plan supported by the Ministry of Environment and Forest – GOI and the United Nations Development Programme. October 12, 2000 at Centre for Ecological Science, Indian Institute of Sciences, Bangalore, p. 9 –14.

Orient–Published by Fundacao,Oriente Lisboa. Editor Carle Alferes Pinto, Agosto 2002, Yiva indusion, p. 70–80.

Palioor, Anita, 2007. Info change India Newa features Save the appemidi. p. 1–4. http:/www.infochangeindia.org/features426.jsp

Plantscultures–Mango in sacred and spiritual life. Mango–spiritual, Mango and religion. p. 2. http://www.plantcultures.org/plants/mango_spiritual.html

Popnoe, W., 1920. *Manual of Tropical and Subtropical Fruits*. Macmillan, New York.

Radha, T. and Nair, S.R. Status of Mango Cultivation in Kerala. *ISHS Acta Horticulturae,* 509: VI International Symposium on Mango. http://www.actahort.org/books/509/509_10.htm

Ramphius, G.E. 1741. Herbarium amboinense, plurimas, Complectens arbores, fruitces, herbas, plantas terrestres et equates, quae in Amonia et adjacentibus reperiuntur insulis (6 vols. In four). Amsterdam.

Randhava, G.S., 1965. The mango varieties of Goa. *Indian Horticulture,* 10(1).

Ravishankar, K.V., Chanrashekara, P., Sreedhara, S.A., Dinesh, M.R., Lalita Anand and Saiprasad, G.V.S., 2004. Diverse genetic bases of Indian polyonic and monoembrionic mango (*Mangifera indica* L) cultivars. *Current Science,* 87(7): 870–871.

Rego, 2002. Editor Carle Alferes Pinto, Agosto 2002, Yiva inclusion.

Rego, Fernandino do, 1979. *Collection, Maintenance and Evaluation of Mango Germplasm in Goa.* Mango Workers Meeing, Panajim.

Rego, Fernando do, 1974. Mango in Goa: Cultivation and industry. In: *All India Seminar on Mango Industry,* Lucknow.

Saeed, Salman. 2002. The south asian mango: Regional variety. *The Fruit of Kings and the King of Fruits: Other Popular Varieties from India*. p. 1–3. http://www.the–south–asian.com/February2002/mangoes–sout per cent 20Asian4.htm.

Sampaio, Joao de Melo, 1902. *A Mangueira*. Nova Goa: Imprensa Nacional.

Singh, Daljit, 1967. *The Mango: Varieties and Nomenclature*. Indian Council for Agricultural Research, New Delhi.

Singh, S.P., 2004. Production potential and biodiversity of mango varieties of Goa. In: *Paper Abstracted in First Indian Horticulture Congress* "On improving productivity, quality post harvest management and trade in horticulture crop" organized by horticulture society of India on 6–9th November, New Delhi, Poster session and genetic resources, 1.24: 7.

Singh, S.P., 2004. Crops from the beach. *Times Agriculture Journal,* September–October, 3(6): 35–37.

Singh, S.P., Devi, Priya and Ramchandradu, K., 2005. Biodiversity in horticulture crops and organic farming for sustainable development of agro ecotourism in Goa. In: *Paper Presented along with Proceeding in National Seminar on Agro Ecotourism* organized by association for coastal agricultural research, I.C.A.R. Res. Complex for Goa, Ela Goa, on Jan 19th–20th, Old Goa Biodiversity and Organic farming in Relation to Agro Ecotourism 19: 38 167–171.

Singh, S.P., Gaonkar, V.Y. and Dhandar, D.G. 2000–2001. *Survey, Collection, Conservation and Evaluation of Fruit Crops*. Annual Report of I.C.A.R. Research Complex for Goa Ela Old Goa.

Singh, L.B., 1960. *The Mango*. World Crop Books, London Leonard Hill (Books) Limited, Interscience Publishers Inc. New York.

Singh, R.N., 1964. Goa and its varieties: My recollections. In: *Goa Today*, May.

Swamy, K.R.M., 2004. *Science Looks at the Mango*. The Sunday, Navhind Times, Paranama, May 16th, p. 5.

Tribhuwan, Pant, Pandey, Vandana and Das, Subhash Chandra. *Agro–Biodiversity Strategy*. Defence Agricultural Research Laboratory, Pithoragarh Uttaranchal), p. 1–11.

Vavilov, N.I., 1949–50. The origin, variation, immunity and breeding of cultivated plants. *Chorn. Bot.*, 13, 26.

Wester, P.J., 1920a. A descriptive list of mango varieties in India. *Phillipp. Agric. Rev.*, 4: 256.

Woodrow, G. Marshall, 1904. *The Mango: Its Culture and Varieties*. Madras: Higgin Botham and Co.

Yadav, L.S. and Rajan, Shailendra, 1993. Genetic resources of mangifera. In: *Advance in Horticulture Vol. 1: Fruit Crops, Part 1*, (Eds.) K.L. Chadha and O.P. Pareek. Malhotra Publishing House, New Delhi, p. 77–91.

Yadav, L.S. Germplasm conservation and utilization in breeding of mango. Part III. PGR conservation and use: Germplasm, introduction, exchange evaluation. p. 9–17. http://www.ipgri.cgiar.org/publications/HTMLPublications/510/ch06.htm.

Yadav, L.S. Rootstock evaluation, conservation and use in tropical fruits. Part III. PGR conservation and use: Germplasm introduction, exchage evoluation. p. 43–44. http://www.ipgri.cgiar.org/publications/HTMLPublications/510/ch06.htm

2011, Biodiversity in Horticultural Crops Vol. 3 *Pages* **204–228**
Editor: **Professor K.V. Peter**
Published by: **DAYA PUBLISHING HOUSE, NEW DELHI**

Chapter 8

Noni (*Morinda citrifolia* L.)

D.R. Singh
Central Agricultural Research Institute
Port Blair Andaman and Nicobar Islands – 744 101
E-mail: drsingh1966@yahoo.com

We the human beings are an integral part of the Mother Nature. Our fate is tightly linked with biodiversity, the huge variety of animals and plants, the places they live and their surrounding environments, around the world. We rely on this diversity of life to provide us with the food, fuel, medicine and other essentials things. Yet this rich diversity is being lost at a greatly accelerated rate because of human activities. This impoverishes us all and weakens the ability of the living systems, on which we depend, to resist growing threats such as climate change. According to WWF the biodiversity has declined by more than a quarter in the last 35 years. The Living Planet Index (LPI), which tracks nearly 4,000 populations of wildlife, shows an overall fall in population trends of 27 per cent between 1970 and 2005. The ever increasing population and our lifestyle are the reasons for this enormous loss and population decline in species. The loss of biodiversity is, arguably, the greatest threat to world stability and security today.

Biodiversity of flora is much affected due to indiscriminate cutting of trees and clearing of forest for agricultural and other developmental purposes. Introduction of high yielding varieties of agricultural crops is slowly depleting the treasure of indigenous varieties high in nutrients and other qualities. The impact of this development is by-late recognized and efforts are being made for conserving the diversity of agriculture crops for future use.

The ever increasing demand of Noni products in global markets in recent years has led to commercial farming of Noni in several western Pacific islands as well as in India. The traditional healers use to give much importance to the plant with some particular morphological characteristics. They believe that the plants with certain phenotypical specifications like shape of the leaf, size of the fruit etc. are of great importance and are effective in healing diseases. They consider a varietal form of

Noni with narrow leaves as more potent without knowledge of their potential medicinal advantages. This information has been at large, is a traditional knowledge without any scientific validation. The fact that leaves do seem to vary implies that traditional healers and other observers of plants, are indeed recognizing a feature that needs scientific scrutiny. Since, Noni is making its transition from wild-harvested to commercial cultivation in India, researchers and farmers are looking for desirable qualities to improve the crop. It is important to understand the diversity within *Morinda citrifolia* and its close relatives. The available intra-varietal diversity and chemical differences will be a potential marketing tool.

The genetic diversity of the genus *Morinda* and the evolution of large number of species having different varieties can be attributed to its wide adaptation and its world wide distribution. The plant is believed originated in India and South East Asia and to be precise in Andaman and Nicobar Islands (Singh, 2009). From here it was naturalized in many parts of Asia, Australia, South America, Caribbean and Polynesian Islands through sea and ocean currents. The tough seed coat, its long dormancy and its ability to germinate even after long exposure to sea water as well as its bouyancy have helped it to spread throughout the tropical regions of the globe. Though, today the plant is found in tropical and subtropical regions, the environmental, soil and other factors such as long isolation must have played an important role in such diversity in both species as well as variety wise. There are several reports of variation among various cultivars within a particular variant. The Nicobarese, one of the aboriginal tribes of Andaman and Nicobar Islands belongs to Mongoloid race and believed to be inhabitant of these isolated Islands for thousands of years (Justine, 1990). They are very healthy race with medium height and yellowish complexion. *Morinda citrifolia* is known as Lurange and is very popular among them. They have very long association with this plant and used different parts as medicine to cure different ailments involving different organ systems and of various etiologies (Singh *et al.*, 2009). The traditional healers are keen observers of the plant morphology and gained enormous knowledge about the genus, species and even its variants. The scientific information available in this regard are confusing and still incomplete. This chapter is an effort to compile the available information and cover some aspects of the diversity of the genus *Morinda* as well as the species of *Morinda citrifolia* covering divergence between different accessions.

For better understanding the diversity of any particular species or genus, it is essential to know the taxonomical classification. This helps in better identification of plant species and to know their relationship with other similar species or variants. The Genus *Morinda* is classified as follows:

Systematics of Genus *Morinda*

Domain	:	Eukaryota
Kingdom	:	Plantae
Sub-kingdom	:	Viridaeplantae
Phylum	:	Tracheophyta
Sub-phylum	:	Euphyllophytina
Infra-phylum	:	Radiatopses
Class	:	Magnoliopsida
Sub-class	:	Lamiidae
Super-order	:	Gentiananae

Order	:	Rubiales
Family	:	Rubiaceae
Sub-family	:	Rubioideae
Genus	:	*Morinda L.*

Intra Genus Variation

The genus *Morinda* is distributed world wide and presence of more than 140 different species are already recorded and some of the websites claim even presence of 172 species (Singh, *et al.*, 2009). But doubts exist on this claim as several variants of some species are listed as a separate species. The central attention world wide is given to the species *Morinda citrifolia* L. Morphological features which unite the species into one genus include syncarps or capitula (aggregated fruits) forming from ovaries and parts of conate calyces and tri corporate pollen grains with sexine usually finely reticulate (Johansson 1994). The syncarps may be fleshy, as with *Morinda citrifolia* or may be dry as with many other species. Each syncarp is probably derived from condensation of umbels of separate flowers with further condensation within some species of *Morinda* to form a range of flower/fruit configurations. Morphological characteristics which are variable between species include: leaf size, shape, and detail; umbel aggregation position and flower pedicel length; and odor, size and shape of fruit.

Several workers identified different species of genus *Morinda* based on morphological features. Hooker (1880) recorded four species *viz. Morinda aspera* Wight and Arn., *Morinda multiflora* Roxb., *Morinda tinctoria* Roxb. and *Morinda tomentosa* Heyne ex Roth. Gamble and Fischer (1921) proposed three species *Morinda pubescens, Morinda stenophylla* and *Morinda tomentosa.* Johansson (1994) recorded presence of 19 species of *Morinda* in New Caledonia. Mabberley (1997) studied the genus *Morinda* extensively and renamed *Morinda tinctoria as Morinda pubescens.* Peter (2007) recorded presence of different variants in *Morinda citrifolia* and *Morinda pubescens* and proposed two variants of *Morinda citrifolia* and four variants of *Morinda pubescens* and also recorded three other species *Morinda stenophylla, Morinda tomentosa* and *Morinda umbellata* in South India. But many of the species are renamed or accepted as synonyms of other species. The list of available species with its synonyms and distribution is given elsewhere.

Morinda citrifolia L., the perennial bush or small tree, 3 to 12 m tall with medium stem diameter and abundant wide elliptical leaves belongs to the class Magnoliosida, order Rubiales and family Rubiaceae. The species name indicates resemblance of plant foliage to that of citrus species. (Abu Saleha et al., 2007). This plant has much variability in overall plant form, fruit size, leaf size, morphology, palatability, odour of ripe fruit and number of seeds/fruit. Surendran and Mathivanan (2006) described the *Morinda citrifolia* L. having leaf with opposite pinnately veined, blades membranous, broadly elliptical and are 20-45 cm long and 7-25 cm wide. The flowers are perfect, bisexual, small tubular white grouped together and inserted on peduncle with tubular, truncated rim, extremely reduced, with small lobes or short tubular calyx. Corolla is white, 5 lobbed, the tube greenish white, 7 to 9 mm long, margin entire, obtuse or acute at apex and stamen five, scarcely exerted, alternate to the corolla lobes, shorter than corolla tube, equal and glabrous. The lobes are oblong, deltate, approximately 7 mm long. The style exerts just beyond the corolla about 15 mm long with stout 1.5 to 2 cm long petioles and having ovary as inferior, 2 or 4 locular, narrowly obovoid. Stipules are connate or distinct, 1 to 1.2 cm long, the apex entire or 2 to 3 lobed. The small tubular white flowers are grouped together and inserted on the peduncle. The petioles leave ring-like marks on the stalks and the corolla is greenish white (Morton, 1992; Elkins, 1998; Dixon *et al.*, 1999; Ross, 2001; Cardon, 2003, Singh, *et al.*, 2006).

Sapwood is yellow-brown and soft. The bark is gray or brown, smooth to slightly rough. It is possible to find fruits at different stages of maturity on the same plant at the same time. The fruits have densely clustered globose, syncarp, fleshy, yellowish white about 5 to 10 cm long and 3 to 4 cm in diameter, fetid when ripe with seeds ovoid to obovoid or reniform having distinct air chamber and viable even after floating in water for months. Howard (1989), Liogier (1997), Little and Wadsworth (1964), Nelson (1996) and Singh, *et al.* (2009) reported fruits (3-10 cm length, 3-6 cm width) as oval and fleshy in appearance. They observed that the fruits were green until maturity and rapidly changed to light yellow, then translucent white at the time of picking. In the reddish-brown pits spread on the fruit surface contain seeds. Number of seeds/fruit varies 150-200. The shrub has a rooting habit similar to that of citrus and coffee, with an extensive lateral root system and a deep taproot. Each seed has an attached air sac that allows it to float for months in the ocean. (Scot Nelson, 2001). Seeds have a distinct air chamber and can retain viability even after floating in water for months. Seed color is tan dark brown. The fruits are found around the year. Usually the fruits or flowers sprout from alternate nodes and bunching of the fruits with two fruits on the same node is rarely encountered. Usually only one fruit is found in every alternate node.

Morinda trimera is probably the most closely related to species of either the South Pacific or Malaysia. *Morinda trimera* is more commonly found at seashore, within low land whereas *Morinda citrifolia* is found in the coastal strand, moist low and high land. It resembles *Morinda citrifolia* but has smaller leaves and fruits. Some people incorrectly identify the two species of *Morinda* described above as so-called "female" (*Morinda citrifolia*) and "male" (*Morinda trimera)* Noni plants. *Morinda trimera* trees are 4-16 m tall; branches pale brown, terete or sometimes 4-angled, with numerous warts and lenticels. Leaves chataceous, oblanceolate to elliptic-oblanceolate, 12-26 cm long, 3.5-9 cm wide, upper surface glossy, glabrous or very sparsely puberulent, lower surface dull, usually glabrate, occasionally puberulent, especially along veins and in axils of secondary and sometimes tertiary veins, domatia of conspicuous pubescent tufts in the axils of lateral veins and midrib, petioles 0.7-4 cm long, stipules connate, the tube 3-8 mm long, the lobes 2-6.5 mm long. Flowers perfect, 5-12 in globose heads, peduncles 6-60 mm long, bracts 3-10(-30) mm long; calyx 2-3 toothed, persistent in fruit, the tube 2.5-3 mm long, the teeth deltate, minute; corolla green, (2)3-4 lobed, the tube 4-8 mm long, the lobes 4-7.5 mm long; stamens (2)3-4, included; style 5-8 mm long. Syncarp dark green and very firm at maturity, becoming somewhat woody, 2-3.5 cm in diameter, the persistent calyx conspicuous.

Morinda pubescens is an evergreen shrub or small tree growing to 5-10 m tall and have 8-25 cm long, broadly elliptical to narrow elliptical leaves depending on the varieties, acute at both ends, not shining. The flowers are tubular, white, scented, about 2 cm long. The stipules are bifid and fruits are irregular in big fruited varieties and regular in small fruited varieties. The plant is extensively used in India to make the morindone dye sold under the trade name "Suranji". Morindone is used for the dyeing of cotton, silk and wool in shades of red, chocolate or purple. The colouring matter is found principally in the root bark and is collected when the plants reach three to four years of age. If the trees are allowed to mature then hardly any colouring substance remains. The small roots yield most of the dye and those above 1 cm diameter are discarded. The active substance is extracted as the glucoside known as morindin that upon hydrolysis produces the dye. Morindone is a mordant dye giving a yellowish-red colour with an aluminium mordant, chocolate with a chromium mordant, and dull purple to black with an iron mordant.

Morinda umbellata is reported from Tamil Nadu and Andaman and Nicobar Islands having small irregular fruits and long narrow, glossy leaves (Singh, *et al.*, 2009, Peter, 2007). According Drury (2007), it is a climber, glabrous, corolla short infundibuliform, leaves from oblong-lanceolate to cineate oblong, pointed; stipules membranaceous, united in a truncated sheath; peduncles terminal, 3-7 in a sessile terminal umbel about half the length of leaves; capituli globose; calyx margin entire; limb 4 (occasionally 5) cleft; filaments short, inserted into the bottom of the dilated part of the tube; ovary many hairs; anthers exerted; flower white. The roots yield dye of permanent yellow and with addition of sappan-wood, red dye is prepared.

In *Morinda stenophylla,* the leaves are 10-15 cm long, narrow oblong in shape, with variable stipules sometime longer. The leaves are not shining and have small sized fruits (Peter, 2007).

Morinda reticulata or Mapoon Bush is a shrub with a thick, woody rootstock and grows 1.2-1.5 m. The leaves are pale green, 5–10cm long, thick and leathery in texture. It has a white flower with broad white sepals which look very showy in dark green background of leaves. Mapoon Bushes growing on seleniferous soils are Selenium tolerant and accumulate very high concentrations of this element. Cattle, sheep and horses can be poisoned from eating Mapoon Bush containing excess amount of the trace element Selenium (Morton,1992).

Morinda villosa is a climbing shrub with branches more or less hirtuse, longitudinally ribbed. Leaves 2-10 x 2.5 -3.5 cm, to elliptic-lanceolate, pubescent above, hirsute beneath, pale greenish when dry; lateral nerves prominent, 10-15 on either half; base narrowed into the petiole. Flowers usually 4-5-merous in terminal, long pedunculate umbels. Corolla white, tube short, Syncarp deep orange,1.5 cm across (Britto, 2008).

Morinda royoc is a woody vine occasionally found in the coastal hammocks and rarely inland, blooms all round the year (Morton, 1992).

Table 8.1: Genetic Diversity of *Morinda* species

Sl.No.	Species	Distribution	References
1.	*Morinda ammitia*	Australia	Halford and Ford,A.J., Austrobaileya 6: 898 (2004).
2.	*Morinda angolensis*	Zaire, Angola, Zambia	(Good).White,F, Forest Fl. N. Rhodes.: 412 (1962).
3.	*Morinda artensis*	New Caledonia	Montrouz., Mém. Acad. Roy. Sci. Lyon, Sect. Sci. 10: 219 (1860).
4.	*Morinda asperula*	Belize to Guatemala	Standl., Publ. Field Mus. Nat. Hist., Bot. Ser. 22: 54 (1940).
5.	*Morinda asteroscepa*	Tanzania, Malawi	K.Schum., K, Bot. Jahrb. Syst. 34: 340 (1904).
6.	*Morinda aurantiaca*	Brazil (Roraima)	(Krause) Steyerm., Mem. New York Bot. Gard. 23: 385 (1972).
7.	*Morinda badia*	South-Eastern China	Ruan,Y.Z., in Fl. Reipubl. Popul. Sin. 71(2): 327 (1999).
8.	*Morinda bartlingii*	Philippines	Elmer, Leafl. Philipp. Bot. 4: 1340 (1912).
9.	*Morinda batesii*	Cameroon	Wernham, J. Bot. 57: 282 (1919).
10.	*Morinda billardierei*	New Caledonia	Baill., Adansonia 12: 231 (1879).
11.	*Morinda brachycalyx*	Suriname	(Bremek.) Steyerm., Mem. New York Bot. Gard. 23: 385 (1972).
12.	*Morinda brevipes*	Hainan	Hu,S.Y., Arnold,J. Arbor. 32: 399 (1951).
13.	*Morinda buchii*	Haiti	Urb., Symb. Antill. 1: 480 (1899).

Contd...

Table 8.1–Contd...

Sl.No.	Species	Distribution	References
14.	*Morinda bucidifolia* *Morinda nandarivatensis* *Morinda tripetala*	Fiji, Western Samoa	Gray, A. Proc. Amer. Acad. Arts 4: 41 (1860).
15.	*Morinda calciphila*	Malaysia	Wong,K.M., Malayan Nat. J. 38: 97 (1984).
16.	*Morinda callicarpifolia*	China	Ruan,Y.Z., in Fl. Reipubl. Popul. Sin. 71(2): 325 (1999).
17.	*Morinda calycina*	Guyana	(Benth.) Steyerm., Mem. New York Bot. Gard. 23: 385 (1972).
18.	*Morinda candollei* *Morinda ilicifolia*	New Caledonia	(Montrouz.) Beauvis., Ann. Soc. Bot. Lyon 26: 57 (1901).
19.	*Morinda canthoides,* *Morinda acutifolia*	Australia	(Muell.) Halford &.Hend., R.J.F., Austrobaileya 5: 731
20.	*Morinda capitellata*	-	-
21.	*Morinda celebica*	Philippines, Borneo, South Sulawesi	Miq., Fl. Ned. Ind. 2: 246 (1857).
22.	*Morinda cinnamomea*	Thailand	Craib, Bull. Misc. Inform. Kew 1932: 433 (1932).
23.	*Morinda cinnamomifoliata*	China (Guangxi)	Ruan,Y.Z., in Fl. Reipubl. Popul. Sin. 71(2): 336 (1999).
24.	*Morinda citrifolia* *Morinda angustifolia* *Morinda aspera* *Morinda bracteata* *Morinda chachuca* *Morinda chrysorhiza* *Morinda elliptica* *Morinda ligulata* *Morinda littoralis* *Morinda macrophylla* *Morinda mudia* *Morinda multiflora* *Morinda nodosa* *Morinda quadrangularis* *Morinda squarrosa* *Morinda stenophylla* *Morinda teysmanniana* *Morinda zollingeriana* *Morinda tomentosa*	Tropical and East Asia to South West Pacific Central and Eastern Himalaya to Myanmar	Sp. Pl.: 176 (1753).
25.	*Morinda citrina*	South-East China	Ruan,Y.Z. in Fl. Reipubl. Popul. Sin. 71(2): 331 (1999).
26.	*Morinda cochinchinensis* *Morinda trichophylla*	South-East China to Vietnam	DC., Prodr. 4: 449 (1830).
27.	*Morinda collina* *Morinda elongata*	Central and South-East New Calidonia	Schltr., Bot. Jahrb. Syst. 39: 265 (1906).
28.	*Morinda coreia* *Morinda exserta* *Morinda tinctoria* *Morinda pubescens*	India, Sri Lanka to Jawa	Buch.-Ham., Trans. Linn. Soc. London 13: 537 (1822).

Contd...

Table 8.1–Contd...

Sl.No.	Species	Distribution	References
29.	*Morinda coriacea*	Philippines (Mindanao)	Merr., Philipp. J. Sci. 17: 435 (1920 publ. 1921).
30.	*Morinda corneri*	Malaysia	Wong,K.M., Malayan Nat. J. 38: 96 (1984).
31.	*Morinda costata*	New Guinea	Merr. and Perry, J.M., Arnold,J Arbor. 26: 263 (1945).
32.	*Morinda debilis*	Guyana	(Sandwith) Steyerm., Mem. New York Bot. Gard. 23: 385 (1972).
33.	*Morinda decipiens*	North-West and Central New Caledonia	Schltr., Bot. Jahrb. Syst. 39: 266 (1906).
34.	*Morinda deplanchei* *Morinda alyxioides* *Morinda gatopensis* *Morinda podocarpifolia*	North-West and Central New Caledonia	(Hook.f.) Baill. ex.Schum,K. in.Engler,H.G.A and Prantl,K.A.E. Nat. Pflanzenfam. 4(4): 138 (1891).
35.	*Morinda elliptica*	North Western part of India	Ali *et al.* (2000)
36.	*Morinda elliptifolia*	Philippines	Quisumb. and Merr., Philipp. J. Sci. 37: 204 (1928).
37.	*Morinda elmeri*	Borneo	Merr., Univ. Calif. Publ. Bot. 15: 297 (1929).
38.	*Morinda fasciculata*	Ecuador	Benth., Pl. Hartw.: 123 (1844).
39.	*Morinda ferruginea*	Cuba	Rich.A in R.de la Sagra, Hist. Fis. Cuba, Bot. 2: 19 (1850).
40.	*Morinda geminata*	North western parts of India, West and West Central Tropical Africa	DC., Prodr. 4: 447 (1830).
41.	*Morinda gjellerupii*	New Guinea	Valeton, Bot. Jahrb. Syst. 61: 152 (1927).
42.	*Morinda glaucescens* *Morinda schumanniana*	New Caledonia	Schltr., Bot. Jahrb. Syst. 39: 267 (1906).
43.	*Morinda glomerata* *Morinda gemella*	New Guinea, Solomon Island	(Blume) Miq., Fl. Ned. Ind. 2: 247 (1857).
44.	*Morinda grayi*	Fiji	Seem., Fl. Vit.: 130 (1866).
45.	*Morinda guatemalensis* *Morinda mesochora*	Mexico to Honduras	(Donn.Sm.) Steyerm., Mem. New York Bot. Gard. 23: 385 (1972).
46.	*Morinda hainanensis*	Hainan	Merr. and How, Sunyatsenia 5: 188 (1940).
47.	*Morinda hirtella*	New Guinea	Merr. and Perry,L.M. Arnold,J. Arbor. 26: 264 (1945).
48.	*Morinda hispida*	Malaysia (Johor)	Wong,K.M. Malayan Nat. J. 41: 269 (1987).
49.	*Morinda hoffmannioides*	Colombia, Venezuela	Standl., Publ. Field Mus. Nat. Hist., Bot. Ser. 22: 193 (1940).
50.	*Morinda hollrungiana*	New Guinea	Valeton, Bot. Jahrb. Syst. 61: 153 (1927).
51.	*Morinda howiana*	China(Guangdong), Hainan	Hu,S.Y J. Arnold Arbor. 32: 400 (1951).
52.	*Morinda hupehensis*	South China	S.Y.Hu, Arnold,J. Arbor. 32: 400 (1951).
53.	*Morinda hypotephra*	Queensland (Cook)	Muell,F., Victorian Naturalist 6: 55 (July 1889).

Contd...

Table 8.1–Contd...

Sl.No.	Species	Distribution	References
54.	*Morinda jackiana*	Malaysia	Korth., Ned. Kruidk. Arch. 2(2): 227 (1851).
55.	*Morinda jasminoides*	New Guinea, North and East Australia	Cunn,A., Bot. Mag. 61: t. 3351 (1834).
56.	*Morinda kanalensis*	Central and South-East New Caledonia	Baill. ex Guillaumin, Arch. Bot. Mém. 3(5): 43 (1930).
57.	*Morinda lacunosa*	China, Malaysia, Borneo	King and Gamble, J. Asiat. Soc. Bengal, Pt. 2, Nat. Hist. 73: 87 (1904).
58.	*Morinda lastelliana*	Madagascar	Baill., Adansonia 12: 231 (1879).
59.	*Morinda latibractea*	Caroline Island	Valeton, Bot. Jahrb. Syst. 63: 321 (1930).
60.	*Morinda latibracteata*	-	-
61.	*Morinda laxa*	-	-
62.	*Morinda laxiflora*	-	-
63.	*Morinda leiantha*	South-Central China to Indo-China	Kurz, J. Asiat. Soc. Bengal, Pt. 2, Nat. Hist. 41(2): 313 (1872).
64.	*Morinda leichhardtii*	-	
65.	*Morinda leparensis*	Lesser Sunda Island	Valeton, Icon. Bogor.: t. 362 (1913).
66.	*Morinda leptocalama*	New Guinea	Wernham, J. Bot. 56: 77 (1918).
67.	*Morinda litseifolia*	South China	Ruan,Y.Z., in Fl. Reipubl. Popul. Sin. 71(2): 335 (1999).
68.	*Morinda longiflora*	West and West-Central Tropical Africa	G.Don, Gen. Hist. 3: 545 (1834).
69.	*Morinda longifolia*	Thailand	Craib, Bull. Misc. Inform. Kew 1932: 434 (1932).
70.	*Morinda longiloba*	Guyana	(Steyerm.) Steyerm., Mem. New York Bot. Gard. 23: 385 (1972).
71.	*Morinda longipedunculata*	South-East Colombia to South Venezuela	Steyerm., Ann. Missouri Bot. Gard. 75: 340 (1988).
72.	*Morinda longipetiolata*	Venezuela	Steyerm., Ann. Missouri Bot. Gard. 75: 340 (1988).
73.	*Morinda longissima*	China (Yunnan)	Ruan,Y.Z. in Fl. Reipubl. Popul. Sin. 71(2): 323 (1999).
74.	*Morinda lucida*	Tropical Africa	Benth. in.Hooker,W.J., Niger Fl.: 406 (1849).
75.	*Morinda micrantha*	New Guinea	Valeton, Bot. Jahrb. Syst. 61: 153 (1927).
76.	*Morinda microcephala* *Morinda microphylla*	Philippines	Bartl. ex DC., Prodr. 4: 449 (1830).
77.	*Morinda moaensis*	East Cuba	Alain, Contr. Ocas. Mus. Hist. Nat. Colegio *De La Salle* 17: 6 (1959).
78.	*Morinda mollis*	New Caledonia, Fiji	Gray,A. Proc. Amer. Acad. Arts 4: 41 (1860).
79.	*Morinda montana*	North-West New Caledonia	Johanss., J.T., Opera Bot. 122: 49 (1994).
80.	*Morinda morindoides* *Morinda confusa*	West Tropical Africa to Sunda and Angola	(Baker) Milne-Redh., Kew Bull. 2: 31 (1947).
81.	*Morinda multiflora*	Northern and Central India	Royle (1970)

Contd...

Table 8.1–Contd...

Sl.No.	Species	Distribution	References
82.	*Morinda myrtifolia* *Morinda choriophylla* *Morinda forsteri* *Morinda glandulosa* *Morinda velutina*	West and South-Central Pacific	Gray,A, Proc. Amer. Acad. Arts 4: 41 (1860).
83.	*Morinda nana*	Thailand	Craib, Bull. Misc. Inform. Kew 1932: 434 (1932).
84.	*Morinda nanlingensis*	South-East China	Ruan,Y.Z. in Fl. Reipubl. Popul. Sin. 71(2): 329 (1999).
85.	*Morinda naudia*	-	-
86.	*Morinda neocaledonica*	North-West and West Central New Caledonia	(Moore,S) Guillaumin, Arch. Bot. Mém. 3(5): 44 (1930).
87.	*Morinda nitida*	Philippines (Bucas Grande)	Merr., Philipp. J. Sci. 17: 435 (1920 publ. 1921).
88.	*Morinda oblongifolia*	New Guinea	Valeton, Bot. Jahrb. Syst. 61: 153 (1927).
89.	*Morinda odontocalyx*	Bolivia	(Sandwith) Steyerm., Mem. New York Bot. Gard. 23: 385 (1972).
90.	*Morinda officinalis*	Shivalik Hills (Lower belt of H.P.) India, South-East China, Hainan	How,F.C. Acta Phytotax. Sin. 7: 326 (1958).
91.	*Morinda oligocephala*	New Guinea	Merr. and Perry,L.M. Arnold,J. Arbor. 26: 265 (1945).
92.	*Morinda palawanensis*	-	-
93.	*Morinda panamensis*	Central America to Colombia	Seem., Bot. Voy. Herald: 136 (1854).
94.	*Morinda pandurifolia*	Indo-China	Kuntze, Revis. Gen. Pl. 1: 289 (1891).
95.	*Morinda parvifolia* *Morinda cumingiana*	Vietnam, South-East China, Hainan, Taiwan, Philippines	Bartl. ex DC., Prodr. 4: 449 (1830).
96.	*Morinda peduncularis*	Venezuela to North Brazil	Kunth in F.W.H.von Humboldt, Bonpland A.J.A and.Kunth,C.S. Nov. Gen. Sp. 3: 380 (1819).
97.	*Morinda pedunculata*	Caroline Island	Valeton, Bot. Jahrb. Syst. 63: 322 (1930).
98.	*Morinda persicaefolia*	-	-
99.	*Morinda persicifolia* *Morinda lanceolata* *Morinda talmyi*	Bangladesh and West Malaysia	Buch.-Ham., Trans. Linn. Soc. London 13: 535 (1822).
100.	*Morinda philippinensis*	North Borneo to Philippines	Elmer, Leafl. Philipp. Bot. 3: 1044 (1911).
101.	*Morinda phyllireoides* *Morinda fallax* *Morinda ligustrina* *Morinda vieillardii Baill.*	New Caledonia	Labill., Sert. Austro-Caledon.: 49 (1825).
102.	*Morinda piperiformis*	Lesser Sunda Island	Miq., Ann. Mus. Bot. Lugduno-Batavi 4: 135 (1869).
103.	*Morinda platyphylla*	Philippines	Merr., Philipp. J. Sci., C 11: 33 (1916).
104.	*Morinda podistra*	North-East Queensland	Halford and Ford,A.J. Austrobaileya 6: 895 (2004).
105.	*Morinda polyneura*	Jawa	Miq., Fl. Ned. Ind. 2: 244 (1857).

Contd...

Table 8.1–Contd...

Sl.No.	Species	Distribution	References
106.	*Morinda pubiofficinalis*	China (Guizhou, Guangdong)	Ruan,Y.Z. in Fl. Reipubl. Popul. Sin. 71(2): 334 (1999).
107.	*Morinda pulchella*	New Caledonia	Schltr., Bot. Jahrb. Syst. 39: 267 (1906).
108.	*Morinda pumila*	Thailand	Craib, Bull. Misc. Inform. Kew 1932: 435 (1932).
109.	*Morinda reticulata*	New Guinea, Queensland	Benth., Fl. Austral. 3: 424 (1867).
110.	*Morinda retusa*	Madagascar	Poir. in J.B.A.M.de Lamarck, Encycl. 4: 316 (1797).
111.	*Morinda ridleyi*	South Malaya	(King and Gamble) Ridl., Fl. Malay Penins. 2: 118 (1923).
112.	*Morinda rigida*	Thailand to Malaysia	Miq., Fl. Ned. Ind. 2: 246 (1857).
113.	*Morinda rojoc* *Morinda yucatanensis*	Florida, South-East Mexico to Ecuador, Caribbean	Greenm., Publ. Field Mus. Nat. Hist., Bot. Ser. 2: 262 (1907).
114.	*Morinda rosiflora*	China (Yunnan)	Ruan,Y.Z. in Fl. Reipubl. Popul. Sin. 71(2): 324 (1999).
115.	*Morinda royoc*	Brevard to Monroe Key	L., Sp. Pl.: 176 (1753).
116.	*Morinda rugulosa*	China (Guangxi, Hunan)	Ruan,Y.Z. in Fl. Reipubl. Popul. Sin. 71(2): 328 (1999).
117.	*Morinda salomonensis*	Papua New Guinea	-
118.	*Morinda sarmentosa*	Jawa	Blume, Bijdr.: 1006 (1826).
119.	*Morinda scabrida*	Thailand	Craib, Bull. Misc. Inform. Kew 1932: 435 (1832).
120.	*Morinda scabrifolia*	China (Yunnan, Guangxi)	Ruan,Y.Z. in Fl. Reipubl. Popul. Sin. 71(2): 332 (1999).
121.	*Morinda scandens*	-	-
122.	*Morinda schultzei*	New Guinea	Valeton, Bot. Jahrb. Syst. 61: 151 (1927).
123.	*Morinda scortechinii*	Malaysia (Perak, Pahang)	(King and Gamble) Ridl., Fl. Malay Penins. 2: 118 (1923).
124.	*Morinda seibertii*	-	-
125.	*Morinda senegalensis*	-	-
126.	*Morinda sessiliflora*	Windward Island	Bertol., Syll. Pl. Hort. Bot. Bonon.: 5 (1827).
127.	*Morinda siebertii*	Central America to Ecuador	(Standl.) Steyerm., Mem. New York Bot. Gard. 23: 386 (1972).
128.	*Morinda sparsiflora*	Guyana	(Steyerm.) Steyerm., Mem. New York Bot. Gard. 23: 385 (1972).
129.	*Morinda subcaudata*	New Guinea	Valeton, Bot. Jahrb. Syst. 61: 154 (1927).
130.	*Morinda surinamensis* *Morinda venezuelensis*	Suriname, Venezuela	(Bremek.) Steyerm., Mem. New York Bot. Gard. 23: 386 (1972).
131.	*Morinda tenuiflora*	South Tropical America	(Benth.) Steyerm., Mem. New York Bot. Gard. 23: 386 (1972).
132.	*Morinda ternifolia*	India	Royle, 1970

Contd...

Table 8.1–Contd...

Sl.No.	Species	Distribution	References
133.	*Morinda titanophylla*	Gabon, South-West Uganda, East Zaire	Petit,E.M.A Bull. Jard. Bot. État 32: 188 (1962).
134.	*Morinda tonkinensis*	-	-
135.	*Morinda triandra*	New Guinea	Moore,S. J. Bot. 61(Suppl.): 25 (1923).
136.	*Morinda trimera* *Morinda lanaiensis* *Morinda sandwicensis*	Hawaiian Islands, India, Andaman and Nicobar Islands	Hillebr., Fl. Hawaiian Isl.: 177 (1888).
137.	*Morinda triphylla*	North Brazil	(Ducke) Steyerm., Mem. New York Bot. Gard. 23: 386 (1972).
138.	*Morinda truncata*	New Caledonia	JohanssJ.T., Opera Bot. 122: 41 (1994).
139.	*Morinda turbacensis*	Colombia	Kunth in Humboldt, F.W.H. von.A.J.A.Bon pland and Kunth, C.S.Nov. Gen. Sp. 3: 380 (1819).
140.	*Morinda umbellata* *Morinda boninensis* *Morinda padavara* *Morinda shuanghuaensis* *Morinda tetrandra*	Tropical and Sub-tropical Asia to Pacific	Sp. Pl.: 176 (1753).
141.	*Morinda undulata*	China (Yunnan)	Ruan,Y.Z. in Fl. Republ. Popul. Sin. 71(2): 321 (1999).
142.	*Morinda vestita*	-	-
143.	*Morinda vieillardi*	-	-
144.	*Morinda villosa*	India (Assam) to China (Yunnan), Vietnam	Hook.f., Fl. Brit. India 3: 158 (1880).
145.	*Morinda vitrifolia*	-	-
146.	*Morinda waikapuensis*	Hawaiian Islands	H.St.John, Pacific Sci. 33: 376 (1979 publ. 1980).
147.	*Morinda wallichii*	Myanmar, Thailand	Kurz, J. Asiat. Soc. Bengal, Pt. 2, Nat. Hist. 41(1): 313 (1872).

Source: http://zipcodezoo.com/Plants/names/Plants-SN-M-0111.html.

Diversity of *Morinda* species in India

India has a vast biodiversity in flora and fauna. For thousands of years, India showed and highlighted the world with the importance of medicinal plants in well being of humanity and spread its knowledge throughout the world. The *Morinda* plant is not an exception to this fact. On reference of literature till now, 15 different species of *Morinda viz. Morinda angustifolia, Morinda citrifolia, Morinda elliptica, Morinda geminate, Morinda leiantha, Morinda lucida, Morinda multiflora, Morinda officinalis, Morinda pandurifolia, Morinda pubescens, Morinda reticulata, Morinda ternifolia, Morinda trimera, Morinda umbellata, Morinda villosa* are reported in India, distributed throughout the Western Ghats, foot hills of Himalayas, Himachal Pradesh, Odisha, Tamil Nadu, Andaman and Nicobar Islands etc. But on scrutiny, it seems that some of the species reported are synonyms of other species or cannot be authenticated. The distribution of *Morinda citrifolia* L. is rare or nil in higher elevation of hilly regions. The survey of South India indicated that different species or varieties of *Morinda* are distributed

Figure 8.1: Diversity in Fruit Shape and Size

Figure 8.2: *Morinda* as Alley Crop

Figure 8.3: ***Morinda*** **Flower**

Figure 8.4: Insect Pollination in ***Morinda***

Figure 8.5: Peat on Noni

Figure 8.6: *Morinda* Seeds

throughout Tamil Nadu and Kerala. However, the species of *Morinda pubescences* (*Morinda tinctoria*) is present abundantly in most parts of Tamil Nadu and some parts of Kerala. *Morinda citrifolia* is profusely distributed in most parts of Andaman and Nicobar Islands, Kerala especially coastal region and also in the Mangalore of Karnataka but not recorded in Tamil Nadu. Recently an unidentified *Morinda* species with large and leathery leaves was reported in the Dhaandakaranya forest area of Malkanagiri district in Odisha (Singh *et al.*, 2006 and Peter, 2007). More recently, presence of *Morinda* was also recorded in Gujarat. The genus earned itself a respectable position by virtue of its healing properties and positive energy in Indian tradition and culture.

Table 8.2: Distribution of *Morinda* species in India

Morinda Species	*Distribution*	*Reference*
Morinda citrifolia L.	South India, Andaman and Nicobar Islands, Punjab (Peshawar valley), Darjeeling, Bihar, Odisha, Uttar Pradesh	Royle (1970), Singh *et al.* (2005), Peter, (2007)
Morinda elliptica	North Western part of India	Ali *et al.* (2000)
Morinda leiantha	Part of Indo-China	Kurz, J. Asiat. Soc. Bengal, Pt. 2, Nat. Hist. 41(2): 313 (1872).
Morinda multiflora	Northern and Central India	Royle (1970)
Morinda officinalis	Shivalik Hills (Lower belt of H.P)	F.C.How, Acta Phytotax. Sin. 7: 326 (1958).
Morinda pubescences	North Central and Southern India, Bihar, Orissa, U.P.	Royle (1970), Peter (2007)
Morinda umbellata	Andaman and Nicobar Islands, East west peninsula through Assam to Burma, Sri Lanka, India (for dye)	Peter (2007), Drury, (2007), Singh *et al.* (2009)
Morinda reticulata	Endemic to Western Ghats	Britto, J.S. (2008)
Morinda ternifolia	India	Royle (1970)
Morinda trimera	A&N Islands, Tamil Nadu	Peter (2007), Singh *et al.* (2009)
Morinda villosa	Khasi Hills Cachar, Kameng, Tirap	Hoof. F., Fl. Brit. India 3:158 (1880), Britto, J.S. (2008)

Diversity and Distribution of *Morinda* species in Andaman and Nicobar Islands

Biodiversity wise, Andaman and Nicobar Islands are one of the most sought after spots in the world. The genus *Morinda* originated in these islands and spread throughout the tropical regions through sea routes (Singh *et al.*, 2009). The aboriginal tribes of these islands have huge knowledge on the medicinal properties of this genus and are using it in traditional healings and also as food. Survey of the genus in these Islands revealed distribution of three different species throughout the Andaman and Nicobar Islands *viz. Morinda citrifolia, Morinda trimera* and *Morinda umbellata* (Singh, *et al.*, 2005, 2006, 2009).

Intra-Species Variation

McClatchey (2002) postulated two possible hypotheses of intra-species variations. The first hypothesis is that there is only one species of *Morinda citrifolia* and that it contains distinct genetic populations which can be recognized at the varietal level. Presumably these populations do not interbreed very often and therefore are maintained as relatively distinct units. These populations could be distinguished based upon both morphological and chemical differences. Thus, more than one variety of *Morinda citrifolia* exist.

The second possibility is that the current circumscription of the species, *Morinda citrifolia* is inclusive of two or more separate species which do not usually interbreed. If this is the case, then there is reason to believe that the different species would have differing medicinal values for people as well as differing rates of growth, disease resistance, fruit production, etc. Thus, more than one species of *Morinda* exist within the plants which are scientifically circumscribed as *Morinda citrifolia.*

There is a relatively high degree of genetic variability of fruit and leaf within the species *Morinda citrifolia* L. Waki *et. al.* (2007) assessed genetic diversity based on morphological variation and for cultivar identification. He examined the three botanical varieties *Morinda citrifolia* var. *citrifolia; Morinda citrifolia* var. *bracteata* and *Morinda citrifolia* var. *potteri.* Search of literature for different variants of this species shows five different varieties. There are relatively limited literature on these five variants.

1. *Morinda citrifolia* var. *citrifolia*
2. *Morinda citrifolia* var. *bracteata*
3. *Morinda citrifolia* var. *potteri*
4. *Morinda citrifolia* var. *sativa*
5. *Morinda citrifolia* var. *typica*

Morinda citrifolia var. *citrifolia*

Morinda citrifolia var. *citrifolia,* the primary focus of this profile, is of the greatest cultural, economic and medicinal value and is in greater abundance in India, Andaman and Nicobar Islands and Pacific region. This is a morphologically diverse species and with no clear sub-populations bearing unique characteristics. There exist large and small fruited members of this group. Leaves are usually 15-20 cm long, broadly elliptic, acute and shining. Stipules are broad, seeds are wringed, pyrenes with empty ventral cavity and fruits are large with diameter ranging from 3-6 cm.

Morinda citrifolia var. *bracteata*

Morinda citrifolia var. *bracteata* is a small-fruited variety with conspicuous bracts. It is found in Indonesia and other parts of the region between the Indian and Pacific Oceans. It has calyx-limbs with 12 leaf-like, linear-lanceolate lobes 1-1.5 cm long; the stem is straighter and the leaves are smaller than var. *citrifolia* (Nelson, 2006).

Morinda citrifolia Cultivar 'Potteri'

Morinda citrifolia cultivar 'Potteri' is an ornamental type, with green and white leaf variegation, distributed throughout the Pacific (Nelson, 2006).

According to Singh *et al.* (2005, 2008), in the Andaman and Nicobar Islands, particularly in South Andaman, the genus *Morinda* includes trees and shrubs. However, the *Morinda citrifolia* and *Morinda trimera* are profusely distributed in Andaman and Nicobar Islands. As per the record available in herbarium at Botanical survey of India, Port Blair, the species *Morinda umbellata* is mainly confined to some parts of these Islands. Morphological features which unite the species into one genus include syncraps or capitula (aggregated fruits) forming from ovaries and parts of conate calyces, and tricolporate pollen grains with sexine usually finely reticulate (Johansson, 1994). The syncarps may be fleshy, as with *Morinda citrifolia* or may be dry as with many other species. Each syncarp is probably derived from condensation of umbels of separate flowers with further condensation within some species of *Morinda* to form a range of flower/fruit configurations. One of the distinctive adaptations of *Morinda citrifolia* seems to be its ability to colonize new islands and new terrain. It is frequently identified as apart of primary forests or shrub vegetation on small atolls, new lava flows, and other newly emergent surfaces near the ocean. The trees are not uncommon in both dry and wet coastal forests.

In addition to the wide spread varieties or kinds of *Morinda citrifolia*, two unusual varieties were identified by CARI, Port Blair. The first *is M. citrifolia* var *citrifolia* with big sized fruit. Because of the assignment of the first variety, there is an implied typical variety, *Morinda citrifolia* var *citrifolia* cultivar with small fruit and both are present in South Andamans. The different kinds of *M. citrifolia* are distinguished based upon characteristics: fruit, leaf size/shape and fruit odor. The later is likely to be related to fruit chemistry/pharmacology.

Genetic Diversity of *Morinda citrifolia* L. Accessions (RAPD)

Genetic diversity among species or cultivars are usually determined with morphological or biochemical markers. Identification based on morphology alone can not correctly designate different accessions of *Morinda citrifolia*. Extensive use of RAPD-PCR technique and other molecular characterization methods of *Morinda citrifolia* germplasm are envisaged to identify markers of commercial traits which will also help in planning future germplasm collection and selection of parents for breeding programmes.

However, these markers have their own limitations. In the last fifteen years, techniques based on DNA markers are used to detect variation at DNA level and are proven to be very effective for distinguishing between closely related genotypes. DNA markers have several advantages over phenotype markers and are not affected by age, physiological conditions and environmental factors. The RAPD technique provides a convenient and rapid assessment of the differences in the genetic composition of the related individuals and is employed in a large number of plants including medicinal plants. The application of RAPD PCR profiling for the determination of differences among fourteen accessions *viz.* GAH-1, SPG-1, SPG-2, MEM-1, ABF-1, ABF-2, LH-1, ABH-1, AHD-1, JGH-1, GAH- 2, TRA-1, TRA-2 and MAN-1 of *Morinda citrifolia* collected from different locations of Andaman and Nicobar Island was carried out at Central Agricultural Research Institute, Port Blair.

The genomic DNA was isolated from *Morinda citrifolia* leaves by CTAB method. DNA quantification was estimated through UV-VIS spectrophotometer and agarose gel electrophoresis with known standard (Lambda DNA). DNA quality was chequed under UV-light and the purity was calculated from O.D.260/O.D.280 ratio, and also by electrophoresis on 0.8 per cent agarose gel.

The resuspended DNA was then diluted in sterile distilled water to 25ng/µl concentration for use in amplification reactions. A set of 48 random decamer oligonucleotides (Bangalore Genei) was used as single primers for the amplification of RAPD fragments. All the 48 primers were tested at least twice for reproducibility of banding pattern and 28 exhibited amplification. Polymerase chain reactions (PCR) were carried out in a final volume of 25 µl containing 25 ng template DNA, 100 µM of each of the four deoxynucleotide triphosphate, 20 ng of decanucleotide primer, 1.5 mM $MgCl_2$, 10X Taq buffer (10mM Tris HCl pH9.0, 50 mM KCl) and 0.5 U Taq DNA polymerase. PCR amplification was carried out on thermal cycler well blocks (M.J. Research Inc. USA) under the following conditions, initial denaturation of 4 minutes at 94°C, followed by 40 cycles of denaturation at 94°C for 1 minute, primer annealing for 1 minute at 32 °C, extension for 1 minute at 72°C followed by a final extension for 7 minutes at 72°C. Amplification products were maintained at 4°C until electrophoresis.

The reaction products were separated along side a molecular weight marker (1 kb ladder and 100bp ladder) by electrophoresis on 1.5 per cent agarose gel using 1X TBE buffer at 8V/cm for 3 hours. The gel was visualized and photographed with gel documentation system (Vilber Loubmet, France, Cat.No. Bio ID++ver.99.04). The image profiles of banding patterns were recorded and molecular weight of each band was determined by Molecular Analyst (version 1.5) software.

Data were recorded as presence (1) or absence (0) of band products from the examination of photographic negatives. Using Dice coefficient, a similarity matrix involving 14 accessions was generated with NTSYS-pc (Numerical Taxonomy System, Applied Biostatistics, Inc., New York, USA, software version 2.02e). Dendrogram was created by UPGMA cluster analysis (Figure 8.7).

The accessions were grouped in to two major clusters A and B having 74 per cent similarity. The first major cluster A as grouped into two sub clusters A1 and A2 having 79 per cent similarity, sub cluster A1 was further subdivided in two groups A1X and A1Y having 82 per cent similarity. The A1Y had a solitary collection from AHD-1 while A1X was further subdivided into two Subgroups $A1X^1$ and $A1X^2$. $A1X^1$ had two accessions JGH-1 and GAH-2 which are closely related and showed 88.5 per cent similarity.

$A1X^2$ was grouped in to $A1X^2A$ and $A1X^2B$ having 86 per cent similarity. $A1X^2A$ comprises collection from SPG-2, ABF-1, ABF-2, LH-1 and ABH-1, in which accessions belonging to LH-1and ABH-1 had 93 per cent similarity while $A1X^2B$ had two accessions JGH-1 and GAH-2 having 88 per cent similarity. The second sub cluster A2 was grouped into A2X and A2Y. A2Y represented by only accession MAN-1 while A2X had two accessions TRA-1 and TRA-2 having 83 per cent similarity. The second major cluster B represented by only one accession MEM-1.

Out of 811 bands, 335 bands were polymorphic. The polymorphism percentage was calculated as per the method suggested by Blair *et. al.* (1999). The study showed 41 per cent polymorphism among the 14 accessions of *Morinda citrifolia.* The result indicated that despite of their morphological identity, substantial polymorphism was observed among *Morinda citrifolia* accessions collected from different Islands.

Table 8.3: Amplified Primer Code and Sequences for RAPD Profiling

Sl.No.	*Primer code*	*Sequance 5'-3'*	*Sl.No.*	*Primer code*	*Sequance 5'-3'*
1.	OPH-1	AATCGGGCTG	15.	OPH-17	AAGCAGCAAG
2.	OPH-2	CAATCGCCGT	16.	OPH-19	AGCATTCGG
3.	OPH-3	TCTGTGCTGG	17.	OPH-20	CACCGTTCTG
4.	OPH-4	GACCGCTTGT	18.	OPH-21	ACTCCGCAGT
5.	OPH-5	GTTGCGATCC	19.	OPH-31	TAGACAGTCG
6.	OPH-6	TCGCCGCAAA	20.	OPH-33	AGGCCGTATC
7.	OPH-7	AGCGTCACTC	21.	OPH-34	ATGAGTCCAC
8.	OPH-8	GTCCGTACTG	22.	OPH-36	TCAAACTCGG
9.	OPH-9	GGTGCTCCGT	23.	OPH-39	TAGCCGTCAA
10.	OPH-10	GACCGACCCA	24.	OPH-41	ATTTGATCGC
11.	OPH-11	CGGTTCCCCC	25.	OPH-42	ACGCTGATCA
12.	OPH-13	GTCTGACGGT	26.	OPH-43	AACCGACGGG
13.	OPH-14	CAGCTCAAGT	27.	OPH-44	TGCCCTGCCT
14.	OPH-15	CGATCGAGGA	28.	OPH-45	CTTGCCTCCC

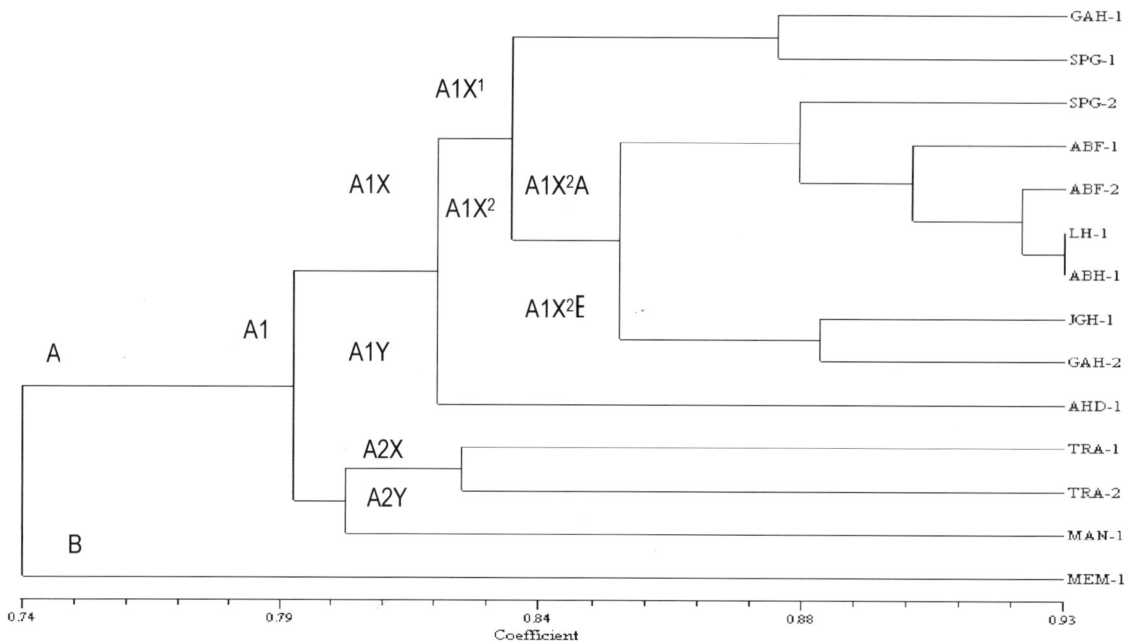

Figure 8.7: Dendrogram Showing Diversity of *Morinda citrifolia* Accessions Based on RAPD

The phenotypic diversity of fruit shape and size among different accessions of *Morinda citrifolia* collected from different location of Andaman and Nicobar Islands were significant. This may be due to the genetic diversity, soil condition of the location and other factors which play important role in growth and production of the plant. The average fruit weight varied from 248 g to 66 g and length and diameter varied from 11.6 cm to 6.9 cm and 7.1cm to 4.9 cm respectively in different accessions. This indicated clear phonotypical variation between different accessions. The other physicochemical parameters also varied between different accessions. The specific gravity ranged from 0.98 to 0.92 and ascorbic acid content from 139.87 to 90.50 mg/100g in fruits of different accessions evaluated.

Conservation Strategies

There are more than 140 species of *Morinda* are already reported worldwide, many of them are medicinally and economically important. The traditional healers around the world are using one or another species for well being of humanity. The importance of this genus and its wide diversity calls for conservation of each species and its genotypes *in-situ* and *ex-situ* conditions for future research and availability. Due to the alarmingly growing population and urbanization, there is high chance of loosing species diversity leading to extinction. Once the diversity is lost, there is no chance to revive it. Hence, steps must be taken to conserve each and every genotype and the information for each should be shared for getting a clear idea of the genetic and morphological diversity as well as its importance for exploiting its full potentials through commercial cultivation.

The first and foremost conservation should be *in-situ* which needs creating awareness and educating the local people about the importance of the species. This should be followed by *ex-situ*

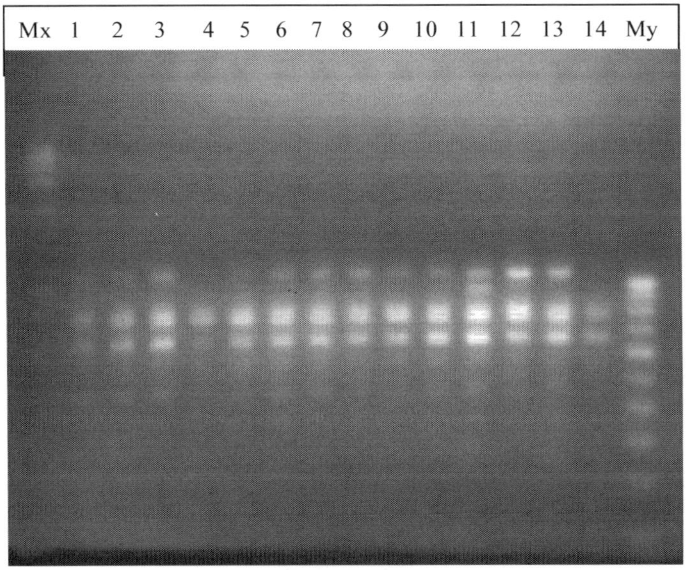

Figure 8.8

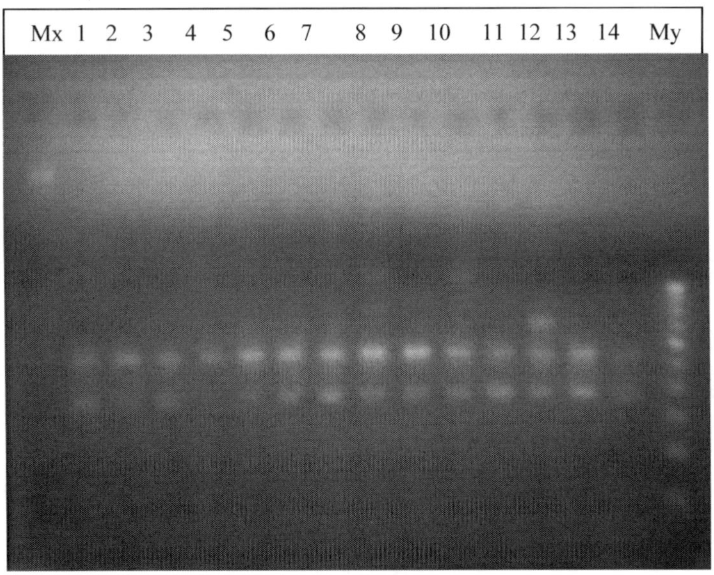

Figure 8.9

Figures 8.8 and 8.9: The Amplification of Primer OPH 31& OPH-11 Respectively

RAPD Profile of 14 accessions of *Morinda citrifolia* collected from Andaman and Nicobar Islands Mx-1Kb ladder, My- 100 bp ladder 1- GAH-1, 2- SPG-1, 3- SPG-2, 4- MEM-1, 5- ABF-1, 6- ABF-2, 7- LH-1, 8- ABH-1, 9- AHD-1, 10- JGH-1, 11- GAH-2, 12- TRA-1, 13- TRA-2, 14- MAN-1

conservation through development of gene-garden, gene-bank, botanical garden, seed-bank, cryopreservation of seeds etc.

Future Strategies

There exist doubts on the total number of species and their variants in the genus *Morinda*. This creates confusion which needs to be cleared and a complete data base has to be made available for this species.

Considering the positive discoveries made with the *Morinda citrifolia* L. and other members of the genus *Morinda* so far, there is excellent reason to anticipate that further studies will prove the fruit and its preparations beneficial to health in numerous ways. Noni is a valuable medicinal plant and it is likely to become an increasingly sought-after dietary supplement in the near future. Yet we have more to learn about what the plant contains and how it works. Further phytochemical investigation into it will likely lead to the discovery of other new compounds with amazing properties. Though the *Morinda citrifolia* L. and other members of the genus *Morinda* is having wide health stimulating properties, it needs much more research to establish its scientific authenticity for the proclaimed healing properties. The need of the hour is to identify the active compound, its isolation and understanding its mode of action. New compounds with high levels of pharmacological activity are urgently required for a wide range of human disorders and diseases. A number of scientific publications showed that this plant contains several nutritional and functional compounds but the current state of knowledge is still far from satisfactory. Based on the level of important nutritional and active compounds contents, varieties should be identified and released. It is found that while leaves of one accession are rich in iron whereas another is rich in calcium. If isolated, they can serve an inexpensive source of iron and calcium. And this should be used for preparation of fortified Biscuits for meeting challenge of malnutrition in the world.

The most important compounds identified are damnacanthal and some other phenolic compounds. However, its activity as chemopreventive and anti-HIV needs further investigation. A drug development programme should be undertaken to develop modern drugs with the compounds isolated, but before that, a comprehensive phytochemical profiling, extensive investigation of its bioactivity, mechanism of action, pharmacotherapeutics, toxicity and clinical trials are needed to provide sufficient data. In addition, research should be initiated for identification of elite chemotypes, quality control of produces through marker development techniques such as TLC, HPTLC, standardization of cell culture techniques for production of bioactive compounds and Identification of pathway related to the production of potent bioactive compounds. In order to realize these goals, a number of projects have been and should be initiated by academic Institutions and pharmaceutical companies.

Conclusion

The queen of the genus Morinda: *Morinda citrifolia* var. *citrifolia* is gaining ever increasing global demand for its wide range of properties and uses. Other than this more than 140 species are already reported with several synonyms and variants. But there is lack of complete information of this genus and its species. For exploiting it commercially, different characteristics and medicinal properties should be compiled and made available to the interested groups or individuals. Steps must also be taken for conserving the wide diversity in this genus. Understanding the biodiversity of this genus with its beneficial properties will help the mankind immensely to combat malnutrition, poverty, unemployment, starvation etc. and will bring health and wellbeing to the whole Humanity.

References

Ali, A.M., Ismail, N.H., Mackeen, M.M., Yazan, L.S., Mohamed, S.M., Ho, A.S.H. and Lajis, N.H., 2000. Antiviral, cytotoxic and antimicrobial activities of anthraquinones isolated from the roots of *Morinda elliptica. Pharm. Biol.*, 38(4): 298–301.

Blair, M.W., Panaud, O. and McCouch, S.R., 1999. Inter simple sequence repeat (ISSR) amplification for analysis of microsatellite motif frequency and finger printing in rice (*Oryza sativa* L.). *Theor. Appl. Genet.*, 98: 780–792.

Britto, S.J., 2008. A review of taxonomy of *Morinda* L. (Rubiaceae). *Intl. J. Noni Res.*, 3(1–2): 60–74.

Cardon, D., 2003. *Le Monde des Teintures Naturelles*, Berlin, Paris.

Dixon, A.R., McMillen, H. and Etkin, N.L., 1999. The transformation of Noni: A traditional Polynesian medicine (*Morinda citrifolia*, Rubaiceae). *Economics and Botany*, 53(1): 51–68.

Drury, C.H., 2007. *Ayurvedic Useful Plants of India*, 2nd edn. Asiatic Publishing House, Delhi.

Elkins, R., 1998. *Hawaiian Noni (Morinda citrifolia): Pleasant Grove*. Woodland Publishing, UT.

Gamble, J.S. and Fischer, C.E.C., 1921. Pavetta. In: *Flora of the Presidency of Madras*, Part IV, London, pp. 632–634.

Hooker, J.D., 1880. Pavetta. In: *Flora of British India*, Vol. 3. L. Reeve and Co., London, pp. 149–153.

Howard, R.A., 1989. *Flora of the Lesser Antilles, Leeward and Windward Islands: Dicotyledoneae*, Part 3. Vol. 6. Arnold Arboretum, Harvard University, Jamaica Plain, MA, 658 p.

http://culturesheet.org/rubiaceae:morinda.

http://www.catalogueoflife.org/search_results.

http://zipcodezoo.com/Plants/names/Plants–SN–M–0111.html.

Johansson, J.T., 1994. The genus Morinda (Morindae, Rubiodeae, Rubiaceae) in New Caledonia: taxonomy and phylogeny. *Opera Botanica*, 122:1–67.

Justin, A. 1990. *The Nicobarese* (The ASI Andaman and Nicobar Island Tribe Series). Seagull Books, Calcutta.

Liogier, H.A., 1997. Descriptive flora of Puerto Rico and adjacent islands. Vol. 5. Editorial de la Universidad de Puerto Rico, San Juan, PR, 436 p.

Little, E.L., Jr. and Wadsworth, F.H., 1964. *Common Trees of Puerto Rico and the Virgin Islands.* Agriculture Handbook 249. U.S. Department of Agriculture, Forest Service, Washington, DC, 548 p.

Mabberley, D.J., 1997. *The Plant Book : A Portable Dictionary of the Vascular Plants*, 2nd edn. Cambridge University Press, Cambridge, U.K.

McClatchey, W., 2002. From polynesian healers to health food stores: changing perspectives of *Morinda citrifolia* (Rubiaceae). *Integral Cancer Therapy*, 1: 110–120.

Morton, J.F., 1992. The ocean-going Noni, or Indian mulberry (*Morinda citrifolia*, Rubiaceae) and some of its "colourful" relatives. *Ecological Botony*, 46(1992): 241–256.

Nelson, G., 1996. *The Shrubs and Woody Vines of Florida*. Pineapple Press, Sarasota, FL, 391 p.

Nelson, S.C., 2001. Noni cultivation in Hawaii. *Fruit and Nuts*, 4: 1–4.

Nelson, S.C., 2006. *Morinda citrifolia* (Noni). In: *Species Profiles for Pacific Island Agroforestry*, www.traditionaltree.org (http://www.agroforestry.net/tti/Morinda–noni.pdf).

Peter, P.I., 2007. *Monograph on Noni (Morinda citrifolia L.).* WNRF., Reliance Printers, Chennai, India, pp 852.

Ross, I.A., 2001. *Medicinal Plants of the World: Chemical Constituents, Traditional and Modern Medical Uses.* Humana Press, New Jersey.

Royle, J.F., 1970. *Illustrations of the Botany and Other Branches of the Natural History of the Himalayan Mountains and of the Flora of Cashmere,* Vol. 1. Today and Tomorrow's Printers and Publishers, New Delhi.

Saleh, Abu, Kamnath, Manju and Peter, P.I., 2007. Indian Noni (*Morinda citrifolia*): A unique fruit crop with high medicinal value. In: *Underutilized and Underexploited Horticultural Crops, Vol. 1,* (Ed.) K.V. Peter. New India Publishing Agency, New Delhi, p. 297–300.

Singh, D.R., Medhi, R.P., Manju, S. and D'Souza, Alvira, 2006. Seed Germination studies on *Morinda citrifolia. Intl. J. Noni Res.,* 1(2): 23–28.

Singh, D.R., 2006. Commercial production of Noni possible in A&N Islands. *UTS Voice,* 1–15: 6.

Singh, D.R., 2007. Underutilized fruit crop of Andaman and Nicobar Islands. In: *Underutilized and Underexploited Horticultural Crops,* (Ed.) K.V. Peter. New India Publishing Agency, New Delhi, pp. 11–54.

Singh, D.R., George, Z., Manju, S. and Srivastava, R.C., 2009. Genetic Diversity of *Morinda citrifolia*: Genus, species and its cultivars. In: *National Symposium on Recent Advances in Biodiversity of Indian Subcontinent,* 11–13th March, at ZSI, Port Blair.

Singh, D.R. and Sundar, Jai, 2007. Peptide and mineral profile of *Morinda citrifolia* leaves and fruits (Abstract). In: *Proceedings of 2nd National Symposium on Noni Research,* October 27– 28, pp. 34.

Singh, D.R., Medhi, R.P., Manju, S. and Sikdar, P.K., 2006. Floral characters of *Morinda citrifolia* in Bay Islands. *International. J. Noni. Res.,* 1(2): 1–3.

Singh, D.R., Medhi, R.P., Manju, S. and D'Souza, Alvira, 2006. Morphological and compositional studies of *Morinda citrifolia* grown in Andaman and Nicobar Islands. *International J. Noni Res.,* 1(2): 14–17.

Singh, D.R., Medhi, R.P., Manju, S. and D'Souza, Alvira, 2006. Seed germination studies in *Morinda citrifolia. International J. Noni Res.,* 1(2): 23–28.

Singh, D.R. and Rai, R.B., 2005. Effect of growth regulators in rooting of stem cuttings of *Morinda citrifolia* var. *citrifolia* in Bay Islands. *International. J. Noni. Res.,* 1: 17–22.

Singh, D.R. and Rai, R.B., 2007. *Morinda citrifolia*: An important fruit tree of Andaman and Nicobar islands. *Nat. Pro. Rad.,* Vol. 6(1), National Inst.of Science. Comm. and Inform. Reso. CSIR., New Delhi, pp. 62–65.

Singh, D.R., Rai, R.B. and Singh, L.B., 2005. Great Morinda: A potential underutilized fruit grown in Bay Islands. *The Daily Telegrams,* Port Blair, pp. 2.

Singh, D.R., Rai, R.B. and Singh, B., 2005. The Great Morinda: A potential underutilized fruit for tsunami affected areas in Bay Islands. *The UTS' Voice,* April 16–30, Port Blair, pp. 21.

Singh, D.R., Srivastava, A. and Srivastava, R.C., 2008. Genetic diversity and analysis of *Morinda citrifolia* accession across Andaman and Nicobar using RAPD primer and marker. *Noni Search: Third National Symposium*, 18–19 October, pp. 57.

Singh, D.R., Srivastava, A., Srivastava, A.K. and Srivastava, R.C., 2008. *Morinda citrifolia L.* (Noni): A review of the scientific valediction for its nutritional and therapeutic properties. *Noni Search: Third National Symposium*, 18–19 October, pp. 79.

Singh, D.R. and Srivastava, R.C., 2007. Studies on growth and development stages of *Morinda citrifolia.* In: *2nd National Symposium on Noni Search*, October 27– 28, pp. 53.

Singh, D.R. and Srivastava, R.C., 2008. *The Noni*. Central Agricultural Research Institute, Port Blair, pp. 1–40.

Singh, D.R., Srivastava, R.C. and Subhash, C., 2008. Noni *Morinda citrifolia* L.: A review of agricultural research. *Noni Search: Third National Symposium*, 18–19 October, pp. 81.

Singh, D.R., Srivastava, R.C., George, Z. and Subhash, C., 2008. Development of protocol for packaging and shipment of *Morinda citrifolia. Noni Search: Third National Symposium*, 18–19 October, pp. 68.

Singh, D.R., Srivastava, R.C., Raja, R. and Manju, S., 2008. Salt tolerance of *Morinda citrifolia* by using Sodium chloride as irrigation water during seedling growth. *Noni Search: Third National Symposium*, 18–19 October, pp. 81.

Singh, D.R., Srivastava, R.C., Srivastava, A.K. and Manju, S., 2008. Estimation of physicochemical and post harvest losses of *Morinda citrifolia L* accessions grown in South Andaman. *Noni Search: Third National Symposium*, 18–19 October, pp 88.

Singh, D.R., Srivastava, R.C., Subhash, C. and Srivastava, A.K., 2006. *Morinda citrifolia*: An evergreen plant for diversification in commercial horticulture. In: *1st National Symposium on Noni Search*, October 7–8, pp. 9–27.

Singh, D.R., Srivastava, R.C., Subhash, C. and Srivastava, A.K., 2007. *Morinda citrifolia*: An evergreen tool for diversification in commercial horticulture *Monograph on Noni*. WNRF, Chennai, pp. 18–33.

Singh, D.R., Sudha, R., Pandey, V.B., Jayakumar, V., Srivastava, R.C. and Damodaran, T., 2007. Influence of Noni (*Morinda citrifolia* L.) herbal extracts on growth and production of chillies, *Capsicum annum* var. Arka Lohit. *2nd National Symposium on Noni Search,* October 27– 28, pp. 39.

Singh, D.R., Srivastava, R.C., Chand, S. and George, Z., 2009. Healing properties of *Morinda citrifolia*: ITK of Nicobarese. In: *Proceedings of the National Workshop-cum-Seminar on Status and Future Strategies of Hort. Dev. in A&N Islands.* 23rd to 25th January, CARI, Port Blair.

Singh, D.R., Srivastava, R.C., Sudha, R., Banu, Saira and Singh, L.B., 2009. Underutilized fruits: An approach towards nutritional security. In: *Proceedings of the National Workshop-cum-Seminar on Status and Future Strategies of Hort. Dev. in A&N Islands.* 23rd to 25th January, CARI, Port Blair.

Singh, D.R., Srivastava, A.K., Srivastava, A. and Srivastava, R.C., 2009. Ethno Medicinal Plants of Andaman and Nicobar Islands: Current Status and Future. In: *Proceedings of the National Workshop-cum-Seminar on Status and Future Strategies of Hort. Dev. in A&N Islands.* 23rd to 25th January, CARI, Port Blair.

Singh, D.R., Srivastava, R.C., Singh, S., Saira Banu, Sudha, R. and Manju, S., 2009. Tapping Biodiversity of Medicinal Plants for Livelihood Improvement in Islands. In: *Proceedings of the National Workshop-cum-Seminar on Status and Future Strategies of Hort. Dev. in A&N Islands.* 23rd to 25th January, CARI, Port Blair.

Singh, R.S. and Singh, D.R., 2006. Agro-climatic requirements and potential for Noni (*Morinda citrifolia* L.) cultivation in Andaman and Nicobar Islands of India. In: *Proceedings of 1st Symposium on Noni Search*, October 7–8.

Surendiran, G. and Mathivanan, N., 2006. An overview of the potentials of *Morinda citrifolia.* In: *First National Symposium on Noni Search*, WNRF, Hyderabad, 7–8th October.

Waki, J., Okpul, T. and Komolong, M.K., 2007. A descriptor list for morphological characterisation of Noni (*morinda citrifolia* L). *The South Pacific Journal of Natural Science*, 10: 61–66.

2011, Biodiversity in Horticultural Crops Vol. 3 *Pages* 229–250
Editor: Professor K.V. Peter
Published by: DAYA PUBLISHING HOUSE, NEW DELHI

Chapter 9

Pomegranate (*Punica granatum* L.)

*V.T. Ram Chandra, P. Kumar Jadhav and K. Dhinesh Babu**
National Research Centre on Pomegranate
NH-9 By Pass Road, Shelgi, Solapur – 413 006, Maharashtra
**E-mail: dh_babu@yahoo.co.in*

Pomegranate (*Punica granatum*. L.) is a favourite dessert fruit of tropical and sub tropical regions (Jalikop 2003). It is capable of growing in a variety of climates ranging from tropical to temperate and as a result, a large number of diverse forms have evolved over the years in nature (Levin 1995). Pomegranate is becoming an export oriented crop for the last one decade and its area and production are increasing with a faster pace. Its demand for internal consumption and export is a driving force for promotion of its cultivation in the country. Consequently, Maharashtra is a leading state for production of pomegranate which is contributing more than 80 per cent of total production. At present, major produce is consumed as fresh fruit in India and hardly 2 per cent of the produce is processed and 1 per cent exported which is very low compared to other pomegranate growing countries like Spain, Israel, Iran and Egypt. (NRCP. 2007). Hostile agro-climate of arid ecosystem makes fruit culture a difficult enterprise, but even under such environment, pomegranate has proved to be a successful fruit crop. The built in capacity to withstand heat, drought and salinity, besides high profitability have made its cultivation increasingly attractive. The unique plasticity of this fruit crop is evident from the threshold limits, it exhibits for higher (44° C) and lower (-12 °C) temperatures (Westwood 1978). The plant habit, fruiting and flowering physiology are, however, altered with the habitat (Pareek and Sharma 1993). In the sub-temperate regions of the country (Western Himalaya), seedling types of trees locally known as *Daru* come up naturally in abundance. *Daru* genotype is very common and gregarious in gravel and boulder deposits of dry ravines in the outer Himalaya. These hardy deciduous seedling trees growing since long time possess better climatic adaptability and resistance to pests and diseases owing to natural selection (Sharma and Sharma 1990). The tropical varieties are evergreen and produce sweet fruits mainly consumed fresh, but the fruits of *Daru,* though highly acidic, are also commercially important as they are extensively used in the preparation of *anardana,* an acidulant condiment product (Jalikop *et al.,* 2005).

The pomegranate is cultivated around the Mediterranean region for several millennia. In Georgia, to the east of the Black Sea, there are wild pomegranate groves outside of ancient abandoned settlements. The ancient city of Granada in Spain was renamed after the fruit during the Moorish period. It is also extensively grown in South China and in Southeast Asia, whether originally spread along the route of the Silk Road or brought by sea traders. The pomegranate was introduced by Spanish colonists into the Caribbean and Latin America.

The genus name *Punica* is named for the Phoenicians, who were active in broadening its cultivation, partly for religious reasons; consequently in classical Latin, the fruit's name was *malum punicum* or *malum granatum*, where "malum" was broadly applied to many apple-like fruits. The name "pomegranate" derives from Latin *pomum* (apple) and *granatus* (seeded). This has influenced the common name for pomegranate in many languages (*e.g.* German: *Granatapfel*, seeded apple; French: grenade *f;* Greek: (rodi); Hindi: (anar); Italian: melagrana *f;* Turkish: nar; Spanish: granada).

The fruit has wide consumer preference for its attractive juicy, sweet-acidic and refreshing arils and there is a growing demand for good quality fruits both for fresh use and processing into juice, syrup and wine. Since time immemorial, it is being used as a symbol of fertility because of innumerable seeds. It is a food medicine of great importance. All parts of the tree, roots, bark, leaves, flower, rind and seeds are used for medicine. Since ancient time, it is used in folk medicine in the Middle East, Iran and India as medication for antiviral, antifungal and antibacterial activity. The *Charaka*, the great physician of ancient India, prescribed a large number of formulations using different parts of pomegranate in the treatment of diseases like dyspepsia, stomach ache, dysentery, diarrhoea, tapeworm, inflammations, bronchitis, cardiac disorders *etc.* Presently, scientists are rediscovering its immense medicinal potentials to cure several diseases.

Origin and Distribution

The pomegranate is one of the oldest known edible fruits. This fruit crop is known domesticated in the Middle East more than 5000 years ago (Adsule and Patil 2005). This fruit tree is one of the species mentioned in the Bible and the Koran and is often associated to fertility. It is native to Iran (Persia) and perhaps some surrounding areas. It was cultivated in ancient Egypt and early in Greece and Italy. The fruit was very popular in Iraq. In time, it spread into Asia (Turkmenistan, Afghanistan, India, China, etc.), North Africa and Mediterranean Europe. The domestication process took place independently in various regions and not only in the Mediterranean region (Evreinoff 1949, Zukovskij 1950, Melgarejo and Martínez 1992). In ancient Egypt, the pomegranate received the name " *Arhumani*". The Romans first called this species "malum punicum" (Punic apple or apple of Carthage) that evolved to "punicum granatum" and C. Von Linne, finally, gave the name *Punica granatum*. It is considered an excellent tree for growing in arid zones for its resistance to drought conditions. It is now widely cultivated in Mediterranean, in tropical and subtropical areas. It can be encountered as regular plantations in Cyprus, Egypt, Morocco, Spain, Tunisia and Turkey. A high number of scattered trees on the borders or within other fruit orchards are reported in many Mediterranean countries where the fruit is very popular in local markets. It is cultivated in Central Asia and to some extent in the USA (California), Russia, China and Japan for fruit production and is also developed as an ornamental tree in East Asia (Mars 1996, Tous and Ferguson 1996). Earlier pomegranate was considered a minor fruit in India and its regular cultivation started after 1986 on commercial scale. Now, Maharashtra has become the main pomegranate producing state in India followed by Karnataka and Andhra Pradesh.

Botany, Cytogenetics and Description

Scientific Classification of Pomegranate

Kingdom	:	Plantae
Division	:	Magnoliophyta
Class	:	Magnoliopsida
Sub-class	:	Rosidae
Order	:	Myrtales
Family	:	Lythraceae
Subfamily	:	Punicoideae
Genus	:	*Punica*

The genus *Punica* belongs to family Lythraceae, which was previously placed in family Punicaceae. Recent phylogenic studies show that Punica belongs to family Lythraceae and is classified in that family by the Angiosperm phylogeny group (Graham *et al.*1998). Its better known species is the pomegranate (*Punica granatum* L.). The only other species in the genus, the Socotra pomegranate (*Punica protopunica*), is endemic on the island of Socotra (Yemen). It differs in having pink flowers and smaller and less sweet fruit. The species *granatum* has two subspecies *viz.* Chlorocarpa and Prophyrocarpa. Dwarf pomegranate (*P. granatum* L. var. *nana*) is a miniature variety of the pomegranate, the most suitable for pot plant production. In out door plantation, it reaches a maximum height of 2 m (Beach 1980). Further, Levin (1985) also reported dwarf fruiting form of pomegranate. Some authors classified the ornamental dwarf pomegranate as a distinct species, *Punica nana* (Melgarejo and Martínez 1992). According to Smith (1976), *P. granatum* has 2n = 2x =16, 18 chromosomes. The somatic chromosome number of cvs. Dholka, Ganesh, Kandhari, Muscat White and Patiala was found 2n= 16, while the cultivar Double Flower has 2n =18 (Nath and Randhawa 1959d). The chromosome number in Vellodu and Kashmiri varieties was 2n = 18 with 1 or 2 quadrivalent associations at meiosis (Raman *et al.*, 1963). The former is found in Transcaucasia region while the latter in Central Asia (Patil and Karale 1985).

The pomegranate is a neat, rounded shrub or small tree that grows from 6 to 10m., but more typically from 4 to 5.25m. in height. Dwarf varieties are also reported. It is usually deciduous, but in certain areas (Tropical climate) the leaves persist on the tree. The trunk is covered by a red-brown bark which later becomes gray. The branches are stiff, angular and often spiny. The plant has a strong tendency to sucker from the base and considered long-lived. There are specimens in Europe known to be over 200 years of age. The vigour of a pomegranate, however, declines after about 15 years. The pomegranate has glossy, leathery leaves which are narrow and lance-shaped.

The attractive scarlet, white or variegated flowers are over 3cm across and have 5 to 8 crumpled petals and a red, fleshy, tubular calyx which persists on the fruit. The flowers may be solitary or grouped in twos and threes at the end of branches. It is self-pollinated as well as cross-pollinated by insects. Cross-pollination increases the fruit set. However, wind pollination is insignificant.

Flowering time varies according to geographical situation and cultivars. In tropical climate, pomegranate flowers almost throughout the year whereas in subtropics, it flowers once a year. In areas of low winter temperature, the tree is deciduous, but in tropical conditions, it is evergreen or partially deciduous. Under tropical condition, flower bud differentiation takes place at varied times.

Time span between start of flower bud elongation and anthesis varies between 14 to 28 days depending on the variety and climatic conditions (Gur 1986). Thus, in subtropical climates of the northern hemisphere, flowering occurs from the last week of March till the second week of May (Singh *et al.*, 1978, Fouad *et al.*, 1979). Several distinct flushes of flowering on the same tree are also seen commonly.

In subtropical central and western India, there are 3 distinct seasons of flowering. In major pomegranate areas of deccan plateau, the flowering is regulated in three distinct seasons *i.e. Ambe Bahar* (January-February flowering), *Mrig Bahar* (June - July flowering) and *Hasth Bahar* (September–October flowering). *Ambe Bahar* is the most commonly preferred by the growers because of high yield as compared to other flowering seasons (Singh *et al.*, 1967, Prasanna Kumar 1998). In Karnataka, flowering was observed in June-August for 80-87 days but flowering in March and September extended over 22 to 30 days only (Nalawadi *et al.*, 1973). In Punjab, only one flowering season is observed from April–June (Josan *et al.*, 1979a). Under Delhi conditions, depending on cultivar, the flowering may occur once a year or twice (Nath and Randhawa 1959d). In temperate climate of Himachal Pradesh, the flowering takes place during middle of April (Parmar and Kaushal 1982.). In Bihar, flowering occurs twice during February- March and July-August. The flowers are short peduncled or sub-sessile. Hermaphrodite and male flowers, as well as intermediate forms occur on the same tree. The calyx of the hermaphrodites is urceolate (pitcher-like) with a broad, well developed ovary. Male flowers are smaller with a companulate (bell- shaped) calyx and a rudimentary ovary. The intermediate forms exhibit various degrees of ovary degeneration. Fruits arising from such flowers drop early, or if they reach maturity they are mis-shaped (Nath and Randhawa 1959c). Heterostyly is common in pomegranate flowers. Hermaphrodite flowers are usually homostylous or pin eyed (*i.e.* the stigmas are on the same level or higher than the anthers) and male flowers are thrum eyed (i.e the stigmas are beneath the level of the anthers). However, many exceptions to this rule apparently exist (Nath and Randhawa 1959a, Josan *et al.*, 1979a,Gur 1986). The per cent of hermaphrodites out of the total number of flowers on a pomegranate tree depends on the cultivar, the flowering season and other unknown environmental factors. In the beginning of the main flowering season, this per cent is higher than at the end of the season (Gur 1986). In several cultivars, 25 to 60 per cent hermaphrodite flowers, 20 to 47 per cent male flowers and 14 to 24 per cent intermediate forms were reported in India but their ratio varied according to cultivars (Nath and Randhawa 1959a, Nalawadi *et al.*, 1973, Singh *et al.*, 1978). Fruit yield, however, depends on number of hermaphrodite flowers as positive correlation was found between the bearing capacity and the per cent of perfect flowers (El Sese 1988, Chaudhari and Desai 1993).

Pomegranate has a very specific flowering habit that the flower buds do not open at night. Maximal opening rates were between 11 am and 2 pm, through the exact time depends on climate, flowering period and cultivars (Nath and Randhawa 1959b, Nalawadi *et al.*, 1973, Josan *et al.*, 1979b, Game 1987). Like other fruit crops, an optimum temperature for the opening of flower buds is required in pomegranate that varies with cultivars. In "Muscat White", the optimum temperature was between 37 to 38°C (Gur 1986).

Dehiscence of anther starts a few hours prior to anthesis or somewhat later depending on the cultivars. It is completed 7 to 15 hours later or may take 24 hours for completion (Gur 1986). The exact time span depends on the cultivar and climatic conditions. High atmospheric humidity and temperature may hasten dehiscence to a certain degree (Nath and Randhawa 1959d, Josan *et al.*, 1979b). Number of pollen grains/anther vary between 16,000 to 23, 000 (Nath and Randhawa 1959c). On an average, 27,00,000 pollen grains were counted in a flower (Rao and Abdul Khader 1962). Pollens stored at 9° C remain viable for 92 days and germinate better than those kept at room temperature *i.e.* 25° C (Saoji and Rudra 1991). Stigmas are receptive a day before anthesis. The peak receptivity coincides with time of anthesis. Later the receptivity, as determined by pollination test, declines gradually and ceases altogether at the end of the 3rd day (Nath and Randhawa 1959b, Nalawadi *et al.*, 1973) or 5th day (Gur 1986). Fruit retention increased from about 30 per cent during early flowering to about 80 per cent during full bloom and/or late flowering (El Sese, 1988, Hussein *et al.*, 1994). Flowers produced 4 to 5 weeks after the onset of blooming gave the highest fruit set (90 per cent) with the best fruit quality.

The pomegranate is described variously as self-pollinated, self and cross-pollinated, highly cross-pollinated or often cross-pollinated. The use of a marker gene, governing the pigmentation of bud flower and petiole base, confirmed the fact that pomegranate ("Ganesh" and "Kabul Yellow") is self and, in low proportion (13 per cent), cross-pollinated (Jalikop and Sampath Kumar 1990). The trials conducted with several cultivars showed that fruit set was 79 per cent and 43.3 per cent, respectively for intact open and self pollinated flowers or 26.4 per cent and 66.2 per cent for the same after emasculation (Karale *et al.*, 1993).

The fruit is a fleshy berry. The nearly round, 7.5 to 15cm wide fruit is crowned at the base by the prominent calyx. The tough, leathery skin or rind is typically yellow overlaid with light or deep pink or rich red. The interior is separated by membranous walls and white, spongy, bitter tissue into compartments packed with sacs filled with sweetly acid, juicy, red, pink or whitish pulp or aril. In each sac,there is one angular, soft or hard seed. High temperatures are essential during the fruiting period to get the best flavor. The pomegranate may begin to bear in 1 year after planting, but 2.5 to 3 years are more common. Under suitable conditions, the fruits mature in 5 to 7 months after bloom. The fruit is non-climacteric and needs be harvested when reached at full maturity.

Not much work is done on the inheritance pattern of pomegranate. Manohar *et al.* (1981) reported that economic characters like rind weight, acidity per cent, fruit weight, arils/fruit, yield/tree and fruits/tree exhibit high heritability and high genetic advance. Thus, satisfactory benefits can be derived if these characters are used for selecting individual with better yield potential. Use of a soft seeded cultivar as male parent slightly decreased seed hardiness in hard seeded cultivars (Purohit 1987). Other selection criteria which may affect yield include improved plant architecture, photosynthesis and insect and disease resistance. Yield is not controlled by a single gene. It is considered a polygenic trait. Desai *et al.* (1994) observed that desired characteristics of bigger fruits, higher aril, juice, total soluble solids and vitamin C contents, coloured arils and soft seeds were favourably associated with each other. Jalikop and Kumar (1990) noticed that the dominant gene (R) present in "Ganesh" confers red and the recessive gene (W) present in "Kabul Yellow" imparts yellow pigmentation of the petiole base, leaf margin, flower bud and fruit rind. Recently, comparison of soft, semi-soft and hard-seeded types showed that for many fruit attributes, soft and semi-soft groups were similar where as hard seeded was a class apart, having significantly higher fruit weight and volume than the former two groups, but they possess as much aril as soft and semi-soft seeded types with significantly low juice per cent. Despite smaller fruit size, soft-type can be an appropriate parent in developing varieties for juice production. Hybridization within soft types was expected to generate restricted variability as revealed by low co-efficient of variation in this group. The hard seeded types, on the other hand, could be good donors for improving some fruit attributes (Jalikop and Kumar 1998).

The description/evaluation of pomegranate germplasm is always performed using self-devised list of descriptors with special emphasis on pomological (physico-chemical) characters (Mars and Sayadi, 1992, Melgarejo and Martínez 1992, Mars and Marrakchi, 1998a) and sometimes on technological properties (Al Kahtani 1992). Recently, a descriptor for pomegranate was prepared in the framework of the "CIHEAM Collaborative Working Group on Underutilized Fruit Crops" (Mars *et al.*, 1997). Melgarejo *et al.* (1995) demonstrated the intervarietal differences in fatty acid composition. These differences correlate well with the sweetness of the juice. The sour varieties contained the highest amounts of myristic and linolenic acid. Results also showed that fatty acid composition of Mediterranean pomegranates can differ from those of Oriental varieties. Ben Nasr *et al.* (1996) classified 8 Tunisian pomegranate cultivars into 4 groups using the pro-anthocyanidins analysis. Great attention must be paid while using morphological or physiological characters since some of them may be strongly influenced by the environment like fruit size, rind and juice colour, ripening date, sugars and acidity in the juice, etc. (Shulman *et al.*1984, El Kassas *et al.*, 1992, Ben Nasr *et al.*1996). Varieties are often classified as sweet, sweet-sour and sour, early, mid season and late, juicy and table fruit, soft-seeded and hard-seeded or major and minor. The names originate frequently either from the place of cultivation or from colour of fruit (Mars and Marrakchi 1998a). Despite the large number of local varieties, in almost all countries, very a few are commercially utilised and propagated in nurseries as in Spain (Llacer *et al.*, 1995), Tunisia (Mars 1995) or Turkey (Aksoy 1995).

Methods of Propagation

Pomegranate seeds germinate easily and are considered the oldest method of propagation. In general, 60-75 per cent seed germination is observed in most of the cultivated varieties. In dwarf pomegranate (*P. granatum* L. cv. nana), seed germination is very low (Cervelli and Belleth 1994, Jalikop 2007) and poor germination was due to presence of a water soluble inhibitor in seeds. Seedlings show high degree of variability, thus, propagation by seeds is discouraged for commercial cultivation.

Among vegetative methods, hardwood, semi-hardwood and softwood stem cuttings were used for propagation of pomegranate. In "Ganesh", IBA (5000ppm) with quick dip method (one minute dip) was optimum for better rooting (73.3 per cent), field survival and growth (Panwar *et al.*, 2001). Ghosh *et al.* (1988) compared the performance of IBA and NAA using hardwood, semi hardwood and soft wood cuttings. They reported that IBA was better for inducing roots in stem cuttings and IBA (5000ppm) gave very high rooting success (83.33 per cent). As far as rooting is concerned, hardwood cuttings were better than semi-hard and soft wood cuttings. Further, Panda and Das (1990) noted higher rooting success (76.10 per cent) with hardwood cuttings using IBA (5000ppm) in quick dip method under intermittent mist condition. They also reported that IBA was better than IAA for induction of rooting. With lower concentration of IBA (100ppm) in prolonged dip method (24 hrs), hardwood cuttings of 20cm length gave high rooting success under Punjab condition (Sandhu *et al.*, 1991). Reddy and Reddy (1989) treated hardwood cuttings (25-30cm) with IBA (2500ppm) + Paclobutrazol (2500ppm) and found higher rooting success. In a separate study, they also found that hardwood cuttings (25-30 cm) with IBA (2500ppm) and NAA (2500ppm) after basal wounding gave higher rooting success in pomegranate (Reddy and Reddy 1990). Use of p-hydroxybenzoic acid (PHB) at 1000ppm + 2500ppm NAA gave higher rooting success when the shoots were ringed and etiolated 15 days prior to making cuttings (Hore and Sen, 1993). Recently, Tripathi and Shuka (2004) also found better rooting and survival in hardwood cuttings using PHB (1000ppm) + IBA (5000ppm).

It was observed that cutting thickness also influenced the rooting success. Dhillon and Sharma (2002) found that cutting thickness of 0.8–1.2 cm was the most appropriate for getting high rooting success. They recorded November as the most suitable time for raising cuttings under Punjab condition. Position of cuttings (basal, sub-apical and apical) responded differently when used for raising of stem cuttings. Basal cuttings (1.0–1.25 cm thickness) treated with IBA (5000ppm) gave higher rooting success than other types of cuttings taken. It was recorded that high C:N ratio and carbohydrate reserves of the cuttings was responsible for higher rooting success in basal cuttings (Purohit and Shekharappa 1985). Rooting media also play vital role in enhancing the rooting success of stem cuttings. Hardwood cuttings gave better rooting success when planted in river silt (Baghel and Saraswat 1989, Deol and Uppal, 1990).

In Maharashtra, mostly layered plants are used for pomegranate cultivation.Very high rooting success (94 per cent) with layering method using IBA (1500ppm) was recorded when layering done between June and August (Hegde and Sulikeri 1989, Hore and Sen, 1994). The most appropriate ratio of soil, sand and FYM as rooting media was 2:1:1. It was observed that ferulic acid (1000ppm) + NAA (10,000) was the most effective for rooting in layers. Further, it was registered that ferulic acid(1000ppm) and IBA(5000ppm) induced about 100 per cent rooting in layers (Hore and Sen, 1995).

Propagation of pomegranate is also possible by root cuttings. Root cuttings taken from young trees (<4 yrs old) gave better result particularly when planted in February (Mendilcioglu 1968). Except, stem cutting, layering and root cutting methods, other propagation methods like budding and grafting have not been tried in pomegranate. Kar *et al.* (1989) explored the possibility for top working in wild pomegranate by budding and grafting methods. They reported that May, June and July could be optimum time for top working. Side veneer grafting gave 100 per cent success when grafting was done between July 1 and 15.

For regeneration of pomegranate through tissue culture techniques, works were initiated in India and aboard. Omura *et al.* (1987a) could achieve organogenesis using leaf explants. They also tried to induce tetraploidy with colchicine treatment during the course of *in vitro* culture of dwarf pomegranate cultivars (Omura *et al.,* 1987b). Later on, Mahishni *et al.* (1991) were successful in culturing shoot-tip explants on supplemented MS medium and then transferring them to Lloyd and Mc Crown woody plant medium. They achieved 80 per cent success in establishment of plantlets in 1:1:1(v/v) peat, perlite and sand mixture. Kanatharajah *et al.* (1998) tried to standardize tissue culture techniques. Little attempt appears to have been made on using isozymes or molecular (genetic) markers to identify or measure genetic variability among existing genotypes or hybrids resulted due to innumerable crosses. Further, somatic embryos were induced from petal explants on MS medium supplemented with various plant growth regulators and nutrients (Nataraja and Neelambika 1996).

Biodiversity

Some parts of the Mediterranean area are considered as native lands of pomegranate. Almost all of the varieties in the region are of local type selected by unknown persons and maintained by vegetative propagation. Authors from different countries listed many local cultivars/varieties (Levin, 1995a, Ozguven, 1996, Mars, 1996, Mars and Marrakchi 1998a). The locally grown material may be considered as the pomegranate primary gene pool. It grows wild in the Near east, Transcaucasia, Dagestan and also in Asia Minor. In these regions, hybridization between cultivated and wild forms is, probably, still taking place (Zukovskij 1950). Wild forms (populations) would be the secondary gene pool. The tertiary pomegranate gene pool would consist of forms of the pomegranate wild relative (*P. protopunica*). It was presumed that *P. protopunica* played a part in the origin of the cultivated pomegranate. Zukovskij (1950) affirmed that the Socotran pomegranate has not played any part in the origin of the cultivated one.

Genetic studies are rare or lacking entirely because pomegranate was not subject of much scientific investigation in the past. Some studies based on morphometric criteria are performed to determine the degree of polymorphism within local material (Mars and Marrakchi 1998b). Results show a considerable phenotypic (presumably genetic) diversity among accessions. High degree of cross pollination exists in pomegranate. This leads to heterozygosity among cultivars. At present over 70 cultivars, both from indigenous or exotic sources, are available in India. Mars and Marrakchi (1998b) reported morphometric studies on 63 pomegranate accessions in Tunisia as part of the variety evaluation process. Since majority of them are of seedling origin, they provide wide variability with respect to shape and size of fruit, mellowness of seeds, aril colour, rind colour, sweetness and sensitivity of fruits to bursting or cracking (Keskar *et al.*, 1993, Mars and Marrakchi 1998b). Presence of a wide variation for organic acids content in pomegranate juice is recently reported from Portugal, Spain and Turkey and Melgarejo. Artes (2000) reviewed the differences in acidity levels reported by different workers in Egypt, Israel Saudi Arabia Yugoslavia, Indian, Russia and USA. Similarly, a great diversity in organic acid composition is detected in pomegranate juice. Citric and malic are the major acids in the pomegranate juice (Melgarejo and Artes 2000, Poyrazoglu *et al.*, 2002, Miguel *et al.*, 2004). Oxalic and tartaric acids dominated in pomegranate varieties analyzed by Melgarejo *et al.* (2004). Presence of small amounts of acetic, fumaric, lactic, quinic and succinic acids have also been reported by Melgarejo and Artes (2000) and Poyrazoglu *et al.* (2002). Further, according to Melgarejo and Artes (2000), sweet varieties sometimes contain more of malic than citric acid. In sour varieties on the other hand, citric acid is always the major acid. In India, wild pomegranate is found growing in Western Himalaya (Jammu and Kashmir, Himachal Pradesh and Uttarakhand). Recently, diversity in wild pomegranate is also reported from Himachal Pradesh by Singh and Singh (2006). Remarkable variability was observed among clones collected from the forest. High variability was found in respect to fruits/tree (215-769), weight (54.28-85.12 g), arils/fruit (48-165.7), TSS (17.5-20.5 °brix) and acidity (2.46-6.60 per cent). Similarly, variability in plant height, leaf and fruit size, stem, rind and aril colour, number of thorns and their size, TSS, acidity etc. are recorded in accessions collected from Uttarakhand (NRCP, 2007).

Uses

Nutrition and Other Utility

Pomegranate fruit is a source of essential nutrients and is easy to digest. It is a diet full of energy for children, elderly sick and pregnant ladies. It has the power to cure hunger, quench thirst and refresh the drunkard. Nutritive value depends on variety, climatic condition, cultural practice stage of fruit etc. The essential nutrients available in a pomegranate are given in Table 9.1.

The fruits are mainly used for direct consumption. A few processed products like juice, jams, jellies, syrups, carbonated beverages, wine etc. are also prepared from good and damaged fruits with cracks, cuts, or bruises in the rind. Wild pomegranate seeds are sometimes used as a spice, known as *anardana* most notably in Indian and Pakistani cuisine but also as a replacement for pomegranate syrup in Persian and Middle Eastern cuisine. The seeds are separated from the flesh, dried for 10-15 days and used as an acidic agent for chutney and curry production. The seeds may also be ground to avoid seeds becoming stuck in the teeth when eating dishes prepared with them. Seeds of the wild pomegranate *daru* from the Himalayas, are considered the highest quality source for this spice. And *anardana* is the main source of income in some parts of Jammu and Kashmir, Himachal Pradesh and Uttarakhand.

The arils are rich in sugars, vitamins, polysaccharides, polyphenols and minerals. They have low oil content and are rich in polyunsaturated (n-3) fatty acids. The pomegranate seed extract possesses high antioxidant activity. The antioxidant capacity of commercial pomegranate juice is three times higher than those of red wine and green tea. The antioxidant qualities of pomegranate juice make it appealing for the production of health supplements and nutraceuticals (Singh *et al.*, 2002).

Table 9.1: Nutritive Value of Pomegranate Fruit/100g Edible Portion

		Vitamins	
Energy	65-78 kcal	Thiamine (B1)	0.003 mg-0.06
Moisture	72-78 per cent	Riboflavin (B2)	0.012- 0.03 mg
Protein	0.05–1.6 g	Niacin (B3)	0.18-0.3 mg
Carbohydrates	15.4-19.6 g	Pantothenic acid	0.596 mg
Fiber	3.4- 5.1 g	Folic acid (B9)	0.03 mg
Ash	0.36-0.73 g	Ascorbic acid (c)	5.3-16 mg
Pectin	0.27g		
Minerals		**Sugars**	
Sulphur	12 mg	Total	14.5 g
Manganese	0.77	Glucose	5.46 g
Copper	0.34	Fructose	6.14
Chlorine	2 mg	Sucrose	Nil-0.1 per cent
Zinc	0.82 mg	**Acids**	
Iron	0.3-1.8 mg	Citric Acid	142-2317 mg
Sodium	0.9 mg	Malic Acid	135-176 mg
Potassium	133–259 mg	Oxalic acid	14 mg
Magnesium	4-44 mg	Boric acid, succinic acid and tartaric acid	Traces

In some countries, fruit is used for preparation of various food items. In Turkey, pomegranate (Turkish: *nar*) is used in a variety of ways, notably as pomegranate juice (Turkish: *nar eksisi*), used as a salad dressing to marinate meat, or simply to drink straight. The arils are also used in salads, in Muhammara (Turkish Walnut Garlic Spread) and in Gullac, a famous Turkish dessert. In Greece, it is used in many recipes; like *kollivozoumi*, a creamy broth made from boiled wheat, pomegranates and raisins; legume salad with wheat and pomegranate; traditional Middle Eastern lamb kabobs with pomegranate glaze; pomegranate eggplant relish; avocado and pomegranate dip; are just some of the dishes, it is used in culinary. Pomegranate is also made into a liquour and popular fruit confectionery that can be used as ice cream topping, or mixed with yogurt, and even spread as jams over toast for breakfast. *Grenadine*, a reduced juice from fresh pomegranate arils, is common in Northern India not only for desserts, but also to marinate meat. Due to its content of proteolytic enzymes, it acts as a meat tenderizer. Either as fresh-extracted juice or in the more durable form of syrup (*dibs ar-rumman*). *Anardana* is a common souring agent in Western Asia and may be used in the Turkish salad *kisir* made from precooked cracked wheat (*bulgur*), parsley and possibly raw vegetables. A similar product in Georgia is called *masharabi*. Pomegranate concentrate is particularly common in Iran, where it suits the local preference for spicy yet sweet-fruity flavours; a famous example of its use is *khoresht fessenjan*, duck or chicken pieces braised in a thick sauce made of pomegranate and walnuts. Dried pomegranate arils (*Anardana*) make an interesting alternative for raisins in cakes and other European sweets.

The fruit has a fairly long shelf-life at room temperature and hence was carried on long journyes through desert climates as a source of water and nourishment. Today, pomegranates make nice ornaments for fruit bowls or Christmas wreaths, as the fruits are marketed around the holiday season.

All parts of the tree are utilized as source of tannin for curing leather. The trunk bark contains 10–25 per cent tannin and was formerly important in the production of Morocco leather. The root bark has a 28 per cent tannin content, the leaves 11 per cent and fruit rind as much as 26 per cent. The latter is a by-product of the *Anardana* industry.

Both the rind and the flowers yield dyes for textile. The bright red juice was often used as an ink in ancient times. Ink can be made by steeping the leaves in vinegar. In Japan, an insecticide is derived from the bark. The pale yellow wood is very hard while available only in small dimensions, is used for walking sticks and in wood crafts.

Therapeutic Value

An age old proverb, '*Ek anar sau bimar*', clearly indicates the importance of pomegranate for a sick person. It is a miracle fruit full of health tonics and several other ingredients having potential to cure diseases. The benefit largely comes from the antioxidants in juice. It is an important source of phenolic compounds, with anthocyanins being one of the most important, especially the 3-glucosides and 3, 5-diglucosides of delphinidin, cyanidin, and pelargonidin (Du *et al.*, 1975). These components along with gallagyl-type tannins, ellagic acid derivatives and other hydrolysable tannins could contribute in some way to the antioxidant activity of pomegranate juice.

More recently Chinese researchers reported that pomegranate peel yielded double the amount of antioxidants than pulp and has more potential as a health supplement. The therapeutic properties of pomegranate are not only due to various antioxidants but perhaps the alkaloids and numerous trace elements present in the fruit also play a medicinal role. The elements cobalt, copper, manganese and zinc were detected in pomegranate. These elements are associated in several enzymes which play significant role in the metabolic processes. The other trace elements like molybdenum, vanadium, chromium, nickel, tin, selenium and silicon also occur in pomegranate, known essential for human beings. Although each of these nutrients may play protective role in the prevention of chronic diseases, their combined effect with other components in whole food may be greater due to additive and synergistic action. Such additive and synergistic effects are not available through supplementation with isolated or synthesized nutrients. It will, therefore, be prudent not to replace pomegranate fruit with dietary supplements of nutrients contained in it with the hope to derive the desired health benefits quickly. The various health benefits that we can derive from pomegranate fruit or its products are summarized here.

Tonic

Pomegranate fruit being source of energy, several important minerals, vitamins, sugars and acids is a tonic for sick as well as healthy people. It tones functions of all body parts *e.g.* skin, brain, digestive, respiratory, circulatory and excretory system. It increases body resistance against infectious diseases specially tuberculosis.

Blood Pressure and Heart Problems

The anthocyanins in red/purple fruit are most abundant in pomegranate. Darker the red colour, richer it is in anthocyanins. It reduces systolic blood pressure and hence, cures hypertension, which is the cause of heart and kidney failure. It is reported that inhibiting serum angiotension converting enzyme (ACE) present in the fruit has a power to reduce systolic blood pressure. Hardening of arteries (atherosclerosis), due to deposition of cholesterol and other substances in the arteries and hence, heart strokes, are reduced in people taking pomegranate fruit or juice regularly (Aviram *et al.*, 2004). Some authors reported that anthocyanins decreased cadmium accumulation in liver and kidney, the concentration of bilirubin and urea in blood serum, and aspartate aminotransferase and alanine aminotransferase activities (Gil *et al.*, 2000, Kowalczyk *et al.*, 2003). In several human clinical trials, the juice of the pomegranate is effective in reducing several heart risk factors, including LDL oxidation, macrophage oxidative status, and foam cell formation, all of which are steps in atherosclerosis and heart disease. Tannins are identified as the primary components responsible for the reduction of oxidative states which lead to these risk factors (Kaplen *et al.*, 2001). Extracts of bark, leaves and immature fruit rind are used as an astringent to halt blood hemorrhages.

Skin Diseases

Anthocyanins are anti-ageing. Punicalagin found exclusively in pomegranate breaks down to ellagic acid which is a highly potent chemical for skin care. It reduces freckles and bad effects of sun rays on skin. Inclusion of pomegranate juice in diet reduces bad effects of sun rays and is more effective than sunscreen creams and lotions.

Cancers

A miracle medicine for lungs and prostrate cancer, pomegranate juice not only prevents but even cures prostrate cancer (NGS 2006). In UK involving 93 men of proven prostate cancer showed considerable prevention of the disease and even reversed its progress. Researcher suggested that pomegranate may not only have cancer protective effect but also chemotherapeutic effects against prostate cancer in humans. Research done in Korea indicated that pomegranate juice and seed oil have chemo-preventive action against breast cancer (Horra 2003). It has prevented 47 per cent inhibition of cancer lesion formation. Seed oil found abundantly in pomegranate seeds, contains pucinic acid which reduces risk of skin cancer and breast cancer by 47 per cent in ladies (Kim 2002). Further, similar report was made in America indicated that pomegranate juice was a potential anticarcinogenic agent with preventive potentials against breast cancer.

Brain Problems

Brain injury in unborn babies (Alzhelmers disease) and premature births may be reduced, if pregnant ladies regularly use pomegranate (Halliday 2005). Hypoxia ischemia, a condition brought about by decreased blood flow and oxygen to the baby's brain, is linked to premature birth and other complications during pregnancy. It is indicated that the brain injury was reduced by more than 60 per cent with pomegranate juice, when tested on mice. It also refreshes overworked and tired brains.

Diabetes

Pomegranate juice is good for diabetics as it has no or negligible sucrose. Glucose and fructose present are attached to anthocyanins which reduce or keep diabetics under check (Esmaillzadeh *et al.*, 2004). Australian researchers found that the pomegranate flower extract improved hyperglycemia and obesity in type–II diabetic patients.

Digestive Problems

Flower and fruit extracts of pomegranate are used effectively since ancient times to cure digestive disorders like bilious vomiting appetite, indigestion, diarrhoea and dysentery. The extracts have antibacterial properties against *Escherichia coli* and *Salmonella typhi*. Decoction of bark is used for deworming. It is highly effective against tapeworms. Pomegranate showed good activity against *Staphylococcus aureus* and *Candid* sp. Decoction of pomegranate peel showed the best antibacterial activity *in vitro* against *Vibrio cholerae.*

Other Medicinal Uses

Pomegranate juice along with little saffron is beneficial in recovery from typhoid, gastric and asthamatic fevers. Paste of pomegranate seeds dissolves kidney and bladder stones. Juice of pomegranate has been found beneficial in curing leprosy. Dry pulverized flower buds are an excellent remedy for bronchitis. Decoction of flowers gargled relieves oral and throat inflammation. Powders and paste prepared from dry rind are good tooth medicines. These cure teeth and gum disorders *viz.* bleeding gums, pyorrhea etc. Many food and dietary supplement makers have found the advantages of using pomegranate extracts (which have no sugar, calories, or additives), instead of the juice, as healthy ingredients in their products. The juice can also be used as an antiseptic when applied to cuts. Pomegranate seed is the richest source of estrogen-the female hormone, hence, pomegranate juice is very helpful to ladies for menopause related problems.

Conservatories

Plant genetic resources are basic raw materials for improvement of any crop plant. A wider genetic diversity is fundamental for the development of new variety with good quality and higher yield (Frankel and Hawkes, 1975, Holden and Williams 1984). The plant diversity could be observed in land races, local selections, elite cultivars, wild relatives, obsolete farmer's named and old varieties, commercial and modern bred cultivars, special genetic stock as parental lines and genetic stock with known attributes. Although genetic base in pomegranate is low, but still variability is available in the form of seedlings, landraces, primitive cultivars etc. in nature owing to cross pollination and their high adaptability in diverse situations ranging from the tropical to temperate climates as reported by Jalikop and Kumar (1990) and Levin (1995b), which could be exploited scientifically for the crop improvement and commercial utility. For the preservation of local pomegranate germplasm, prospections are realized and collections are already established in several Mediterranean countries like Spain, Morocco, Tunisia, Greece, Turkey, Egypt, etc.(Mars 1996). Introduction of foreign material is limited. Gulick and Van Sloten (1984) reported that in India there are three principal collections containing, each one, at least 30 accessions and that a collection was established in the USA (Georgia) with 35 accessions. Levin (1995b) reported that 8 countries of the former USSR have collections of pomegranate germplasm. Azerbaijan, The Ukraine, Uzbekistan and Tajikistan have relatively large collections of 200-300 accessions. The collection of the Turkmenistan Experimental Station (TES) of Plant Genetic Resources was established in 1934 and is the largest in the world containing 1117 accessions and recently, it was enhanced to 1157 (Levin 1995b). Samples are from 26 countries on 4 continents, consisting of cultivars received from commercial firms and through interchange with other scientific institutions and material collected from wild populations and landraces (Levin 1994, 1995a). However, TES has a core collection which is 10 per cent of the size of the main collection (Levin 1996). For the unique wild relative of the pomegranate (*P. protopunica*), an expedition during 1989 located the species at 5 sites in the Socotra island (Yemen) and seeds were collected and successfully germinated (Guarino *et al.*, 1990). Collected material is maintained as living plants. According to Levin (1994) experiments have shown that crypreservation is suitable for storage of seed and pollen of pomegranate cultivars. Also, natural reserves are used in Transcaucasus and Central Asia for *in situ* conservation of wild pomegranate population. Interchange of plant material is not frequent. Guidelines for its safe movement are not available. But it seems not to be problematic and it depends on the general policies for genetic resources for each country. Pathological constraints which may impede the movement of germplasm are not reported. Recently the "Hop Stunt Viroid" (HSVd) was detected in pomegranate, which is not previously described as host of this pathogen (Astruc *et al.*, 1996). Nevertheless, Levin (1994) reported that since 1964, material from the collection of the TES has been sent to scientific institutes to form working collections. In India, at present 7 SAU's and 4 ICAR Institutes are maintaining germplasm of pomegranate in their repositories.

Table 9.2: Status of Germplasm of Pomegranate in India

Sl.No.	Centers	Number of Collections
1.	PAU, Abohar, Punjab	19
2.	ANGRAU, Anantpur, Andhra Pradesh	29
3.	TNAU, Arrupukottai, Tamil Nadu	24
4.	CCSHAU, Bawal, Haryana	09
5.	CIAH, Bikaner, Rajasthan	152
6.	IIHR, Bangalore, Karnataka	64
7.	RAU, Jobner, Rajasthan	09
8.	CAZRI, Jodhpur, Rajasthan	34
9.	MPKV, Rahuri, Maharashtra	52
10.	S.K. Nagar, Gujarat	52
11.	NRCP, Solapur, Maharashtra	187

Research and Development Organizations

The pomegranate research in India was first started at College of Agriculture, Pune in 1905 (Keskar *et. al.* 1993). Genetic improvement in pomegranate has, for centuries, depended on the selection from seedling variability and their clonal propagation (Pareek and Sharma 1993). Seedlings from selected open pollinated fruits of "Alandi" were raised at Ganeshkhind fruit Experiment Station, Pune in 1932 from which a promising type, GBG-1, was identified and released for commercial cultivation in 1936. This was renamed as "Ganesh" in 1970. This cultivar has pinkish and sweet arils and soft seeds unlike the deep pink and sour aril and hard seeds in "Alandi" (Keskar *et al.*, 1993). This variety was very popular till the last decade of the 20th century. After release of cultivar 'Bhagwa' by MPKV, Rahuri, the major cultivated area of pomegranate was occupied by this cultivar and the area under "Ganesh" started declining. Earlier pomegranate was considered as a minor fruit crop, therefore, not much emphasis was given on crop improvement. Some of the institutions in India or abroad initiated work on pomegranate in the past and released promising varieties.

Introduction of deciduous varieties namely Sellimi, Roman, Chokab, Suffami, Wellingi, Ras-el-Baghi from Baghdad, Palestine and other Mediterranean countries failed to establish in warm climate of India. Similarly, introductions like Gulsha Rose Pink, Shirin Anar, Gulsha Red from erstwhile USSR showed poor performance owing to their deciduous habit. Some introductions like Gulsha Red Pik, Appuli, Gulsha Red Lupania, Bedana Sedana, Kandhari and Khog are being used in breeding programmes to induce dark red aril pigment in the popular cultivars (Pareek and Sharma 1993, Keshar *et al.*, 1993). In recent years, a few commercial cultivars including soft seeded dark red aril types *viz.* Wonderful from USA; Males, Be Hastah, Alah, Agha Mohammad Ali, Post Sephid, Sirin from Iran and Rannyiz G-1-8-23, Rannyiz G-1-3-34, Cherenyj G-1-8-7 and few cultivars from Tunesia were introduced (Singh and Rana 1993) and they showed variability in their morphological and qualitative traits.

Ramu *et al.* (1996) reported a new seedless selection "RCR-1" from cv."Alandi." It gave 267 fruits/tree in 10th year with average fruit weight of 220 g/fruit and average yield of 58.7 kg fruit/tree. A critical evaluation of vegetative propagated plants of "Ganesh" resulted in identification of five promising clones, *viz.* G-107, G-132, G-133. G-134 and G-137 (Sawant 1973). The performance of these promising selection was further studied by Keskar *et al.*(1990). They reported that G-137 was distinctly superior in performance than the parental Ganesh. Selection from the seedlings of cv. Muskat was attempted by Choudhury and Shirsath (1976) and over 47 high yielding individuals superior in fruit quality (soft seeded with high TSS) were identified for further studies. Later, a progeny of over 4000 seedlings were developed from the selected 47 types and further selection was made (Patil 1976, Bhapkar 1976, Karale 1977, Kolhe 1980). On the basis of yield and fruit quality L26P39, L20P5, L26P46, L4P20, L3P38 were selected owing to better performance (Karale 1995). Naik (1975) surveyed orchards originated from Muskat and identified five better types *viz.* P-13, P-16, P-23, P-26 and SK-1. Later, P-23 and P-26 were released for commercial cultivation in 1986 (Keskar *et al.*, 1993).

"Jyothi", a pomegranate with soft seeds and pink aril is released from University of Agricultural Sciences, Bangalore. This is a selection from the seedlings of Bassein Seedless and Dholka varieties (Sulladmath 1985). From TNAU, Coimbatore, a soft seeded superior pomegranate cultivar CO-1 was selected from the assemblage of 28 genotypes (Khader *et al.*, 1982). Sayed *et al.* (1985) reported that a clone Acc. No. 455 had outstanding performance in Tamil Nadu. Later on this was released as "Yarcaud- 1". Fruits of this selection are of medium size with easily peelable rind. The seeds are soft with attractive deep purple aril. Liang and Cheng (1991) selected 3 high yielding, better quality pomegranates and named them as Duo Hong I, Duo Quing II and Duo Bai 2. Shi (1991) could identify a chance seedling as a top quality pomegranate in China and named it as "Taishan Dahongshiliok". On an average, single fruit of this variety weights 400-450 g but can reach up to 750 g. Its fruits have bright red rind, globose or oblate shape and soft seeds. They are very juicy with 17-19 °bx TSS and resistant to cracking. Recently, Liu *et al.* (1997) discovered a sport of the "Qingpitian-87 Qing 7" which is early bearing, productive and highly adaptable. Its fruits have large weight 365-650 g, on an average, but reaching 1365g, oblate round and yellowish green with a red blush. The arils are large, juicy and very sweet. According to Dong (1997)"Huashu Dahongshiliu" is another highly promising, self-fertile variety resistant to pests and diseases. It produces average yield of 130-180 kg/tree. Its fruits are of good quality round bright red with pink-fleshed seeds and 15.5-16.5 °birx TSS. At CCSAU Hisar, cultivar Shirin Anar and Russian seedling were found resistant to bacterial leaf spot. *Punica protopunica* Balf is found resistant to diseases and can be used in breeding programmes or as a rootstock (Pareek and Sharma 1993).

In order to incorporate blood red colour of Russian types into "Ganesh" several crosses were made at Rahuri in 1976. Out of 122 F_1 hybrids, seven had deep red aril colour but hard seeds and inferior taste than "Ganesh" (Kale 1986). Back crossing with Ganesh also did not result in improvement in quality attributes. Some desirable recombinants in F_2 progeny raised from open pollinated fruits of these hybrids were identified (Keskar *et al.*, 1989, 1993). A promising selection from the F_2 population (No.61) combining all desirable quality attributes was released under the name "Mridula".

Breeding programme for improvement of the pomegranate is being done at Indian Institute of Horticultural Research, Bangalore since 1984. The major emphases are given to develop vigorous growing plants having attractive fruits, deep red, bold arils, soft, small seeds, sweet juice and resistance towards bacterial blight. More than 2900 hybrids of single, double, three way and other complex crosses including F_2's were evaluated (Prasanna Kumar 1998). The promising progenies are selected. They not only exhibit heterosis but combine such attributes like superior fruit quality of "Ganesh " and "Kabul," deep red colour of "Gulsha Rose Pink" and vigour of "Yarcaud". Hybrid No. 15-9-94 has dark red, non-sticky arils and soft seeds with high sweetness and low tannin (Pareek 1996). This is named "Ruby". Levin (1990) reviewed achievements of 25 years of pomegranate breeding at Turkmenistan Academy of Agricultural Sciences, Garrygala (Turkmenistan) and reported that the newly bred "Sverkhrannil" is notable being very early and small seeded.

Attempts are made to develop varieties using mutagens. Levin (1990) treated pomegranate seeds with chemical mutagens and γ- rays and obtained forms with good fruit and juice quality and good keeping quality among mutant seedlings. The best result was obtained with γ- rays at 10-20 kR. Promising mutants were also bred using N, N-dimethyl N-dimethly N-nitrosourea treatment. Among the most promising forms produced, was the soft seeded seedling "Sverkhrannil", a super-early variety with fruit ripening in August. It contains 11-13 per cent sugars, 15 per cent dry matter and 7.2-14.7 mg ascorbic acid/100g of juice. Earlier Memedov (1984) treated pollens and seeds with chemical mutagens and observed significant variation in size of the treated pollens and plants raised from the treated seeds.

At present, the main objectives in pomegranate breeding are aimed at to obtain types which produce small soft seeds with attractive reddish (pink) arils and fruit skin. Priority is given to a variety having thicker, sweet juicy aril and large number of seeds with an easily manageable upright growth having semi tall and dwarf habit of the tree. Thornlessness of the twigs is also a desirable character as it helps in cultural management of the tree. Recently, bacterial blight of pomegranate caused by *Xanthomonas axonopodis* pv. *punicae* is reported to cause enormous yield losses (60-80 per cent) in Maharashtra, Karnataka and Andhra Pradesh (Sharma *et al.*, 2006). Therefore, the most priority area would be to develop bacterial blight resistant variety in India. Besides, wilt is also an emerging problem for which resistant rootstocks and tolerant varieties are to be identified. Among the insect pests and physiological disorders, fruit borer (*Virachola isocrates*), fruit cracking and internal breakdown of arils are also important problems to be solved. Sources of resistance to these problems need to be identified for further breeding programme.

For promotion of pomegranate industry in India, ICAR started work under ad-hoc Scheme entitled, "Research on some selected fruits of the arid and semi-arid areas in India" financed by the AP Cess fund of ICAR in 1976 at 10 centres (Abohar, Jodhpur, Chandanwell, Hisar, Khedoi, Rahuri, Lahaul, Yercaud, Bangalore and Udaipur). The scheme was merged during the Sixth Five Year Plan to form the Cell III of the All India Coordinated Fruit Improvement Project (AICFIP) in 1978. During the Seventh Five Year Plan, the Cell III of the AICFIP was restructured to form an independent Project entitled, "All India Coordinated Research Project on Arid Zone Fruits" which was later strengthened during the VIII Plan. At present, the project has twelve centres located in different parts of the country. Out of twelve centers, three are located in Rajasthan, two in Gujarat and one each in Tamil Nadu, Karnataka, Andhra Pradesh, Punjab, Haryana, Uttar Pradesh and Maharashtra. Looking into its increasing demand for internal consumption and export potential, threat created by diseases and pests, ICAR realised that some more infrastructural facilities for pomegranate research may be created. This resulted in establishment of a full-fledged National Research Centre on Pomegranate in 2005 at Solapur in Maharashtra to solve the problems of pomegranate industry in India and act as a nodal agency.

References

Adsule, R.N. and Patil, N.B., 2005. Pomegranate. In: *Handbook of Fruit Sciences and Technology: Production Composition, Storage and Processing*, (Eds.) D.K. Salunkhe and S.S. Kadam. Marcel Dekker, New York, pp. 455–464.

Aksoy, U., 1995. Present status and future prospects of underutilized fruit production in Turkey. *Cah. Options Mediterranean*, 13: 97–07.

Al Kahtani, H.A., 1992. Intercultivar differences in quality and postharvest life of pomegranate influenced by partial drying. *Journal of American Society of Horticulture Science*, 117(1): 100–104.

Astruc, N., Marcos, J.F., Macquaire, G., Candresse, T. and Vallas, V.P., 1996. Studies on the diagnosis of hop stunt viroid in fruit trees: Identification of new hosts and application of a nucleic acid extraction procedure based on non-organic solvents. *European Journal of Plant Pathology*, 102: 837–846.

Aviram, M., Rosenblatt, M. and Gaitini, D., 2004. Pomegranate juice consumption for 3 years by patients with carotid artery stenosis reduces common carotid intima-media thickness, blood pressure and LDL oxidation. *Clinic and Nutrition*, 23(3): 423–433.

Baghel, B.S. and Saraswat, B.K., 1989. Effect of different rooting media on rooting and growth of hardwood and semi-hardwood cuttings of pomegranate (*Punica granatum* L.). *Indian Journal of Horticulture*, 46(4): 458–462.

Beach, M., 1980. Dwarf and other pomegranate. *Light and Garden*, 17(1): 13–15.

Ben Nasr, C., Ayed, N. and Metche, M., 1996. Quantitative determination of the polyphenolic content of pomegranate peel. *Zeitschrift fur Lebensmittel Untershug und Forshung*, 203(4): 374–378.

Bhapker, S.K., 1976. Seedlings selection in the pomegranate cv. Muskar. *M.Sc. (Agri.) Thesis*, MPKV, Maharashtra.

Cervelli, C. and Belletti, P., 1994. Effect of thermic treatments and seed manipulation on emergence of dwarf pomegranate (*Punica granatum* L. 'Nana'). *Acta Horticulturae*, 362: 189–195.

Chaudhari, S.M. and Desai, U.T., 1993. Effects of plant growth regulators on flower sex in pomegranate (*Punica granatum* L.) *Indian Journal of Agricultural Sciences*, 63(1): 34–35.

Choudhury, K.G. and Shirsath, N.S., 1976. Improvement of pomegranate (*Punica granatum* L.) by selection. *South Indian Horticulture*, 24: 56–59.

Deol, I.S. and Uppal, D.K., 1990. Effect of different rooting media on rooting and growth of hardwood and semi hardwood cuttings of pomegranate (*Punica granatum* L.). *Punjab Horticulture Journal*, 30(1–4): 141–144.

Desai, U.T., Masalkar, S.D. and Chaudhari, S.M., 1994. Association of fruit characters in pomegranate. *Annals of Arid Zone*, 33(2): 157–158.

Dhillon, W.S. and Sharma, K.K., 2002. Rhizogenesis of pomegranate in relation to planting time and cutting thickness. *Indian Journal of Horticulture*, 59(2): 150–152.

Dong, Shunsheng, 1997. Huashu Dahongshiliu: a promising pomegranate variety. *China Fruits*, 2: 54.

Du, C.T., Wang, P.L. and Franics, F.J., 1975. Anthocyanins of pomegranate, *Punica granatum*. *Journal of Food Sciences*, 40(2): 417–418.

El Kassas, S.E., Amen, K.I.A., Hussein, A.A. and Osman, S.M., 1992. Effects of certain methods of weed control and nitrogen fertilization on the yield fruit quality and some nutrient contents of Manfalouty pomegranate trees. I. Flowering and fruit setting. *Assiut Journal of Agricultural Sciences*, 23(3): 183–198.

El Sese, A.M., 1988. Physiological studies on flowering and fruiting habits of some pomegranate cultivars under Assiut conditions. *Assiut Journals of Agricultural Sciences*, 19(4): 320–336.

Esmaillzadeh, 2004. Concentrated pomegranate juice improves lipid profiles in diabetic patients with hyperlipidemia. *Journal of Medicinal Food*, 6(3): 157.

Evreinoff, V.A., 1949. Le grenadier. *Fruits Journals of Agriculture Sciences*, 19(4): 320–336.

Fouad, M., Barkat, M.R. and El-yazal, S.A., 1979. Bud burst activity, flowering and fruit set of de la Grenouiller and Manfalot pomegranate cultivars under Giza conditions. Research Bulletin No. 1089, *Facility of Agriculture and Animals, Shams University*.

Frankel,O.H. and Hawkes, J.G., 1975. *Crop Genetic Resources: Today and Tomorrow*. Cambridge University Press, Cambridge.

Game, R.V., 1987. Studies on floral biology of some cultivars of pomegranate (*Punica granatum* L.). *M.Sc. (Ag.) Thesis*, MPKV, Maharashtra.

Ghosh, D., Bandyopadhyay, A. and Sen, S.K., 1988. Effect of NAA and IBA on adventitious root formation in stem cuttings of pomegranate (*Punica granatum* L.) under intermittent mist. *Indian Agriculturist*, 32(4): 239–243.

Gil, M.I., Tomas-Barberan, F.A., Hess-Pierce, B., Holcroft, D.M. and Kader, A.A., 2000. Antioxidant activity of pomegranate juice and its relationship with phenolic composition and processing. *Journal of Agriculture Food Chemistry*, 48(10): 4581–4589.

Graham, S.A., Thorne and Reveal, 1998. Validation of subfamily names in Lythracae. *Taxon.*, 47(2): 435–436.

Guarino, L., Miller, T., Baazara, M. and Obadi, N., 1990. Socotra: The island of Bilss revisited. *Diversity* 6(3–4): 28–31.

Gulick, P. and Van Sloten, D.H., 1984. Directory of germplasm collections. 6–I–*Tropical and Subtropical Fruits and Tree Nuts*, IBPGR, Rome.

Gur, A., 1986. *Punica granatum*. In: *CRC Handbook of Flowering*, (Ed.) A.H. Halevy. CRC Press, Inc. Boca Raton, Florida, pp. 147–150.

Halliday, Jess, 2005. Pomegranate might fight brain damage in babies. Food navigator.com/Europe. *Health and Nutrition* USA, June 29.

Hegde, N.K. and Sulikeri, G.S., 1989. Effect of indole butyric acid (IBA) on the rooting of air-layers of pomegranate. *Current Research, University of Agricultural Sciences*, 18(11): 161–162.

Holden, J.H. and Williams, J.T., 1984. *Crop Genetic Resources: Conservation and Evaluation*. George Allen and Unwin, London.

Hore, J.K. and Sen, S.K., 1994. Root formation in air-layers of pomegranate with NAA and auxin synergists. *Annals Agriculture Research*, 15(3): 310–314.

Hore, J.K. and Sen, S.K., 1995. Role of non-auxinic compounds and IBA on root regeneration in air–layers of pomegranate. *Current Research, University Agricultural Sciences*, 24(5): 83–85.

Hore, J.K. and Sen, S.K., 1993. Root formation in pomegranate (*Punica granatum* L.) stem cuttings with NAA and auxin synergists under intermittent mist. *Crop Research*, 6(2): 252–257.

Horra, J.J., 2003. Chemopreventive effects of pomegranate seed oil on skin tumor development of CD 1 mice. *Journal of Medicinal Food*, 6(3): 157.

Hussein, M.A., El Sese, A.M., El Mahdy, T.K. and Abd-El-Sabour, 1994. Physiological studies on thinning effects on the yield and fruit quality of Manfalouty pomegranate: A Flowering behaviour and sevin, NAA and hand thinning effects on fruit thinning yield and certain fruit characteristics. *Assiut Journal of Agricultural Sciences*, 25(3): 27–40.

Jalikop, S.H. and Kumar, P.S., 1990. Use of gene marker to study the mode of pollination in pomegranate (*Punica granatum* L.). *Journal of Horticulture Sciences*, 62: 221–233.

Jalikop, S.H. and Kumar, P.S., 1998. Use of soft, semi soft and hard-seeded types of pomegranate (*Punica granatum* L.) for improvement of fruit attributes. *Indian Journal of Agricultural Sciences*, 6(2): 87–91.

Jalikop, S.H., 2003. Rosetted siblings in F_2 of a cross in pomegranate (*Punica granatum* L.) can be useful model for resetting investigations. *Euphytica*, 131: 333–342.

Jalikop, S.H., 2007. Linked dominant alleles or inter-locus interaction results in a major shift in pomegranate fruit acidity of Ganesh x Kabul Yellow. *Euphytica* (In Press).

Jalikop, S.H., Rawal, R.D. and Kumar, R., 2005. Exploitation of sub-temperate pomegranate *Daru* in breeding tropical varieties. In: Proc 7th IS on TZFTS. *Acta Horticulturae* 696, ISHS, p. 107–112.

Josan, J.S., Jawanda, I.S. and Uppal, D.K., 1979a. Studies on floral biology of pomegranate. I. Flowering habit, bud development and floral morphology. *Punjab Horticulture Journal*, 19: 59–65.

Josan, J.S., Jawanda, J.S. and Uppal, D.K., 1979b. Studies on floral biology of pomegranate. II. Anthesis, dehiscence pollen studies and receptively of stigma. *Punjab Horticulture Journal*, 19: 66–70.

Kale, R.H., 1986. Studies on F_1 hybrids of pomegranate. *M.Sc. (Agri.) Theisis*, MPKV, Maharashtra.

Kanatharajah, A.S., Dewitz, I. and Jabbari, S., 1998. The effect of media, plant growth regulators and source of explants on *in vitro* culture of pomegranate (*Punica granatum* L.). *Erwerbsobstbau*, 40(2): 66–68.

Kaplan, M., Hayek, T., Ayelet Raz, Raymond Coleman, Lislie Dornfield, J. Vaya and Aviram, M., 2001. Pomegranate juice supplementation to atherosclerotic mice reduces macrophage lipid peroxidation, celluar cholesterol Accumulation and development of atherosclerosis. *Journal of Nutrition*, 131: 2082–2089.

Kar, P.L., Singh, R.P. and Chadha, T.R., 1989. Studies on the standardization of top working technique in wild pomegranate trees. *Horticulture Journal*, 2(1): 68–70.

Karale, A.R., 1995. Pomegranate cultivation: Advances and achievements. In: *Lecture Notes of Summer Institute on Dryland Horticulture* at MPKV, Maharashtra pp. 30–42.

Karale, A.R., Supe, V.S., Kaulgud, S.N. and Kale, P.N., 1993. Pollination and fruit set studies in pomegranate. *Journal of Maharashtra Agriculture University*, 18(3): 364–366.

Karale, A., 1977. Seedlings selection in the pomegranate (*Punica granatum* I.) cv. Muskat. *M.Sc. (Agri.) Thesis,* MAPU, Maharashtra.

Keskar, B.G., Chaudhuri, K.G. and Karale, A.R., 1989. Varietal improvement in the pomegranate. *Maharashtra Journal Horticulture*, 4(2): 66–68.

Keskar, B.G., Karale, A.R., Dhawale, B.C. and Chaudhuri, K.G., 1990. G– 137: A promising clonal selection (pomegranate). *Journal of Maharashtra Agricultural Univeristy*, 15(1): 105–106.

Keskar, B.G., Karale, A.R. and Kali, P.N., 1993. Improvement of pomegranate. In: *Advances in Horticulture*, Vol. 1, (Eds.) K.K. Chadha and O.P. Pareek. Malhotra Publishing House, pp. 399–405.

Khader Abdul, J.B.M. Md., Kulusekaran, M. and Muthuswami, S., 1982. Co–1: A Soft seeded selection. *South Indian Horticulture*, 30(4): 298.

Kim, N.D., 2002. Chemopreventive and adjuant therapeutic potential of pomegranate for human breast cancer. *Breast Cancer Research and Treatment*, 71(3): 203.

Kolhe, V.H., 1980. 'Pomegranate of Muskat pomegranate by seedling selection. *M.Sc.(Ag) Thesis*, MPKV, Maharashtra.

Kowalczyk, E., Kopff, A. and Fijalkowski, P., 2003. Effect of anthocyanins on selection of biochemical parameters in rats exposed to cadmium. *Acta Biochemistry Pol.*, 50(2): 543–548.

Levin, G.M., 1985. Dwarf fruiting forms of pomegranate. *Referativnyi Zhurnal*, 3/55: 751.

Levin, G.M., 1990. Induced mutagenesis in pomegranate (In Russian). *Referativnyi Zhurnal* 10Y a 3399: 126–128.

Levin, G.M., 1994. Pomegranate (*Punica granatum* L.) plant genetic resources in Turkmenistan. *Plant Genetic Resources Newsletter*, 97: 31–36.

Levin, G.M., 1995a. Aspects of pomegranate culture in Turkmenistan. *Plant Genetic Resources Newsletters*, 102: 29–31.

Levin, G.M., 1995b. Genofund of pomegranate in Turkmenistan: Problems of desert development. *Plant Genetic Resources Newsletter*, 103: 84–89.

Levin, G.M., 1996. Pomegranate (*Punica granatum* L.) collection research in Turkmenistan. *Plant Genetic Resources Newsletter*, 106: 47–49.

Liang, C.C. and Cheng, Y.D., 1991. Selections of elite pomegranate. *Journal of Fruit Science*, 8(10): 59–60.

Liu Hong Xian, Cheng Ya Don and Chen Xue Guei, 1997. Promising pomegranate selection "87–Qing 7". *Journal of Fruit Science*, 14(1): 59–60.

Llacer, G., Martinez Valero, R., Romero, M. and Toribio, F., 1995. Present status and future prospects of underutilized fruit tree crops in Spain. *Cah. Options Mediterranean*, 13: 69–78.

Mahishni, D.M., Muralikrishna, A., Shivashankar, G. and Kulkarni, R.S., 1991. Shoot-tip culture method for rapid clonal propagation of pomegranate (*Punica granatum* L.). In: *Horticulture: New Technologies and Application,* (Eds.) J. Prakash and R.L.M. Pierik. Kluwar Academic Publisher Dordrecht, Netheralnds, pp. 215–217.

Manohar, M.S., Tikka, S.B.S. and Narhulal, 1981. Phenotypic variation and its heritable components in some biometric characters in pomegranate (*Punica granatum* L.). *Indian Journal Horticulture*, 8: 187–190.

Mars, M. and Marrakchi, M., 1998a. Conservation at valorization des resources genetiques du grenadier (*Punica granatum* L.) in Tunisia. *Plant Genetic Resources Newsletters*, 114: 35–39.

Mars, M. and Marrakchi, M., 1998b. Diversity among pomegranate (*Punica granatum* L.) germplasm in Tunisia. *Genetic Resources and crop Evolution* (In Press).

Mars, M. and Sayadi, S., 1992. Etude comparative de la qualite des fruits de cinq varieties de grenadier (*Punica granatum* L.) *Revue des Regions Arides* 4/92: 45–57.

Mars, M., Melgarejo, M., Amoros, A. and Martinez, R., 1997. *Pomegranate Descriptors*. CIHEAM Collaborative Working Group on Underutilized Fruit Crops in the Mediterranean Region.

Mars, M., 1995. La culture du grenadier (*Punica granatum* L.) et du figuier (*Ficus carica* L.) en Tunisie. *Cah. Options Mediterranean*, 13: 85–95.

Mars, M., 1996. Pomegranate genetic resources in the Mediterranean region. In: *Proc. First MESFIN Plant Genet. Research Meeting*, Tenerife, Spain, 2–4 October 1995, pp. 345–354.

Melgarejo, P. and Martínez Valero, R., 1992. *El Granado*. Ediciones Mundi Prensa, Madrid.

Melgarejo, P. and Artes, F., 2000. Total lipid content and fatty acid composition of oilseed from lesser known sweet pomegranate clones. *Journal of Sciences Food Agriculture*, 80(10): 1452–1454.

Melgarejo, P., Salazar, D.M. and Amoros, A., 1995. Total lipids content and fatty acid composition of seed oils from six pomegranate cultivars. *Sciences Food Agriculture*, 69: 253–256.

Melgarejo, P., Martinez, J.J., Hernandez, F., Martinez Font, R., Barrows, P. and Erez, A., 2004. Kaolin treatment to reduce pomegranate sunburn. *Scientia Horticulturae*, 100(1/4): 349–353.

Memedov, G.M., 1984. Meiotic abnormalities in induced forms of pomegranate following treatment of seed with various doses of physical and chemical mutagens. *Referativnyi Zhurnal*, 4: 6526.

Mendilcioglu, K., 1968. Investigations on the propagation of important fruit species by cuttings. *Ege University Ziraat Fakultesti Dergisi*, A.5(2): 171–195.

Miguel, G., Dandlen, S., Neves, A. and Martins, D., 2004. The Effect of two methods of pomegranate (*Punica granatum* L.) juice extraction on quality storage at 4°C. *Biomedical and Biotechnology*, 5: 332–337.

Naik, V.B., 1975. Seedlings selection in pomegranate (*Punica granatum* L.) cv. Muskat. *M.Sc. (Agri.) Thesis*, MPKV, Maharashtra.

Nalawadi, V.G., Farooqui, A.A., Dasgupta, N., Reddy, M.A., Gubbaiah Sulikeri, G.S. and Nalini, A.S., 1973. Studies on the floral biology of pomegranate (*P. granatum* L.). *Mysore Journal of Agriculture Sciences*, 7: 213–225.

Nataraja, K. and Neelambika, G.K., 1996. Somatic embryogenesis and plantlets from pertal cultures of pomegranate (*Punica granatum* L.). *Indian Journal of Experimental Biology*, 34(7): 719–712.

Nath, N. and Randhawa, G.S., 1959a. Studies on floral biology of pomegranate. I. Flowering habit, flowering season, bud development and sex ratio in flowers. *Indian Journal of Horticulture*, 16: 61–68.

Nath, N. and Randhawa, G.S., 1959b. Studies on floral biology of pomegranate.II. Anthesis, dehiscence, pollen studies and receptivity of stigma. *Indian Journal of Horticulture*, 16: 121–135.

Nath, N. and Randhawa, G.S., 1959c. Studies on floral biology of pomegranate. III. Pomegranate Pollination, fruit set and seed formation. *Indian Journal of Horticulture*, 16: 136–140.

Nath, N. and Randhawa, G.S., 1959d. Classification and description of some varieties of (*Punica granatum* L.). *Indian Journal of Horticulture*, 16: 191–201.

NGS, 2006. Pomegranate juice confirmed as miracle medicine for prostrate cancer. In: *News Target*, July 3, National Geographic Society.

NRCP, 2007. *Annual Report*. National Research Centre on Pomegranate, Solapur, Maharashtra, p. 5.

NRCP, 2007. *Vision–2025*. National Research Centre on Pomegranate, Solapur, Maharashtra, p. 15.

Omura, M., Matsuta, N., Moriguchi, T. and Kozaki, I., 1987a. Adventitious shoot and plantlet formation from cultured pomegranate leaf explants. *Horticulture Sciences*, 22(1): 133–134.

Omura, M., Matsuta, N., Moriguchi, T., Kozaki, I. and Sanada, T., 1987b. Establishment of tissue culture methods in dwarf pomegranate (*Punic granatum* L. var *nana*) and application for the induction of variants. Bulletin of the *Fruit Tree Research Station,* A (Yatabe) Japan, 44: 17–44.

Ozguven, A.I., 1996. The genetic resources of pomegranate (*Punica granatum*) in Turkey. In: *Proc. First MESFIN Plant Genet. Res. Meeting* Tenerife, Spain, 2–4 Oct. 1995, pp. 269–284.

Panda, J.M. and Das, R.C., 1990. Regeneration of pomegranate stem cuttings treated with IAA and IBA under intermittent mist. *Orissa Journal of Horticulture,* 18(1–2): 32–37.

Panwar, R.D., Kaushik, R.A., Singh, Surinder, Gupta, R.B. and Singh, S., 2001. Effect of indole butyric acid (IBA) on rooting of hard-wood cuttings in pomegranate (*Punica granatum* L.). *Haryana Journal of Horticulture Sciences,* 30(1–2): 72.

Pareek, O.P. and Sharma, S., 1993. Genetic resources of under exploited fruits. In: *Advances in Horticulture,* Vol. 1, (Eds.) K.L. Chadha and O.P. Pareek. Malhotra Pub., New Delhi, pp. 189–225.

Pareek, O.P., 1996. Indian jujube and pomegranate. In: *50 Years of Crop Science Research in India,* (Eds.) R.S. Paroda and K.L. Chadha. ICAR, New Delhi, pp. 557–537.

Parmar, C. and Kaushal, M.K., 1982. *Wild Fruits of the Sub-Himalayan Region.* Kalyani Publishers, Ludhiana, pp. 74–79.

Patil, A.V. and Karale, M.K., 1985. Pomegranate. In: *Fruits of India: Tropical and Subtropical,* (Ed.) T.K. Bose. Naya Prakash, Calcutta, pp. 537–548.

Patil, B.N., 1976. Seedling selection in the pomegranate cv. Muscat. *M.Sc. (Agri.) Thesis,* MPKV, Maharashtra.

Poyrazoglu, E., Gokmen, V. and Artik, N., 2002. Organic acids and phenolic compounds in pomegranate (*Punica granatum* L) grown in Turkey. *Journal of Food Composition and Analysis,* 15(5): 300–303.

Prasanna Kumar, B., 1998. Pomegranate. In: *A Textbook on Pomology,* Vol. 3, (Ed.) T.K. Chattopadhyay. Kalyani Publication, Ludhiana, pp. 161–188.

Purohit, A.G. and Shekharappa, K.E., 1985. Effect of type of cutting and indolebutyric acid on rooting of hardwood cuttings of pomegranate (*Punica granatum* L.). *Indian Journal of Horticulture,* 42(1–2): 30–36.

Purohit, A.G., 1987. Effect of pollen parent on seed hardness in pomegranate. *Indian Journal Agricultural Sciences,* 57: 753–755.

Raman, V.S., Kesavan, P.C., Manimekalai, G., Alikhan, W.M. and Rangaswami, S.R., 1963. Cytological studies in some tropical fruit plants: Banana, annona, guava and pomegranate. *South Indian Horticulture,* 11(3 and 4): 27–33.

Ramu, B.S.S., Reddy, P.N., Hussain, S.A. and Patil, P.B., 1996. A new seedless selection from pomegranate cv. "Alandi" *Current Research University of Agricultural Sciences,* 25(4): 72–73.

Rao, V.N.M. and Abdul Khader, J.B.M., 1962. Estimation of pollen production in fruit crops. *Madras Agricultural Journal,* 49: 152–166.

Reddy, Y.T.N. and Reddy, Y.N., 1989. Rooting of pomegranate cuttings as influenced by paclobutrazol, polythene covering and basal wounding. *Indian Farming,* 39(4): 27–28.

Reddy, Y.T.N. and Reddy, Y.N., 1990. Effects of basal wounding, growth regulator and polythene covering on rooting of pomegranate cutting. *Journal of Maharashtra Agricultural University,* 15(2): 153–155.

Sandhu, A.S., Minhas, P.P.S., Singh, S.N. and Kambhoj, J.S., 1991. Studies on rhizogenesis in hard wood cuttings of pomegranate. *Indian Journal of Horticulture*, 48(4): 302–304.

Saoji, A.A. and Rudra, J., 1991. Estimation of protein contents in the stored pollen grains of *P. granatum* L. *Advances in Plant Sciences*, 4(2): 316–318.

Sawant, R.B., 1973. Clonal selection in Ganesh pomegranate (*Punica granatum* L.). *M.Sc. (Agri.) Thesis*, MPKV, Maharashtra.

Sayed, S., Ramdas, S., Nanjan, K. and Muthuswami, S., 1985. YCD–1 Pomegranate. *South Indian Horticulture*, 33(1): 67.

Sharma, K.K., Sharma, J. and Kumar, P., 2006. *Important Diseases, Disorders and Insect Pest of Pomegranate and their Management*. Technical Bulletin. NRC on Pomegranate, Maharashtra, 1: 1–16.

Sharma, S.D. and Sharma, V.K., 1990. Variation for chemical characters in some promising strains of pomegranates. *Euphytica*, 49: 131–133.

Shi, C.D., 1991. Taishan Dahongshilliou a top quality pomegranate cultivar. *China Fruits*, 4: 27 (In Chinese English abstract).

Shulman, Y., Berstein, L.F. and Lavee, S., 1984. Pomegranate fruit development and maturation. *Journal of Horticultural Sciences*, 59(2): 265–275.

Singh and Singh, 2006. Diversity of wild pomegranate in Himachal Pradesh. *Progressive Horticulture*, 38(1): 49–52.

Singh, B.P. and Rana, R.S., 1993. Promising fruit introductions. In: *Advances in Horticulture*, Vol. 1, (Eds.) K.L. Chadha and O.P. Pareek. Malhotra Publishing House, New Delhi, pp. 43–66.

Singh, R.P., Kar, P.L. and Dhuria, H.S., 1978. Studies on the behaviour of flowering and sex expression in some pomegranate cultivars. *Plant Sciences*, 10: 29–31.

Singh, R.P., Murthy, K.N.C. and Jayaprakasha, G.K., 2002. Studies on the antioxidant activity of pomegranate (*Punica granatum*) peel and seed extracts using *in vitro* models. *Journal of Agriculture Food Chemistry*, 50(1): 81–86.

Singh, S.S., Krishnamurthi, S. and Katyal, S.L., 1967. *Fruit Culture in India*. ICAR, New Delhi, pp. 207–214.

Smith, P.M., 1976. Minor crops. In: *Evolution of Crops Plants*, (Ed.) N.W. Simmonds, Longman, London, pp. 301–324.

Sulladmath, U.V., 1985. *Progress Report*. Third National Workshop Arid Zone Fruit Research. MPKV, Rahuri, 5–8 July.

Tous, J. and Fergson, L., 1996. Mediterranean fruits. In: *Progress in New Crops*, (Ed.) J. Kanick. ASHS Press, Arlington, VA, pp. 416–430.

Tripathi, S.N. and Shukla, H.S., 2004. Propagation of pomegranate (*Punica granatum* L.) cultivars by stem cuttings with indole butyric acid and p-hydroxybenzoic acid. *Indian Journal Horticulture*, 51(4): 362–365.

Westwood, M.N., 1978. *Temperate Zone Pomology*. WH Freeman and Company, San Francisco.

Zukovskij, P.M., 1950. *Punica*. In: *Cultivated Plants and their Relatives*. State Publishing House Soviet Sciences, Moscow, pp. 60–61.

2011, Biodiversity in Horticultural Crops Vol. 3 *Pages* 251–274
Editor: Professor K.V. Peter
Published by: DAYA PUBLISHING HOUSE, NEW DELHI

Chapter 10

Heliconia

T. Janakiram[1] *and Anuradha Sane*[2]
[1]*Division of Floriculture,*
Indian Agricultural Research Institute, New Delhi – 110 012
E-mail: tolety@iihr.ernet.in
[2]*Division of Plant Genetic Resources,*
Indian Institute of Horticultural Research, Hessaraghatta Lake, Bangalore – 560 089

Heliconias belonging to Heliconiaceae are the most attractive of all exotic tropical flowering plants. They are known as lobster's claws, parrots flower, parrot plantain and false plantains. The heliconias are gaining importance due to their diversity in both colour and form and have good potential as commercial cut flowers. Its brilliant colour, exotic form, long straight peduncles and excellent post harvest characteristics make it an outstanding flower for the florist trade. *H. psittacorum* and *H. latispatha* x. *H. psittacorum* are excellent for floral arrangements. It is well adapted to all major agro climatic zones of the country and can be cultivated upto a height of 3000 to 4000 m.a.m.s.l.

Habitat

Most of the species of *Heliconia* can be found in moist or wet regions, but some are found in seasonally dry areas.Although most heliconias thrive in the humid lowland tropics at elevations below 500m. the largest number of species are found in middle elevations, rain forests and cloudy forest habitats (Tom, 1997). The abundance, population structure and recruitment of a local understorey herb (*Heliconia acuminata*) were compared in four secondary forest sites in Amazonas, Brazil. The density of naturally emerging *H. acuminata* seedlings ranged 0-6 seedlings/quadrat, and there was a highly significant difference among the four stands. Results suggest that secondary forest type may be a poor predictor of *H. acuminata* density or population structure. However, significant differences were observed among forest types in H. *acuminata* seed germination and seedling establishment (Bruna and Ribeiro, 2005).

Morphology and Anatomy

In general, heliconias have banana-like foliage and have erect or pendulous terminal inflorescence composed of two or more boat-shaped bracts arising from a central floral axis. Depending on variety, the height varies from 0.6 to 6.5 m, often with extensive rhizomatous growth.

The morphological and anatomical features of the inflorescence bracts were studied in *H. rivularis, H. spathocircinata* and *H. velloziana* (subgen. *Heliconia*) and in *H. angusta, H. hirsuta* and *H. subulata* subsp. *gracilis* (subgen. Stenochlamys). In the subgen. Heliconia, the bracts were cymbiform and in Stenochlamys, they were lanceolate-conduplicated. All the species had bracts with uniseriate epidermis, with thin-walled cells, tetracytical stomata in *H. hirsuta* and *H. subulata*, and polycytical in others; the vascular bundles and air canals were distributed in arc-shaped and the fibre bundles were situated close to the abaxial surface. The bract shape associated with the distribution of fibre bundles and with size of air canals were different in both subgenera studied. An identification key was made based on morphological and anatomical features (Simao and Scatena, 2004).

Species and Cultivars

A native of tropical America, genus *Heliconia* has more than 250 species mostly native to the New World tropics where they are found in open clearings. Two main types are, erect heliconia with standing straight bracts pointing up and pendent heliconia with hanging bracts pointing down. The great interest for heliconias by the external market has placed this crop in a prominent position in flower production chain. Establishment, whether outdoors in the ground or in containers or in interiorscapes, demands a knowledge of growth habits and environmental requirements. Considerations of maintenance include grooming and pest and weed control. Problems include chlorosis, failure to flower and excessive growth (Criley, 1999).

Thirty species of heliconias (*Heliconia spp.*) were evaluated for the purpose of diversifying the number of species available to the local cut flower market. A point scoring system was used to determine those species most suitable for use as cut flowers. This system was based on the flowering period, shrub production, stem rigidity, post harvest durability, handling, packaging and transport, shipping durability, and market value. The species considered fairly suitable, with scores ranging from 25 to 35 points, were the following, presented in an ascending order of importance: *H. hirsuta, H. bourgaeana, H. forerοi, H. rauliniana, H. angusta* var. *flava, H. caribaea, H. chartacea, H. latispatha, H. angusta cv. Holiday, H. episcopalis, H. rostrata, H. angusta* var. *aurorea, H. sampaioana, H. librata* and *H. velloziana*. The species regarded as quite suitable, with scores higher than 35 points, were the following, presented in a descending order of importance: *H. orthotricha, H. bihai, H. stricta, H. psittacorum, H. aurea* and *H. wagneriana*. The point scoring system used in this study was found an excellent tool to establish the cut flower sales potential of the species studied (Castro *et al.*, 2006)

A nomenclatural study on heliconias (*Heliconia* spp.) based mainly on information obtained from published articles in national and international periodicals, as well as from books published between 1900 and 2002, was conducted. Data analysis revealed 176 species of heliconias occurring in the Neotropical Region and 6 species occurring in the Pacific Islands, giving a total of 182 species. Nearly half of the species described are found in Colombia (94), although a number of species were also found in Ecuador (60), Panama (56), Costa Rica (47), Brazil (37), Peru (32), Venezuela (26), Nicaragua (22), Guatemala (16), Bolivia (15), Honduras and Mexico (14), and Suriname (13). Ninety-four of the 182 species are endemic to certain regions. On account of the high level of endemism, it is of pressing importance to define a national/international programme aimed at the collection and conservation of species germplasm under this genus (Castro *et al.*, 2007).

Figure 10.1: Bihai Yellow Damear

Figure 10.2: Omtotricha velvat

Figure 10.3: Wagemania

Figure 10.4

Figure 10.5

Figure 10.6

Figure 10.7

Figure 10.8

It was observed that short heliconia genotypes- the cultivars and hybrids of *H. psittacorum*- have a larger number of shoots than tall heliconia genotypes, in both planting conditions. Results from the APC, measured 304 DAP, showed that among the genotypes grown in full sun, *H. psittacorum* cultivar 'Red Opal' occupied half of the plot area (11.387 cm^2) while genotype *Heliconia* x *nickeriensis* occupied an area larger than the plot area (22.541 cm^2). The genotype *H. orthotricha* showed the smallest APC (725 cm^2). In shaded condition, APC varied from 750 cm^2 (*H. bihai* 'Nappi Yellow') to 10.201 cm^2 (*H. stricta* 'Fire Bird'). The evaluation of these traits during the first 316 DAP allowed us to conclude that a single plant spacing for different heliconia genotypes is inappropriate and can lead to future problems in plant management (Costa *et al.*, 2006).

Medium flower size, bold coloration, good keeping qualities and a long season of bloom characterize the selections of *H. orthotricha*, a species relatively new to the palette of bold tropical cut flowers. An evaluation of four cultivars, Eden Pink, Garden of Eden, Macas Pink and Candy Cane, was conducted over a 2-year period to determine their seasonality of flowering and productivity, and developmental progression from shoot emergence to flower harvest. The shortest average periods from shoot emergence to flower harvest occurred when shoots emerged during the spring and the longest during the autumn, and the variations were for 'Eden Pink' 170-283 days, 'Garden of Eden' 125-254 days, 'Macas Pink' 156-291 days, and 'Candy Cane' 131-248 days. Approximately 50 per cent of the flowers of each cultivar were harvested during summer months, but all produced flowers throughout the year. Per plant productivity during their first 12 months following planting ranged from a low of 40 for 'Eden Pink' to 54 for 'Candy Cane' and 'Macas Pink' and 57 for Garden of Eden, but all produced more than 200 inflorescences by 2 years after planting from a small number of plants. All four cultivars may be characterized as welcome additions to the export markets for tropical flower producers (Criley *et al.*, 2003).

Floral Biology

Six cultivars of *H. psittacorum* were selected for studies on their natural fruit-bearing ability, pollen formation and pollination under the tropical climatic conditions. Three of them, namely Tay, Andromeda and Lady Di, were partially fertile with a very low rate of fruit set, ranging 2.8 to 4.7 per cent. They were diploids with 2n = 24 chromosomes. The process of pollen formation (microsporogenesis) was normal, and pollen grains were all uniform in size and appeared normal. The poor fruit set of these 3 cultivars was attributed to poor pollen germination on stigmas rather than poor pollination or self-incompatibility. The other 3 cultivars, namely Petra, Sassy and Iris, were completely sterile. Their pollen grains were of variable sizes and appeared to be abnormally fragmented. Over 80 per cent of the pollen grains aborted 1-2 days before pollination. These abnormal features were consistent with irregular distribution of chromosomes during meiosis in microsporocytes. All 3 cultivars were confirmed to be triploid (2n = 3x = 36)(Lee *et al.*, 1994). In Heliconia, thread-like structures connecting the pollen grains are described. Anthers of *H. auraniaca, H. collinsiana, H. psittacorum var. subulata* and, for comparison, *Strelitzia reginae* and *Musa uranoscopus* were examined, following critical point drying, in the SEM. Calcium oxalate raphides and long threads were observed within the extruded pollen mass. The long threads (250-600 micro m) are extremely variable, their shape ranging from very thin and smooth to solid, rough and twisted. These threads are decay products of the walls separating the pollen chambers, and products of the rupture of the mature anthers in the stomium region. The pliable cell threads mix with the pollen and entangle individual grains to form aggregates. This ensures that the pollen becomes embedded in the feathers or attached to the smooth, unsculptured beak of pollinating hummingbirds (Trochilidae). Structure and origin of the Heliconia threads differ from those of the related genus *Strelitzia* (Strelitziaceae). No similar threads were observed in *M. uranoscopus* (Rose and Barthlott, 1995).

A study was conducted to test if two sympatric species of *Heliconia, Heliconia spathocircinata* and *Heliconia laneana* var. *flava*, differ in their reproductive seasons and guilds of flower visitors. Individuals of *H. spathocircinata* and of *H. laneana* var. flava were tagged within a transect of 4000 x 4 m. Reproductive phenophases (flowering, unripe, and ripe fruits) were recorded, and circular statistics were used to test for the regularity and frequency of phenophases. Flower visitors were observed during two flowering periods for each species. Both species had well-defined and synchronous reproductive periods, with a small overlap, suggesting a sequential flowering model. Biotic and abiotic factors seem to affect phenophases of both species. However, herbarium data suggest that the flowering period in *H. spathocircinata* varies lesser than in *H. laneana,* despite its wider geographic distribution. The hummingbird *Phaethornis idaliae* was the main floral visitor of both *Heliconia* species studied. *H. laneana* was also legitimately visited by another hummingbird species, *Glaucis hirsuta*, and by Lepidoptera. Some Lepidoptera, particularly the bee *Trigona spinipes*, were illegitimate visitors ('nectar robbers'). We suggest that the flower of *H. spathocircinata* allowed a wider array of visitors, because it is bigger (Cruz *et al.*, 2006).

The developmental anatomy and morphology of the ovule and seed in several species of *Heliconia* were investigated as part of an embryological study of the Heliconiaceae and to provide a better understanding of their relationships with the other families of the Zingiberales. *Heliconia* species have an ovule primordium with an outer integument of both dermal and subdermal origin. The archesporial cell is divided into a megasporocyte and a single parietal cell, which in turn are divided only anticlinally to form a single parietal layer, disintegrating later during gametogenesis. The embryo sac was fully developed prior to anthesis. In the developing seed, the endosperm was nuclear, with wall formation in the globular stage; a nucellar pad was observed during embryo development, but later became compressed. The ripe fruit contained seeds enveloped by a lignified endocarp that formed the pyrenes, with each pyrene having an operculum at the basal end; the embryo was considered to be differentiated. Most of these characteristics are shared with other Zingiberales, although the derivation of the operculum from the funicle and the formation of the main mechanical layer by the endocarp are unique to the Heliconiaceae. (Simao *et al.*, 2006).

Artificial hybridization among 14 species of neotropical Heliconia was studied. Individuals in cultivation were used as parents in crosses primarily between species with pendent inflorescences which are normally distributed allopatrically. At Finca La Selva normally sympatric species with either pendent or erect inflorescences were crossed in their natural habitats. Observation of pollen tube growth by means of fluorescence microscopy and seed set were used to determine the extent of crossability. Crossability barriers between majority of species were strong and foreign pollen tubes were inhibited at the stigmatic surface, within the stylar tissue or within the ovary. The site of inhibition was consistent for each pair of species, and depended on the parentage and the direction of the cross. Although additional isolating mechanisms, such as pollinator specificity and phenological separation were present, pre-fertilization crossability barriers acted as the ultimate mechanism to prevent hybridization. The type of barrier (stigmatic, stylar or ovarian) that existed between two species was not dependent upon the geographical distribution of the parental species or the specific types of pollinators which visit them, but in some cases might indicate taxonomic relationships (Kress,1983).

Reproductive Biology

Anther development, microsporogenesis and microgametogenesis in several species of *Heliconia* were investigated as part of a complementary embryological study of the Heliconiaceae. All studied

Heliconia species present bithecate and tetrasporangiate anthers with fertile pollen grains; only *H. rivularis,* a natural hybrid, presented sterile pollen grains of variable size and no content. The anther wall has a uniseriate epidermis and endothecium, the latter with helicoidal thickenings, although some cells of the middle layers also showed thickenings; the biseriate tapetum is of amoeboid non-syncytial type, since the tapetum cells did not fuse together forming a true plasmodium. The microsporogenesis is successive leading to isobilateral tetrads. The inaperturate pollen grains had a very reduced exine consisting of a thin, more or less continuous layer with small spines upon; the pollen grain shape is variable among the species, all of them presenting heteropolar pollen, except *H. angusta* with isopolar ones. Most of these characteristics were shared with other studied Zingiberales, although more studies need to be done (Simao *et al.,* 2007)

Propagation

Increased interest in tropical cut flower export in developing nations has increased the demand for clean planting stock. The most popular ones are *Strelitzia,* various gingers and Heliconia. The morphology of seed and vegetative propagation of *H. velloziana,* a rhizomatous herb with a large aerial part, was studied. The seeds took ~4 months to germinate, probably due to the undifferentiated embryo and hard endocarp. The seedlings had 2 first foliar structures as scale leaves and the primary root developed, but the adventitious roots were more conspicuous. Vegetative propagation by rhizomes was faster than sexual propagation. The roots developed after one month, and young shoots started developing after 4 to 6 weeks. The young aerial shoots presented a variable number of cataphylls and two protophylls. The seed of *H. velloziana* is described (Simao and Scatena, 2003). In *Heliconia stricta* cv. Dwarf Jamaican, it was initiated in week 17 and reached 44 per cent; T50 occurred at week 24 and T10-90 at 13 weeks; this was the species with the most irregular pattern of germination.(Maciel and Mogollon,1998). In an experiment to determine their tolerance to high temperatures recommended for eradicating nematodes, rhizomes of Heliconia survived treatment in 48°C hot water for periods up to 1 h and in 50° up to 30 minutes (Criley, 1988).

Micropropagation

Embryo Culture

Embryos from mature fruits of *Heliconia rostrata* were excised and cultured in basal medium containing MS salts, vitamins and sucrose. Sucrose was essential for embryo development. In the medium without sucrose, the embryos died in culture. Sucrose concentrations of 1, 2 and 3 per cent (w/v) were beneficial for embryo development. Sucrose concentrations of 6, 9 and 12 per cent (w/v) inhibited embryo growth. The addition of kinetin, isopentenyladenine and zeatin did not improve embryo growth and development. (Torres *et al.,* 2005). A protocol was developed for the *in vitro* induction of embryogenic callus in *Heliconia chartacea* cultivars Sexy Pink and Sexy Scarlet from ovary sections. Culture media were supplemented with IAA (0, 1 and 2 mg/litre) and 2,4-D (0, 5, 10, 15 and 20 mg/litre). Sexy Pink presented a higher frequency of callus development compared to Sexy Scarlet. Culture media supplemented with 1 mg IAA + 10 mg 2,4-D/litre and 2 mg IAA + 5 mg 2,4-D/litre gave 100 per cent callus formation for Sexy Pink inoculated explants.(Ulisses *et al.,* 2007).

Shoot Tip Culture

Shoot explants from *Heliconia psittacorum* rhizomes were surface-sterilized and cultured in modified MS medium. Treatments comprised: 0.25 mg IAA/litre and benzyladenine (BAP) at 4, 6, 8 or 10 mg/litre. When buds were cultured, it turned green from a creamy-white colour within 10-14 days of culture and subsequently increased in size. Treatment with 6 mg BAP/litre recorded the highest regeneration (per cent) (59.0), followed by treatment with 4.0 and 8.0 mg BAP/litre (both 50.0). In other media, buds remained green, retained their size and thereafter turned brown and died. When the buds were maintained in the medium with 6 mg BAP/litre for 2-3 months, shoots did not proliferate. Treatment with 0.25 mg IAA/litre + 2.0 mg BAP/litre recorded the highest shoot proliferation (per cent) within 4-5 weeks of culture. In other media tested, the buds either turned brown or died before shoot proliferation. Shoot growth was studied in MS medium supplemented with 0.25 mg IAA/litre and 2.0 mg BAP/litre. A maximum shoot length of 8 cm was observed (mean: 5.4 cm). A maximum of 7 leaves/culture was observed (mean: 4.2). For rooting, treatments comprised: 0.25, 1.0 or 1.5 mg IAA/litre and/or 0 or 0.25 mg BAP/litre. No rooting was observed in treatment with only 0.25 mg IAA/litre. Rooting was observed in the medium with either 1.0 or 1.5 mg IAA/litre only. Generally, root initiation started earlier in the medium with 1.5 mg IAA/litre than in the medium with 1.0 mg IAA/litre. Treatment with 1.5 mg IAA/litre recorded a higher rooting average and root length than treatment with 1.0 mg IAA/litre.(Talukdar *et al.*, 2002).

A method of *in vitro* propagation of *H. psittacorum* cv. Rhizomatosa by using terminal shoot tip explants was developed. Various concentrations (25, 50, 100 and 500 mg/litre) of cefotaxime were used in liquid Murashige and Skoog (MS) basal medium to control contamination. Lower rates of contamination were observed in shoot tip explants after pretreatment of explants in liquid MS medium supplemented with 500 mg/litre cefotaxime for 3 days. Well developed shoots were successfully obtained after 45 days of culture of shoot tip explants on MS medium supplemented with 4-8 mg/litre benzylaminopurine (BA) [benzyladenine] and 0.5 mg/litre NAA. Addition of 0.01-0.5 mg/litre thidiazuron (TDZ) to the medium was beneficial for rapid growth of shoot tip explants. The effect of various culture methods on growth of shoot tip explants was also studied. (Shiau–YihJuh *et al.*, 1999).

A rapid and flexible propagation system for *H. psittacorum* was developed using thin sections (1 mm) of the base of *in vitro* plantlets. When cultured on MS medium in the presence of 10 micro M 2,4-D, the explants produced rooted plantlets. 2,4-D at 40-80 micro M induced the formation of highly morphogenic callus. When subcultured on basal medium, this callus produced protocorm-like bodies which developed into plantlets (Goh *et al.*, 1995). A protocol was developed for *in vitro* propagation of *H. psittacorum* by culture of terminal and axillary buds of rhizomes. Cultures were initiated on modified MS medium containing 40 micro M BA, 150 ml coconut water, 30 g sucrose and 2 g Gelrite/litre. Shoot multiplication was achieved on the above medium without coconut water, but supplemented with 10 micro M BA. Shoots were rooted on MS basal medium and successfully acclimatized to greenhouse conditions (Nathan, *et al.*, 1992).

Presence of endophytic microorganisms like *Pseudomonas* sp. represents an obstacle even for the evolution of heliconia tissue culture studies. Stem apices of *Heliconia rauliniana* were inoculated and analyzed *in vitro* to investigate the presence of endophytic microorganisms. An antibiotic analysis was also run for posterior microbiological control. The presence of *Pseudomonas* sp. and *Klebsiella* sp. in stem apices confirmed infection. Antibiotics such as chloramphenicol, cefotaxime and the association of both (chloramphenicol+cefotaxime) at rates of 50, 150, 250 and 500 mg l^{-1}, were applied to the MS medium and supplemented with 3.5 mg l^{-1} of 6-BA. After a 50 day-period, 500 mg l^{-1} of cefotaxime was the most efficient form to control endophytic microorganisms in this species of *Heliconia*. The chloramphenicol, although effective for endophytic microorganism control, was inadequate for *in vitro* growth due to its cytotoxicity to explants.(Rodrigues *et al.*, 2005). Various concentrations (25, 50, 100 and 500 mg/llitre) of cefotaxime were used in liquid Murashige and Skoog (MS) basal medium to control contamination. Lower rates of contamination were observed in shoot tip explants after pretreatment of explants in liquid MS medium supplemented with 500 mg/litre cefotaxime for 3 days. Well developed shoots were successfully obtained after 45 days of culture of shoot tip explants on MS medium supplemented with 4-8 mg/litre benzylaminopurine (BA) [benzyladenine] and 0.5 mg/litre NAA. Addition of 0.01-0.5 mg/litre thidiazuron (TDZ) to the medium was beneficial for rapid growth of shoot tip explants (Shiau-YihJuh *et al.*, 1999).

Callus Culture

Callus initiation in heliconia was carried out using different plant tissues, namely, shoot apices, primordial leaves and rhizome sections as explants. Both shoot apices and primordial leaves produced good callus on MS medium with 2.4-D (0.15 mg/l), NAA (1.96 mg/l) and kinetin (4.3 mg/l), whereas rhizome sections failed to induce callus. However, the calli, thus derived remained non-organogenic on the regeneration medium (Siddique and Paswan, 1997).

Axillary Bud Culture

A three-fold increase in length and more than double in breadth *Heliconia psittacorum* axillary buds were successfully established in MS medium supplemented with 5.0-6.0 mg benzyladenine, 0.05-0.1 mg IAA, and 1.50-2.00 mg GA3/litre. The supplementation of 2.5 mg benzyladenine, 0.2 mg IAA, and 2.5 mg GA3/litre was ideal for initiation of multiple shoots and their multiplication rates. MS medium supplemented with 0.75 mg IAA and 0.05 mg NAA/litre gave the maximum rooting response (85 per cent),the earliest root initiation (within 21 days), and maximum number of primary (3.70) and secondary (3.90) roots (Bora and Paswan *et al.*, 2004)

Heliconia obtained by *in vitro* micropropagation on Murashige and Skoog (MS) medium were inoculated with 3 vesicular arbuscular mycorrhizal (VAM) species, alone and in combination. A control treatment without inoculation was also included. The plantlets were acclimatized in a mixture of soil 30 per cent, sand 20 per cent and "torta de filtro" 50 per cent. The fungal species were *Glomus clarum, G. etunicatum* and *Gigaspora margarita*. At 60 days after inoculation for gerbera and 90 days for heliconia, the fresh and dry weights of the aerial and root parts were recorded and the percentage of mycorrhizal colonization of the roots was determined. Although colonization was high, heliconia did not benefit from this inoculation, however Gerbera benefited from colonization by *G. clarum, G. etunicatum* and the mixed inoculation (Sato *et al.*, 1999).

New buds obtained by dark culture from the rhizomes of ten species or cultivars and hybrids of Heliconia were used as explants which were cultured on MS medium supplemented with 2-10 mg/l benzyladenine and 0.2–1.0 mg/l NAA. The experiments showed that the explants could induce shoots and proliferate. The best result for rooting was observed on MS medium supplemented with 0.5 mg/l NAA plus 0.5 per cent active carbon. The plantlets transferred to pots containing sand/turf mixture (1:1), or sand/perlite (1:1), or sand only, showed a survival rate of more than 90 per cent (Zeng-SongJun *et al.*, 2004).

The effects of 3 organic substrates (carbonized rice hull, and dry and green coir dust) and 2 fertilizers (Vitasolo and refuse from earthworm) at 3:1 (v/v) on the acclimatization of plantlets of *Heliconia bihai* obtained by *in vitro* culture were studied. The following interactions were investigated: coir dust (green or dry) x carbonized rice hull; green coir dust x dry coir dust; and Vitasolo x refuse from earthworm. After 75 days, plant height, diameter of pseudostems, number of leaves, and area of the third leaf were evaluated. Carbonized rice hull had a more beneficial effect on the development of micropropagated plants than coir dust (green or dry). No significant difference between dry and green coir dust was observed. The refuse from earthworm was more efficient in the enhancement of plant growth than Vitasolo (Santos *et al.*).

Micropropagation is used in several ornamental species for the large scale commercial propagation of high quality plants. The *in vitro* establishment of a new species should be followed by field evaluation of its behavior to guarantee the maintenance of agronomic characteristics. *Heliconia bihai* var. Lobster Claw I plants obtained from rhizomes and from *in vitro* multiplication were planted in the field. Two hundred plants of each were planted in alternate rows, 20 plants/row, with 10 replications. The experiment was evaluated by the number of suckers, time for flower initiation and cut flower productivity. Flowering was initially observed after 7 months in the field in rhizome-derived plants. The number of suckers,however,was higher in micropropagation-derived plants, resulting in a higher productivity. In addition to a high cut flower production, the market price/plant derived from micropropagation is approximately one third of that for rhizome-derived plants, which makes establishment of a field crop of this species from *in vitro* propagated plants even more viable (Rodrigues *et al.*, 2006)

Genetics and Breeding

Self Incompatibility

Potential for physiological self-incompatibility in *Heliconia bihai* as a mechanism to promote outcrossing was studied. The *H. bihai* is self-compatible. However, plants vary in their degree of compatibility from full to partial self-compatibility. We found only traces of physiological incompatibility but conclusive determination of its mechanism would require further testing. In contrast to Central American Heliconia species, more pollen tubes are able to grow inside *H. bihai* styles following artificial pollination (Melendez-Ackerman *et al.*, 2008)

Varietal Evaluation

The cultivars Andromeda (a clone of *Heliconia psittacorum*) and Golden Torch (*H. latispatha x H. psittacorum*) were evaluated outdoors. Flower production in both cultivars generally started in April or May with a peak between July and September, and a decline in October and November, ceasing when the temperatures dropped below 10°C. The vase life of inflorescences in deionized water averaged 14-17 days for both cultivars held at 23°C. Both cultivars were propagated by division of rhizomes or by tissue culture (Broschat *et al.*, 1984).

Biometric Studies

Genetic variability was assessed in 12 heliconia cultivars. Data were recorded for duration of flowering, number of leaves, plant height, spike length, stalk length, number of bracts, number of flowers/bract, leaf area, bract size, days to first flowering, vase life and number of flowering shoots/ plot/year. Phenotypic coefficient of variation (PCV) were higher than genotypic coefficient of variation (GCV) for all the characters studied. However, higher PCV and GCV estimates were found for flower yield, bract size and leaf area. High heritability with high genetic advance was observed for flower yield, leaf area and bract size (Sheela *et al.*, 2005).

Twelve Heliconia cultivars were assessed for morphological, flowering and vase life parameters. Analysis of variance revealed significant differences among cultivars for all characters studied, indicating high potential for crop improvement through selection. Estimates of phenotypic coefficients of variation were higher than the genotypic coefficient of variation for all characters studied, indicating the role of the environment on the expression of the genotype. High heritability coupled with high genetic advance was recorded for number of flower shoots, stalk length, plant height, duration of flower, bract size, vase life and days to first flowering, indicating the preponderance of additive effects in the expression of these characters (Sheela *et al.,* 2007).

An experiment was conducted to study genetic variability, heritability and correlations between floral stem characteristics important for the genetic improvement of cultivars and hybrids of *Heliconia psittacorum*. Data were recorded for number of days for the inflorescence emission after shoot formation (NDEI), number of days before harvesting the inflorescence after inflorescence emission (NDCI), flower mass stem (MH), number of leaves in the stem at the moment of inflorescence emission (NFH); stem length (CH), and inflorescence length (CI). The highest heritability and genetic coefficient values were 97.33, 85.05, 84.98, 84.48 and 82.25 per cent for NDEI, CH, CI, NFH and MH, and 27.96, 23.02 and 16.89 per cent for NDCI, MH and CI, respectively. The genetic correlation of NDEI with NFH was 0.52, which may indicate NFH as a marker for flowering. CH demonstrated negative genetic correlation with NDEI and DCI (-0.72 and 0.81, respectively), which indicate that high stem length was observed in genotypes with lesser number of days from harvest (Costa *et al.*, 2007).

The association between various characters and the direction and magnitude of different characters towards number of flowering shoots in 12 genotypes of Heliconia (Golden Torch, Deep Orange, Lady Di, Gayana, Petra Orange, Dirooji Red, Parakeet, Sassy, Pascal, Andromeda, Black Cherry and St. Vincent Red) were studied in Kerala, India. Correlation revealed that number of flowering shoots/ year had a strong positive correlation with vase life, number of flowers/bract and size both at genotypic and phenotypic levels (Sheela *et al.*, 2006).

Chemical Profiling

Leaves, spathes, flowers, fruits, seeds, rhizomes and roots of *Heliconia bihai, H. latispatha* and *H. psittacorum* collected in Sucre State, Venezuela, were assayed. Interspecific differences in the presence or absence of alkaloids, saponins, flavonoids, tannins, polyphenols, essential oils and/or anthraquinones were observed. Alkaloids were detected in all organs of *H. bihai* except the spathes, in leaves of *H. latispatha* and in leaves and seeds of *H. psittacorum*. Saponins, flavonoids and polyphenols were detected in all species, whereas tannins, essential oils and anthraquinones were absent in all species (Sanabria *et al.,* 1998).

Molecular Markers

Many *Heliconia spp*. are polymorphic with a large number of cultivars. Cultivar identification was primarily based on morphological differences of the flowers and inflorescences. A method for DNA extraction from Heliconia leaves was developed and RAPD analysis was used to detect genetic variations and similarities among *Heliconia* species, cultivars and hybrids. Significant differences in RAPD profiles occurred amongst different *Heliconia* spp. and some distantly related plants. *H. rostrata* had 3 prominent bands that were absent in *H. psittacorum* cultivars. All 19 cultivars of *H. psittacorum* tested had similar profiles using 3 different primers. Three cultivars of the hybrid *H. psittacorum* x *H. spathocircinata* (Golden Torch, Red Torch and Alan Carle) had slightly different RAPD profiles which were distinct from that of *H. psittacorum* x *H. marginata*. Of the 3 triploid cultivars of *H. psittacorum* tested, Iris and Petra had identical profiles with all the primers. A prominent band identified in Iris and Petra was absent in Sassy. It is concluded that Iris and Petra are the same genotypes (Goh *et al.*, 1995)

A protocol was developed to extract DNA from Heliconia leaves and to analyse genetic variation using RAPD. The percentage genetic similarities among *Heliconia* species, cultivars and hybrids were determined. Data from 11 primers suggested that the RAPD technique can be used to distinguish species and cultivars of Heliconia. Using a single 10-mer primer (OPA18), distinct RAPD profiles were determined for 16 cultivars of *H. psittacorum* grown at the Jurong BirdPark, Singapore, a Heliconia Society of International depository. It is suggested that the characteristic profiles generated by RAPD may be used as additional DNA markers for classifying different species and cultivars of Heliconia. The phylogenetic tree derived from the RAPD data showed that all of the 16 cultivars examined are closely related to each other, providing the first genetic evidence that this large group of cultivars has a common genetic background. Moreover, 2 triploid cultivars of *H. psittacorum* (Iris and Petra) showed identical RAPD profiles with 10 different primers in agarose and polyacrylamide gels, suggesting that they are of the same genotype.(Kumar *et al.,* 1997). Seventeen *Heliconia* species and varieties were analyzed using RAPD markers. Eight primers, which produced the highest number of bands, were used for DNA amplification. The genetic similarity matrix constructed with Jaccard's coefficient using RAPD marker scores showed that the highest value was between Petra Orange and Parakeet, while the lowest was between Golden Torch and *H. humilis.* The 17 species and varieties of Heliconia formed nine distinct clusters at similarity coefficient value of 0.42, implying a strong parallelism between genetic and morphologic/taxonomic variability of Heliconia genotypes. Petra Orange, Deep Orange, Parakeet, Pascal, and Alan Carle formed a big cluster within which Petra Orange and Parakeet formed a more cohesive entity (Sheela *et al.,* 2006)

Mutation Breeding

The high economic importance of ornamental plants represents a stimulus for breeders in the research for new cultivars. Among the several employed techniques, gamma radiation induced mutation can bring promising results in different *Heliconia* species. The 50 per cent lethal dose (LD 50) determination, in *in vitro Heliconia bihai* cultivation, is the initial step for utilization of that breeding method. Different doses, *i.e.* 0, 10, 20, 30, 40 and 50 Gy gamma rays in 25 shoot apices were tested. Results indicated that the LD 50 for this *in vitro* culture ranged between 30 and 40 Gy (Rodrigues *et al.,* 2005).

Cultivation

Soil and Culture Medium

The effects of culture medium (field soil, yellow soil + peat + mushroom compost, yellow soil + mushroom compost + sand and yellow soil + peat) on growth,cut flower yield, leaf structure and ultramicroscopic structure of *H. psittacorum* were determined. *H. psittacorum* produced green leaves with normally developed cells and high cut flower yield when grown in yellow soil and peat. The mixture of yellow soil + peat + mushroom compost or yellow soil + mushroom compost + sand inhibited growth of the plants and gave plants with leaves exhibiting chlorosis, densely arranged cells and damaged chloroplasts.(Chen-GuoJu *et al.,* 2005).

A study was conducted in a greenhouse to estimate the evapotranspiration and crop coefficient for *Heliconia psittacorum* x *H. spathocircinada* cultivar Alan carle. The evapotranspiration (ETc) was estimated using soil water balance method. A class A pan evaporation was used to estimate the reference evapotranspiration (ETo). It has been used inside the greenhouse and another one outside, from an agrometeorological station located 500 m from the experimental area. Using the ETc and ETo data, the crop coefficient (Kc) was determined for each crop development stage. The crop evapotranspiration average was 2.2 mm/day. The Kc from ETo estimated by class A pan evaporation in the greenhouse was higher than those from the estimated ETo using the class A pan evaporation located at the agrometeorological station. Thus, the average values of crop coefficient for initial crop development and flowering stages were 0.41, 0.78 and 1.26, respectively, when using inside pan evaporation and 0.29, 0.60 and 0.92, when using outside pan evaporation data (Gomes *et al.*, 2006).

Manures and Fertilizers

Golden Torch plants (probably *Heliconia latispatha* x *H. psittacorum*) range in height from 1.0 to 1.8 m, and produce 4 or 5 leaves/shoot followed by a terminal inflorescence with 3 or 4 bracts. The entire inflorescence is uniformly orange-yellow in colour. In trials in Southeastern Florida with outdoor plants, flower production peaked from July to September and, under full sun and with high NPK application (3.6 kg 18-6-12 NPK m2/year), averaged 84 flowers/m2/year. The optimum temperature range for flower production was 21 to 35 degrees C. Growing plants under 63 per cent shade reduced flower production by about 50 per cent. The postharvest life of cut inflorescences placed in deionized water averaged 14 to 17 days at 23°C. (Broschat *et al.*, 1984a). Plants of Andromeda (a clone of *Heliconia psittacorum*) range in height from 1.0 to 1.8 m, and produce 5 leaves/shoot followed by a terminal inflorescence with 3 or 4 bracts. The bracts are red, fading to light orange at their bases. The lower half of each bract and the upper 2 to 3 cm of the peduncle are covered with a waxy white bloom. The florets are orange with black tips and the main axis of the inflorescence is orange. In trials in Southeastern Florida with outdoor plants, flower production peaked from July to September. Under full sun and with high NPK application (3.6 kg 18-6-12 NPK/m^2/year) flower production averaged 130 flowers/m^2/year in the first year and 160 flowers/m^2/year in the second year. The optimum temperature range for flower production was 21 to 35 degrees C. Growing plants under 63 per cent shade reduced flower production to about 35 flowers/m2/year in first-year beds and 65 to 75 flowers/m^2/year in second-year beds. The postharvest life of cut inflorescences placed in deionized water averaged 14 to 17 days at 23°C (Broschat *et al.*, 1984b).

Heliconia psittacorum cultivars St. Vincent Red and *H. psittacorum* x *H. spathocircinata* cv. Golden Torch were supplied with N (200, 400 and 600 kg/ha), P (80, 169 and 240 kg/ha) and K (80, 160 and 240 kg/ha) fertilizers in a field experiment conducted in Goias. Pseudostem length, floral stem length and diameter and flower production in cv. Golden Torch and flower production in cv. St. Vincent Red increased with increasing rates of N. Increasing rates of P and K resulted in decreasing values of the parameters measured. NPK at 600:160:160 and 400:240:160 kg/ha were optimum for flower production of cultivars Golden Torch and St. Vincent Red, respectively (Ferreira and Oliveira, 2003). Containerized plants of *Heliconia psittacorum x H. spathocircinata* (cv. Golden Torch) were grown in a greenhouse for 8 months from early summer to winter under selected combinations of N, P, and K. Fertilizer rates ranged from zero to rates which exceeded those reported in the literature by 50-100 per cent. Biomass variables (vegetative and inflorescence dry weight, and leaf area) were predicted to be maximized at high N and high N:P, and N:K ratios corresponding to N-P-K application rates of 1.2, 0.5, and 0.6 kg/m^3, respectively (~2:1:1). However, the number of shoots and flowers produced/rhizome were maximal at lower N:K ratios (1:1). Flower yield could therefore be optimized with appropriate fertilizer application, provided attention was paid to the N:K ratio so that size of plants and their flowers were not compromised by efforts to increase shoot and flower number. The heavier the rhizome planted, the shorter the time for shoot emergence and flowering to occur, and greater the number of flowers harvested. However, rhizome weight had no effect on number of shoots to emerge. The probability of shoots flowering declined markedly with order of shoot emergence, although this could be increased with appropriate mineral nutrition. The maximum number of leaves subtending the inflorescence (7) was obtained at high N and P rates. Flower production was probably limited by declining solar radiation in autumn, and by within-plant competition for rooting space (Clemens and Morton, 1999).

Plant Density

The effect of two spacings (20 and 40 cm) and 4 sizes of rhizome (10, 20, 30 and 40 g) on number of inflorescences/m^2, growth and other flower characters of *H. psittacorum* were studied. The wider spacing accounted for taller plants and larger inflorescences. The height of plant increased with increase in size of rhizome. The 40 g rhizome exhibited the greatest plant height of 116.80 cm compared to 91.77 cm exhibited by the smallest rhizome (10 g). A similar trend in growth was observed with respect to number of leaves/clump, number of leaves/m^2, number of shoots/clump and number of shoots/m^2. For these characters, the largest rhizome gave the highest mean performance. The largest rhizome also resulted in maximum inflorescence/clump and/metre square, but the inflorescence size was reduced considerably. The smallest rhizome produced the largest inflorescence of 18.56 cm. (Lalrinawani and Talukdar, 2000).

The effect of two spacings (20 and 40 cm) and 4 sizes of rhizome (10, 20, 30 and 40 g) on number of inflorescences/m^2, growth and other flower characters of *H. psittacorum* was studied during 1995-96, in Jorhat, Assam, India. The wider spacing accounted for taller plants and larger inflorescence. The height of the plant increased with increase in size of rhizome. The 40 g rhizome exhibited the highest plant height of 116.80 cm compared to 91.77 cm exhibited by the smallest rhizome (10 g). A similar trend in growth was observed with respect to number of leaves/clump, number of leaves/m^2, number of shoots/clump and number of shoots/m^2. For these characters, the largest rhizome gave the highest mean performance. The largest rhizome also resulted in maximum inflorescence/clump/metre square, but the inflorescence size was reduced considerably. The smallest rhizome produced the largest inflorescence of 18.56 cm (Lalrinawani and Talukdar *et al.*, 2001).

H. stricta and *H. rauliniana* were planted outdoors at 0.5 x 1.0 m, 1.0 x 1.0 m, 1.5 x 1.0 m or 2.0 x 1.0 m. The first inflorescence appeared within approx. 180 days at all spacings. The number of inflorescences/plant was the highest at 0.5 x 1.0 m (8.6 for *H. stricta* and 1.5 for *H. rauliniana*). Stem length was commercially acceptable in *H. stricta* at 0.5 x 1.0 m and 1.0 x 1.0 m and in *H. rauliniana* at all spacings. Inflorescence size was commercially acceptable in both species at all spacings but plants at 1.5 x 1.0 m and 2.0 x 1.0 m had high number of discoloured inflorescences (Ibiapaba *et al.*, 1997). *H. psittacorum* cv. Sassy and Andromeda were grown at spacings of 0.50 x 0.25, 0.50, 0.75 or 1.00 m. In all spacings, the first shoots appeared 20-30 days after transplanting and the first inflorescences after about 120 days. The inflorescences lasted about 14 days. Both cultivars produced the largest number of buds and inflorescences at a spacing of 0.25 x 0.50 m, with Sassy producing more than Andromeda. Treatments did not affect inflorescence length or width. Andromeda showed the best flower stem length at a spacing of 0.50 x 0.50 m, while there was no effect of spacing in Sassy. Flower stems were strong and resistant to handling (Ibiapaba *et al.*, 2000). The effects of rhizome size (10, 20, 30 or 40 g) and spacing (20 x 20, or 40 x 40 cm) on the rhizome production of *H. psittacorum* (a commercial cut flower plant) were studied in Assam, India. The largest rhizomes (40 g) and wider spacing (40 x 40 cm) resulted in the highest values for girth, spread, number of rhizomes/clump and rhizome yield/clump. However, number of rhizomes/m^2 and rhizome yield/m^2 were higher (343 and 7160 g, respectively) for the closer spacing (Lairinawmi and Talukdar, 1999)

Photoperiod

The natural flowering season for *Heliconia* species in their natural habitats is sometimes indicated in the taxonomic literature, but it may be influenced locally by rainfall and drought periods as well as by photoperiod and may not be reliable in indicating production periods elsewhere. Wild-collected plants are often introduced with little or no information on their season of flowering. With more than three dozen species of *Heliconia* grown and exported in international trade, the seasonality of flowering is important to the supply and marketing of this bold tropical flower. In Hawaii, sale records from several commercial growers were collected over several years to identify the peak seasons of flowering as well as duration of flowering season. Such information is helpful in coordinating with the flower markets. *Heliconia* species of commercial interest with strong seasonal flowering periods are noted (angusta, bihai, caribaea, caribaea x bihai, collinsiana, lingulata, rostrata, stricta and wagneriana) as well as species with longer flowering periods such as the *H. psittacorum* cultivars and hybrids (Criley, 2000).

Flowering responses of *H. psittacorum x H. spathocircinata* cv. Golden Torch to temperature and photosynthetic photon flux (PPF) were determined in controlled-environment conditions using a 2 x 2 factorial combination of temperature (32°C day/20°C night and 24°C day/20°C night) and PPF (475 and 710 micro mol/m^2s). Temperature had no significant effect on new shoot production, with an average of 9.3 shoots/plant being produced over the 248 days of treatment. More shoots, however, were produced at the higher PPF level (10.1 shoots, compared with 8.3 shoots at the lower light level). The proportion of shoots that initiated flowers (85 per cent) was similar in all treatments. The time from shoot until inflorescence emergence was significantly shorter at 32/20°C than at 24/20°C (140 and 146 days, respectively) and was unaffected by PPF. This duration was also significantly affected by the interacting effects of order of shoot appearance and number of leaves subtending the inflorescence. The second shoot to emerge had the shortest duration from shoot emergence to inflorescence emergence. The number of leaves subtending the inflorescence increased at the higher temperature and decreased as shoot order increased but was unaffected by PPF. Temperature and PPF levels influenced total leaf area at flowering, with the highest areas being achieved in the high temperature-low PPF combination. Acceptable flower quality with at least 2 opened, well-formed, well-coloured bracts was obtained in all treatments, although flower stems were taller and thicker at 32/20°C and these dimensions increased further with increasing order of shoot appearance. Stems tended to be thinner at the lower PPF level. Overall, temperature was more dominant than light in influencing production and quality of flowers, but developmental factors associated with the order of shoot appearance also played a significant role. Cut flower production of Golden Torch should be feasible in temperature-controlled greenhouses in temperate regions where mean air temperatures can be maintained at ~20 degrees (Catley and Brooking, 1996).

It was previously reported that Heliconia cv. Golden Torch leaves grown in full sunlight exhibited a sustained decrease in PS II efficiency as compared to those grown under shaded conditions. In this study, full sunlight plus low levels of fertilizer application caused a further reduction in photosynthesis, chlorophyll content and Fv:Fm ratio while plants grown at high nutrient levels showed higher values of all these parameters. When plants were grown under intermediate or deep shade, there was no significant difference in any of the parameters irrespective of nutrient supply. In the recovery experiments, plants without fertilizer were refertilized weekly. Maximal photosynthetic rates, chlorophyll content, and Fv:Fm ratio increased gradually after refertilizing the plants grown in full sunlight. However, no significant changes in these parameters were observed in plants grown under intermediate or deep shade over the same period. Total leaf N was measured parallel with all the parameters. Photosynthetic rates, chlorophyll content and Fv : Fm ratio showed a clear linear correlation with total leaf N in plants grown in full sunlight while no clear relationship was observed in plants grown under intermediate or deep shade. These results suggest that acclimatization of Heliconia under full sunlight could be achieved by high nutrient levels.(Jie-He *et al.*, 2000)

Although *Heliconia* is a tropical genus, many species exhibit highly seasonal patterns of flowering. Some cases have been attributed to seasonal rainfall, but research has demonstrated that *H. wagneriana* and *H. stricta* cv. Dwarf Jamaican are short day (SD) species while *H. angusta* initiates its flowers during long days (LD). *H. stricta* cv. Dwarf Jamaican initiates its flowers when its pseudostem has 3 unfurled leaf blades. Anthesis is reached 15-19 weeks after the start of SD. *H. angusta* initiates its flowers when daylengths exceed 13.3 hours and requires 15-17 weeks of long days to reach anthesis. Grower shipping records are used to identify other species with marked seasonality. The results from these studies may be used to manipulate flowering times to advance or extend normal flowering periods (Criley *et al.*, 1999).

Influence of habitat on physiological and structural characteristics was investigated in *H. imbricata*, *H. latispatha*, *H. pogonantha* and *H. wagneriana*, occurring in open sites, receiving full sunlight, *H. mathiasiae*, found at the forest edge under partial sunlight, and *H. irrasa* subsp. *undulata* and *H. umbrophila*, occurring in the forest understorey under deep shade. Light response curves showed a clear gradient with respect to light saturation and rates of maximum net assimilation (Amax). *H. latispatha* showed saturation at higher photon flux densities (PFD, 1400 micro mol/m^{-2} s^{-1}) and higher Amax (14-16 micro mol/m^{-2} s^{-1}) than *H. mathiasiae* (PFD 1000 micro mol/m^{-2} s^{-1}; Amax 7.5-8.5 micro mol m^{-2} s^{-1}) and *H. irrasa* (PFD 250 micro mol m^{-2} s^{-1}; Amax 3.5 mol/m^{-2} s^{-1}). Leaf blade areas were the largest in open sites, and leaf specific mass was also significantly higher, but leaf support efficiency was the highest in understorey species. Species in open sites had thicker leaves with more chlorenchyma, whereas deep-shade species had very thin leaves and low stomatal densities. These rapidly growing herbaceous perennials appear to allocate much of their aboveground biomass to leaf tissues and have a relatively low investment in support tissues, possibly as a result of presence of belowground rhizomes (Rundel *et al.*, 1998)

Seasonal flowering behaviour of *H. wagneriana* was caused by short daylengths (SD) in experiments using artificial short days (8-9 h) and long days as daylength extension or night break lighting with incandescent lamps. The natural time for floral initiation was estimated to be mid- to late October (11 h 40 minutes to 11 h 20 minutes) in Hawaii, and 120-150 days were required from the onset of inductive SD to inflorescence emergence. It is suggested that this information may be used to manipulate flower availability for flower markets (Criley and Sakai, 1997)

Growth and development of *H. bihai* cv. Lobster Claw One and *H. latispatha* (cut flower crops) were studied under 3 shade levels (0, 40 or 60 per cent) for 20 months, during which 5 generations of shoots developed. Plants without shade produced the most pseudostems. The number of shoots/ clump was greater in *H. bihai* than in *H. latispatha*. Abortive and generative shoots developed from all of the first 4 generations. *H. latispatha* showed a higher percentage of aborted shoots than *H. bihai.* Generative shoot height increased with increasing shade and successive generations. The number of leaves/generative shoot decreased as new generations emerged in both species. In *H. bihai* the first 3 generations flowered simultaneously, when the clump reached the age of 12 months; this flowering period lasted 6 months with a peak during Mar.-June, when 95 per cent of the flowers developed. In *H. latispatha* flowering began when plants were 10 months old and showed an irregular pattern during the cycle, with the peak (82 per cent) occurring in July and Aug. The generative stage could begin at the apex when the number of unfurled leaves on the pseudostem was 4 or 5 in *H. bihai* and 5 in with H. *latispatha* (Maciel and Rojas, 1994).

Although *Heliconia* is a tropical genus, many species exhibit highly seasonal patterns of flowering. Some cases are attributed to seasonal rainfall, but research has demonstrated that *H. wagneriana* and *H. stricta* cv. Dwarf Jamaican are short day (SD) species while *H. angusta* initiates its flowers during long days (LD). *H. stricta* cv. Dwarf Jamaican initiates its flowers when its pseudostem has 3 unfurled leaf blades. Anthesis is reached 15-19 weeks after the start of SD. *H. angusta* initiates its flowers when daylengths exceed 13.3 hours and requires 15-17 weeks of long days to reach anthesis. Grower shipping records are used to identify other species with marked seasonality. The results from these studies may be used to manipulate flowering time to advance or extend normal flowering periods (Criley *et al.*, 1994). Although year-round flowering of *H. chartacea* is potentially possible in Hawaii as new shoots develop regularly, flowering is low in the period from late Mar. to early June. The shoots which should flower at this time are produced 40-48 weeks earlier in June-Aug. Floral initiation occurs after shoot emergence when 4 leaves have unfurled with 2-3 still within the enclosing pseudostem. This places flower initiation in the Oct.-Dec. timeframe for these shoots. In contrast, autumn-winter shoots initiate flowers during Jan.-Mar. and average 1 leaf fewer than summer shoots. The greater leaf count and low flowering (per cent) of spring shoots suggest autumn conditions are favourable for leaf initiation and unfavourable for floral initiation. An attempt to promote flower initiation during the autumn with light-break lighting was thwarted by disease (Phytophthora) (Criley and Lekawatana,1995)

Plants of *H. stricta* cv. Dwarf Jamaican were grown in 10-litre containers under full sun or 50 per cent shade for 1 year, and plants of *H. caribaea* cv. Purpurea were grown in an open field for 2.5 years. Rhizomes were soaked for 1 h before planting or plants were sprayed with 30 micro M DCPTA ((3,4-dichlorophenoxy)triethylamine) after 2 leaves had emerged. *H. stricta* plants grown under full sun produced more inflorescences than those grown under 50 per cent shade and DCPTA-treated plants grown under shade produced more pseudostems and were taller than control plants under shade. DCPTA-treated *H. caribaea* plants produced more pseudostems/plant than control plants during their first year, but differences in the number of pseudostems and inflorescences during subsequent years were not significant (Broschat and Svenson,1994).

Vase Life

Tropical flowers, in particular *Heliconia psittacorum,* have a vase life that is limited by inhibition of water uptake. The block to water uptake is mainly due to exudation of mucilage at the cut surface. The effects of benzyladenine on the vase life of heliconia (*H. latispatha*) inflorescences harvested at 2 stages of development were studied. The inflorescences were sprayed with 0, 100, 200 or 300 mg benzyladenine/litre and harvested at stage characterized by the presence of 1-2 fully opened bracts (stage 1) and 3-4 fully opened bracts (stage 2). The experiment was arranged in a random block design consisting of 9 stalks/treatment. The growth regulator was sprayed twice until the stalks were completely wet within the first hour after harvest. The average number of open bracts throughout the vase life was not affected by benzyladenine (2.5 for stage 1 and 4.2 for stage 2). Treatment with benzyladenine extended flower longevity at both stages of development. A positive correlation was observed between benzyladenine concentration and increase in vase life. Spraying benzyladenine(300 mg/litre) increased flower longevity by at least 1.85 fold compared to the untreated flowers.(Moraes, de; *et al.*, 2005).

The sensitivity of cut flowers and foliage to irradiation used for insect disinfestation varies from species to species and to a lesser extent variety to variety. Medium sensitive flowers were heliconia 'Red Stricta' (*Heliconia stricta*). Flowers which showed high sensitivity to 250 Gy were Heliconia 'Keanae'. For the heliconia 'Keanae' inflorescences, a hot water treatment at 45-47.5°C for 40 min effectively reduced irradiation injury (Sangwanangkul *et al.*, 2008).

The optimum harvest time and effective pretreatment solutions and treatment methods of cut *Heliconia rostrata* in relation to the control of spath browning were determined. The middle flowering bud stage was the optimum harvest stage. Whole flower twigs were immersed completely for 1 h in a solution containing 500 mg H< sub>3</sub>BO< sub>4</sub>/litre or 500 mg MgSO< sub>4</ sub>/litre, then placed in a vase solution containing 2 mg benzyladenine + 200 mg Al< sub>2</ sub>(SO< sub>4</sub>)< sub>3</sub> and finally sprayed with 0.1 per cent polyvinyl alcohol. The treatments significantly controlled the browning of spath tissues of cut *Heliconia rostrata*, maintained at a low pH value and reduced the increase of membrane permeability. After the treatments, vase life was 12 and 11.5 days.(Sheng-Aiwu and Liu-Nian, 2006). Studies were conducted to evaluate *Heliconia psittacorum x H. spathocircinata* cv. Golden Torch, first flower stem morphological and physiological characteristics as well as postharvest longevity, under macronutrients omissions. A greenhouse experiment was conducted under the technique of the lacking element. Inflorescences produced under N omission treatment showed a pale orange colour and floral stem deformation. Stem length and diameter as well as inflorescence length were reduced under omission of N, P or K, down to 31.23 per cent when compared to the complete solution treatment. Floral stem dry matter and post harvest longevity were reduced on 67 and 38.46 per cent, respectively, on either N or K omission treatments. These nutrient omission treatments reduced floral stem production from the second shoot emitted. Greater postharvest longevity is to be found at higher floral stem diameter and dry matter. Carbohydrate ratio in underground parts presented positive correlation (0.90) with floral stem dry matter. Floral stems showed greater carbohydrate ratio on the leaves withdrew inflorescence appearance(Castro,-A-C-R; *et al.*, 2007). The longevity of cut flowers is limited by its ephemeral nature and by several stresses, leading to the decrease of water uptake, depletion of stored carbohydrates, increase of respiratory activity and ethylene production, and by enhancing flower sensitivity to ethylene. Wide range of techniques are available to extend flower preservation, including use of flower preservatives, inhibitors of ethylene action, growth regulators, calcium and control of temperature and flower dehydration. For most flowers, the xylem blockage of a cut flower is reduced by recutting the stem, reducing the pH of the vase water to 4 and by applying antibacterial substances to the water. Use of sucrose in pulsing solution or as constituent of vase solution may extend the vase life of flowers by improving water balance, stimulating flower opening or by delaying the senescence due to lower synthesis of ethylene, as observed in carnation. Flowers can be classified as sensitive, low or ethylene-insensitive based on the responses to the growth regulator. Cytokinins and gibberellic acid may have positive affect on the vase life of some flowers, specially in those with low or ethylene-insensitive, like *Heliconia sp*. The reduction of temperature reduces tremendously the CO< sub>2</sub> and ethylene production and its action. However, chilling sensitive flowers such as bird-of-paradise and Heliconia cannot be stored bellow 10 to 13 degrees C due to intense tissue discolouration (Finger and Barbosa, 2006).

Insect Pests and Diseases

Snail (*Achatina fullica*)

Occasionally, snails chew the young leaves of heliconias by scraping, making big irregular holes, Adult and young ones devour plants during nights.

Management

Handpicking and killing by dropping snails in 5 per cent salt solution reduce population. Sprinkling of salt crystals in the paths or around pots kills snails as they come into contact. Spreading Snail kill (3 per cent metaldehyde pellets) in the field effectively reduces snail population. Spray of neem oil (10 ml/l) and soap nut extract (60 g/l) protect foliage from damage.

Harvesting

The flowers are harvested with peduncles of 70 cm or more. Flower stalks should be cut near the ground early in the day while still turgid. It is important to minimize desiccation after harvest as heliconias do not take up water well, and flowers harvested at midday can have as much as a week poorer post harvest life than those harvested in the morning. Inflorescences last longer if harvested from well irrigated plants than from plants under stress. Flowers must be cut at the desired stage of opening since further opening of the bracts does not occur after cutting,

Postharvest

After trimming, the whole inflorescence is given an insecticide dip (5 minutes) and subsequently rinsed. Manual cleaning is necessary on some inflorescence to dislodge insect and remove dead flowers. The stalks are held in water to prevent drying of inflorescence before packing. Heliconias have a post harvest life of 14 to15 days in deionized water and flowers are damaged when stored at temperatures below 10°C.

Grading and Packaging

The large heliconias are usually cut to fit a 150 cm long box while the *H. psittacorum* types are trimmed to lengths of 60 to 90 cm. They are often packed in a bunch of 10, and sleeved in plastic film or open weave netting. Up to 25 such bunches may be packed in a 150 x 50x 25 cm box. Medium sized heliconias such as *H. bihai*, and *H. stricta* may be packed 20 to 50 per box while the large *H. caribaea* are packed as 10 to 5 pieces.

References

Abalo, J.Ev., 1999. Heliconias for the ornamental industry. *Acta Horticulturae*, 486: 313–315

Bora, S. and Paswan, L., 2003. *In vitro* propagation of *Heliconia psittacorum* through axillary bud. *Journal of Ornamental Horticulture* New Series, 6(1): 11–15

Broschat, T.K. and Svenson, S., 1994. EDCPTA enhances growth and flowering of heliconias. *HortScience,* 29(8): 891–892.

Broschat, T.K., Donselman, H.M. and Will, A.A., 1984. 'Andromeda' and 'Golden Torch' heliconias. *HortScience,* 19(5): 736–737.

Broschat, T.K., Donselman, H.M. and Will, A.A., 1984. Andromeda: A red and orange heliconia for cut flower use. *Circular, Agricultural Experiment Stations,* University of Florida, S–309: 5.

Broschat, T.K., Donselman, H.M. and Will, A.A., 1984. Golden Torch: An orange heliconia for cut-flower use. *Circular, Agricultural Experiment Stations,* University of Florida, S–308: 4.

Bruna, E.M. and Ribeiro, M.B.N., 2005. Regeneration and population structure of *Heliconia acuminata* in Amazonian secondary forests with contrasting land-use histories. *Journal of Tropical Ecology*, 21(1): 127–131.

Castro, C.E.F.de, May, A. and Goncalves, C., 2006. Heliconia's species for cut flowers. *Revista Brasileira de Horticultura Ornamental*, 12(2): 87–96.

Catley, J.L. and Brooking, I.R., 1996. Temperature and light influence growth and flower production in Heliconia 'Golden Torch'. *HortScience*, 31(2): 213–217.

Clemens, J. and Morton, R.H. Optimizing mineral nutrition for flower production in Heliconia 'Golden Torch' using response surface methodology. *Journal of the American Society for Horticultural Science*, 124(6): 713–718.

Chen, GuoJu; Chen, RiYuan; Wu, XiaoYing; Huang, DingHua and He ShuiHua, 2005. The effect of different culture mediums on growth and cell structure of *Heliconia psittacorum*. *Journal of South China Agricultural University*, 26(1): 28, 30.

Costa, A.S.da; Loges, V; Castro, A.C.R de; Verona, A.L., Pessoa, C. de O. and Santos, V.F. dos, 2006. Number of shoots and area per clump of heliconia. *Horticultura Brasileira*, 24(4): 460–463.

Criley, R.A., 1988. Propagation of tropical cut flowers: *Strelitzia, Alpinia* and *Heliconia. Acta Horticulturae*, 226(11): 509–517.

Criley, R.A., 2000. Seasonal flowering patterns for Heliconia shown by grower records. *Acta Horticulturae*, (541): 159–165.

Criley, R.A. and Paull, R.E., 1993. Review: Postharvest handling of bold tropical cut flowers *Anthurium, Alpinia purpurata, Heliconia*, and *Strelitzia*. *Acta Horticulturae*. 337: 210–212.

Criley, R.A., Sakai, W.S., Lekawatana, S. and Kwon, E., 1999. Photoperiodism in the genus *Heliconia* and its effect upon seasonal flowering. *Acta Horticulturae*, 486: 323–327.

Criley, R.A., Uchida, J.Y. and Fu, Z.Q., 2003. Productivity and periodicity of flowering in *Heliconia orthotricha* cultivars. *Acta Horticulturae*, 624: 207–212.

Criley, R.A., Sakai, W.S., Lekawatana, S. and Kwon, E., 1999. Photoperiodism in the genus *Heliconia* and its effect upon seasonal flowering. *Acta Horticulturae*, 486: 323–327.

Criley, R.A., 1995. Propagation of Zingiberaceae and Heliconiaceae. *Revista Brasileira de Horticultura Ornamental*, 1(1): 14–21.

Cruz, D.D. da, Mello, M.A.R. and Sluys, M. van, 2006. Phenology and floral visitors of two sympatric *Heliconia* species in the Brazilian Atlantic forest. *Flora Jena*, 201(7): 519–527.

Cutraro, J.L., Ercelawn, A.Y., LeBrun, E.G., Lonsdorf, E.W., Norton, H.A. and McKone, M.J., 1998. Importance of pollen and nectar in flower choice by hummingbird flower mites, *Proctolaelaps kirmsei* (Mesostigmata: Ascidae). *International Journal of Acarology*, 24(4): 345–351.

Ferreira, L.B. and Oliveira, S.A., 2003. Study of fertilization with NPK in the growth variable and inflorescences productivity of *Heliconia* sp. *Revista Brasileira de Horticultura Ornamental*, 9(2): 121–127.

Finger, F.L. and Barbosa, J.G., 2006. Postharvest physiology of cut flowers. In: *Advances in Postharvest Technologies for Horticultural Crops*, p. 373–393.

Goh, C.J., Nathan, M.J., Kumar, P.P., Criley, R.A. and Lekawatana, S., 1995. Year around production with high yields may be a possibility for *Heliconia chartacea*. *Acta Horticulturae*, 397: 95–101

Goh, C.J., Kumar, P.P. and Yau, J.C.K., 1995. Genetic variations detected with RAPD markers in Heliconia. *Acta Horticulturae*, 420: 72–74

Gomes, A.R.M., D' Avila, J.H.T., Gondim, R.S., Bezerra, F.C. and Bezerra, F.M., 2006. Evapotranspiration and crop coefficient of *Heliconia psittacorum* L x *H. spathocircinada* (Arist) under greenhouse conditions. *Revista Ciencia Agronomica*, 37(1): 13–18.

Innecco, R., 2000. Evaluation of spacing in *Heliconia psittacorum* L. cv. Sassy and Andromeda. *Ciencia e Agrotecnologia*, 24(1): 181–186.

Ibiapaba, M.V.B. da, Luz, J.M.Q., Shiau, YihJuh, Wang, TsaiYih and Tsay, HsinSheng, 1999. Studies on tissue culture of *Heliconia psittacorum* cv. Rhizomatosa. III. Effect of cefotaxime and plant growth regulators on growth of shoot tip explants in vitro. *Journal of Agricultural Research of China*, 48(4): 42–51.

Ibiapaba, M.V.B. da and Luz, J.M.Q., 1997. Innecco performance of two *Heliconia* species at different planting spacings in Fortaleza (CE). *Revista Brasileira de Horticultura Ornamental*, 3(2): 74–79.

Jerez, E., 2007. The culture of heliconia. *Cultivos Tropicales*, 28(1): 29–35.

Jie, He, Lay, PhengTan and Chong, JinGoh, 2000. Alleviation of photoinhibition in Heliconia grown under tropical natural conditions after release from nutrient stress. *Journal of Plant Nutrition*, 23(2): 181–196.

Kumar, P.P., Yau, J.C.K. and Goh, ChongJin, 1998. Genetic analyses of *Heliconia* species and cultivars with randomly amplified polymorphic DNA (RAPD) markers. *Journal of the American Society for Horticultural Science*, 123(1): 91–97.

Kress, W.J., 1983. Crossability barriers in neotropical Heliconia. *Annals of Botany*, 52(2): 131–147.

Lalrinawani and Talukdar, M.C., 2000. Effect of spacing and size of rhizome on the flower production of Heliconia (*Heliconia psittacorum* L.). *Journal of the Agricultural Science Society of North East India*, 13(1): 48–51.

Lopez, G.A., Sanabria, C.h.E.M. and Davila, P.M., 2001. Comparative morphology of vegetative structures of cultivars 'Distans' and 'Red Yellow' of *Heliconia latispatha* Bentham. In: *Proceedings of the Interamerican Society for Tropical Horticulture*, 45: 82–86.

Lairinawmi and Talukdar, M.C., 1999. Effect of spacing and size of rhizome on the rhizome production of heliconia (*Heliconia psittacorum L.*). *Horticultural Journal*, 12(1): 45–49.

Lairinawmi and Talukdar, M.C., 2000. Effect of spacing and size of rhizome on the flower production of Heliconia (*Heliconia psittacorum* L.). *Journal of the Agricultural Science Society of North East India*, 13(1): 48–51.

Lee, Y.H., Ng, N.Y. and Goh, C.J., 1994. Pollen formation and fruit set in some cultivars of *Heliconia psittacorum*. *Scientia Horticulturae*, 60(1/2): 167–172.

Lekawatana, S. and Criley, R.A., 1989. Pot culture of *Heliconia stricta* 'Dwarf Jamaican'. *Acta Horticulturae*, 252: 123–128.

Melendez Ackerman, E. and Rojas Sandoval, J., 2008. Self compatibility of microgametophytes in *Heliconia bihai* (Heliconiaceae) from St. Lucia. *S Caribbean Journal of Science*. 44(2): 145–149.

Moraes, P.J. de, Finger, F.L., Barbosa, J.G., Cecon, P.R. and Cesa, L.P., 2005. Influence of benzyladenine on longevity of *Heliconia latispatha* Benth. *Acta Horticulturae*, 683: 369–373.

Maciel, N. and Mogollon, N., 1998. Germination in six ornamental Zingiberales. *Proceedings of the Interamerican Society for Tropical Horticulture*, 41: 55–61.

Maciel, N. and Rojas, E., 1994. Growth and development of *Heliconia bihai* and *H. latispatha* under different levels of shade. In: *Proceedings of the Interamerican Society for Tropical Horticulture*, 38: 257–263.

Nathan, M.J., Goh, C.J. and Kumar, P.P., 1992. *In vitro* propagation of *Heliconia psittacorum* by bud culture. *HortScience*, 27(5): 450–452.

Nathan, M.J., Kumar, P.P. and Goh, C.J., 1993. High frequency plant regeneration in *Heliconia psittacorum* L. f. *Plant Science Limerick*, 90(1): 63–71.

Rodrigues, P.H.V., Ambrosano, G.M.B., Lima, A.M.L.P., Mendes, B.M.J. and Rodriguez, A.P.M., 2006. *Heliconia bihai* var. Lobster Claw. I. Cut flower field production from micropropagated versus rhizome derived plants. *Floriculture, Ornamental and Plant Biotechnology*, p. 558–560.

Rose, M.J. and Barthlott, W., 1995. Pollen connecting threads in Heliconia (Heliconiaceae). *Plant Systematics and Evolution*, 195(1/2): 61–65.

Rundel, P.W., Sharifi, M.R., Gibson, A.C. and Esler, K.J., 1998. Structural and physiological adaptation to light environments in neotropical Heliconia (Heliconiaceae). *Journal of Tropical Ecology*, 14(6): 789–801.

Rodrigues, P.H.V., 2005. *In vitro* establishment of *Heliconia rauliniana* (Heliconiaceae). *Scientia Agricola*, 62(1): 69–71.

Rodrigues, P.H.V., Tulmann Neto, A. and Dutra, M. de F.B., 2005. Gamma radiation LD_{50} determination in *Heliconia bihai* (Heliconiaceae) explants. *Revista Brasileira de Horticultura Ornamental*, 11(1): 79–81.

Santos, M.R.A., Timbo, A.L.O., Carvalho, A.C.P.P. and Morais, J.P.S., 2006. Effects of organic fertilizers and substrates on development of micropropagated heliconia plantlets. In: Developmental anatomy and morphology of the ovule and seed of *Heliconia* (Heliconiaceae, Zingiberales), (Eds.) D.G. Simao, V.L. Scatena and F. Bouman. *Plant Biology*, 8(1): 143–154.

Simao, D.G. and Scatena, V.L., 2003. Morphological aspects of the propagation in *Heliconia velloziana* L. Emygd. (Zingiberales: Heliconiaceae). *Brazilian Archives of Biology and Technology*.

Simao, D.G. and Scatena, V.L., 2004. Morphology and anatomy of the bracts in Heliconia (Heliconiaceae) occurring in the Sao Paulo State, Brazil. *Acta Botanica Brasilica*, 18(2): 261–270.

Sato, A.Y., Carvalho Nannetti, D. de, Pinto, J.E.B.P., Siqueira, J.O. and Blank, M. de F.A., 1999. Application of arbuscular mycorrhiza to micropropagated heliconia and gerbera plants during acclimatization period. *Horticultura Brasileira*, 17(1): 25–28.

Siddique, A.B. and Paswan, L., 1997. Callus initiation in heliconia (*Heliconia psittacorum*). *Journal of the Agricultural Science Society of North East India*, 10(1): 150–152

Sanabria, M.E., Crescente, O. and Caserta, A.E.de, 1998. Secondary metabolites as a chemotaxonomic character for three species of the genus *Heliconia* L. (Heliconiaceae) from the State of Sucre, Venezuela. *Ernstia*, 8(1): 27–39.

Sheela, V.L., George, T.S., Rakhi, R. and Lekshmi, P.R.G., 2007. Variability studies in cut flower varieties of *Heliconias. Indian Journal of Horticulture*, 64(1): 109–111.

Sheela, V.L., Rakhi, R., Nair, C.S.J. and George, T.S., 2005. Genetic variability in Heliconia. *Journal of Ornamental Horticulture New Series*, 8(4): 284–286.

Simao, D.G., Scatena, V.L. and Bouman, F., 2007. Anther development, microsporogenesis and microgametogenesis in *Heliconia* (Heliconiaceae, Zingiberales). *Flora Jena*, 202(2): 148–160.

Sheela, V.L., Lekshmi, P.R.G., Nair, C.S.J. and Rajmohan, K., 2006. Molecular characterization of Heliconia by RAPD assay. *Journal of Tropical Agriculture*, 44(1/2): 37–41.

Sheela, V.L., George, T.S., Rakhi, R. and Lekshmi, P.R.G., 2007. Variability studies in cut flower varieties of Heliconias. *Indian Journal of Horticulture*, 64(1): 109–111.

Sangwanangkul, P., Saradhuldhat, P. and Paull, R.E., 2008. Survey of tropical cut flower and foliage responses to irradiation. *Postharvest Biology and Technology*, 48(2): 264–271.

Sheng Aiwu and Nian, Liu, 2006.Control of browning of spath tissues of cut *Heliconia rostrata. Acta Horticulturae Sinica*, 33(4): 898–900.

Shiau YihJuh, Wang TsaiYih and Tsay HsinSheng, 1999. Studies on tissue culture of *Heliconia psittacorum* cv. Rhizomatosa. III. Effect of cefotaxime and plant growth regulators on growth of shoot tip explants *in vitro. Journal of Agricultural Research of China*, 48(4): 42–51.

Sheela, V.L., Nair, C.S.J., Rakhi, R. and Lekshmi, P.R.G., 2006. Correlation studies in heliconia. *Journal of Ornamental Horticulture*, 9(1): 29–32.

Torres, A.C., Duval, F.D., Ribeiro, D.G., Barros, A.F.F. and Aragao, F.A.D., 2005. Effect of sucrose, kinetin, isopentenyl adenine and zeatin on the development of embryos of *Heliconia rostrata in vitro. Horticultura Brasileira*, 23(3): 789–792.

Talukdar, M.C., Lalrinawmi and Singh, S., 2002. *In vitro* propagation of heliconia. *Journal of Ornamental Horticulture New Series*, 5(1): 45–46.

Ulisses, C., Camara, T., Willadino, L. and Albuquerque, C., 2007. Induction of somatic embryogenesis in *Heliconia chartacea* Lane ex Barreiros cv. Sexy pink and cv. Sexy scarlet from sections of ovaries. *Revista Brasileira de Ciencias Agrarias*, 2(4): 281–284.

Zeng SongJun, Guo ShaoCong, Wu KunLin, Zhang YiQi, Chen GuoHua and Duan Jun, 2004. *In vitro* propagation of *Heliconia* plants. *Journal of Tropical and Subtropical Botany*, 12(2): 153–158.

2011, Biodiversity in Horticultural Crops Vol. 3 *Pages* **275–290**
Editor: **Professor K.V. Peter**
Published by: **DAYA PUBLISHING HOUSE, NEW DELHI**

Chapter 11

Marigold

Namita*, Kanwar P. Singh and V.S. Raju Dantuluri
Division of Floriculture and Landscaping,
Indian Agricultural Research Institute, New Delhi – 110 012
**E-mail: namitabanyal@yahoo.co.in*

Flowers are one of the best creations of nature which symbolize life and happiness. They become an integral part of our culture. Among flowers, marigold 'a versatile crop with golden harvest' is one of the commercially grown flowers all over the world and also in different parts of India. Marigold (*Tagetes* sp.) belongs to family Asteraceae and is native to Mexico. It gained popularity because of its adaptability to various soil and climatic conditions, longer blooming period and beautiful flowers which have good shelf life. It is grown as an ornamental crop for loose flowers and as a landscape plant.

Origin and Distribution

Marigold is native to South and Central America, especially Mexico. From Mexico, it spread to different parts of the world in 16th Century. The generic name *Tagetes* was derived from 'Tagtes', the name of an Estrusch God, known for his beauty. Native to American continent, marigold was discovered by the Spanish traveller Cortz who introduced it to Europe. The devotees in Spain placed this flower at the altar of Virgin Mary and from that day, they came to be known as Mary's gold or marigold. The two important species got the names African marigold and French marigold although the original species from which they have been bred were introduced from America.

In the language of flowers, marigold, in general means, anxiety but despair or grief in particular. African marigold represents the vulgar mind and the French marigold is a symbol of jealousy. In England, it is a common belief that if the flowers of African marigold do not open before 7 a.m., it will rain or thunder that day. The marigold at first was regarded with distrust due to its strong disagreeable smell and was considered a poisonous plant. Marigold is also known as friendship flower in the United States.

African marigold was first introduced into Spain early in the 16th century and became popular in Europe in 1585 under the name Flors Africonus by Emperor Charles V who launched an expedition to free Tunis from the Moors where it was naturalized on the Algerian coast. It continued to be called so until 18th Century, although it was correctly figured as Plantas *Tagetes indica* in Fuch's herbal of 1542.

Another well known type of marigold, the French marigold followed a similar route as that of African marigold. Parkinson mentioned that it was called as *Tanacetum peruvianum* as it grew in Peru. James Justice believed that it was a Chinese plant and it was first sent to the Royal garden at Paris from where it was distributed under the name of French marigold. Baily mentioned that *Tagetes erecta* was put into cultivation in 1596 and *Tagetes patula* in 1573 in Europe. Alice Vonk developed white marigold after 56 years of research.

Uses

Marigold flowers are grown for loose flower production. They are used extensively for making garlands, for beautification of mandaps and decoration of cars in marriages and religious offerings. Marigolds are also used for bedding and potted purposes. Some varieties of African marigold which have straight stalk and globular flowers are used as cut flowers for interior decoration also.

Tagetes: A Novel Source of Phytochemical

Tagetes is one of the largest members of plants which hold promise as a bio resource for a variety of useful phytochemicals. There is high variability in the phytochemicals; well known among them are terpenoids, flavonoids, carotenoids and thiopenes and most of these are biologically active. Both the flowers and leaves of Tagetes contain terpenoids where these act as antioxidants and protect lipids, blood and other body fluids from harmful free radicals like single oxygen, hydroxyl, peroxide and superoxide. Flavonoids of flowers and leaves of *Tagetes erecta* have anti-inflammatory activity.

Recently, marigold emerged as a source of carotenoid pigments, which are widely used as dietary supplements. Carotenoids provide bright colouration, serve as antioxidants and can be a source of vitamin A activity.Petals of *T. erecta* have more carotenoids than seeds or sepals and 200 times more than yellow corn. *Tagetes* extract contains approximately 27 per cent carotenoids with 0.4 per cent β-carotene, 1.5 per cent cryptoxanthin esters and 86.1 per cent xanthophylls esters. Lutein accounts for over 60 per cent of the xanthophylls of marigold flowers and acylated with palmitic and myristic acids. Lutein dipalmitate, the main component of petal xanthophylls, possesses pharmacological activity and is used as opthalmological agents. Thiophenes are found in species *viz., Tagetes minuta* L., *T. coronopifolia* Wild, *T. erecta* L., *T. filifolia* A., Gray, *T. lemmonii* A. Gray, *T. patula* L. and *T. multiflora* HBK. Thiophenes act as toxins activated by sunlight or U. V. radiation (300-400 nm), killing pathogens like nematodes, insects, fungi and bacteria.

Industrial Uses

Carotenoids are the major sources of pigments for poultry feed. Dry petals of marigold flower contain about 90 per cent (W/W) carotenoids used in feed for intensification of yellow colour of egg yolks and broiler skin. The Lutein of marigold flower is the source of carotene and used for colouring food stuff. Tagetes oil has a sweet, fruity almost citrus like smell and is yellow to reddish-amber in colour. The main components of this oil are tagetone, limonene, valeric acid and ocimene, etc. *Tagetes minuta* oil is the most valuable and precious for using in high grade perfumes and cosmetics. Tagetes oil has juvenile hormonal and insect repellent activities against flies, ants and mosquitoes. Therefore, this oil is being used on commercial scale for formulation of insect repellents. Marigold plant is a source of thiophane compounds which are nematicidal and hence, used as a nematode trap crop. Now-a-days, marigold therapy with marigold petals is used to cure podiatric skin conditions.

Botany

Marigold belongs to family Asteraceae (Compositae) which comprises of other cultivated flowers like ageratum, aster, chrysanthemum, cornflower, dahlia, daisy, gaillardia and sunflower. Each flower is in fact composed of a mass of small flowers called florets. What we generally understand to be a flower is a complete inflorescence called head or capitulum. Two forms of florets exist in most of the flowers of the family, Asteraceae. Those on the outside of the disc called ray florets and the other which comprise disc are called disc florets. Ray florets are female and zygomorphic. The five florets are fused to form a ligule or a flat petal. The disc florets are tubular, actinomorphic and bisexual. The double forms are generally male sterile as the disc florets are converted into showy petals. Double forms are ornamentally preferred and have good economic value. In marigold, there are two types of male sterility. Male sterility in some lines is linked to apetalous nature, as such male sterile lines without petals. In petalous male sterility, the male sterile lines will have petals. However, commercially apetalous male sterility is popular as it is stable and can be easily identified.

Species and Cultivars

There are more than 50 species of the genus *Tagetes*. Some of species are *Tagetes alamensis, T. anethina, T. angustifolia, T. anisata, T. apetala, T. arenicola, T. arvensis, T. belendium, T. biflora, T. bonariensis, T. campanulata, T. cancellatus, T. chinchilla, T. congesta, T. coronopifolia, T. corymbosa, T. daucoides, T. dianthiflora, T. dianthifolia, T. dichotoma, T. elliptica, T. elongate, T. erecta, T. ernestii, T.filifolia, T. florida, T.filiformis, T. fragrantissima, T. gigantea, T. gilletii, T. gracilis, T. heterocarpa, T. iltisiana, T. jaliscensis, T. laxa, T. lemmoni, T. lucida, T. macroglossa, T. maculate, T. maxima, T. minuta, T. multifida, T. multiflora, T. patula, T. pectinata, T. pumila, T. pringlei, T. rupestris, T. seleri, T. signata, T. rotundifolia, T. tenuifolia, T. tinctoria, T. verticillata, T. triradiata*, etc. The characters of important species are given below.

Figure 11.1: ***Tagetes erecta***

Figure 11.2: ***Tagetes patula***

Tagetes erecta (African Marigold)

The plant is an annual. It grows to a height of 70-100 cm. The plant type is erect. The leaves are pinnately divided and leaflets are serrated. Flower type varies from single to double. The flowers are globular. Flower colour varies from sulfur yellow to orpiment orange. African marigold can be cultivated throughout the year.

Tagetes patula (French Marigold)

The plant is an annual which grows to a height of 30-45 cm. The stem is reddish and the foliage is dark green. Leaves are pinnately divided and leaflets are serrated. Flowers are yellow to brick red. In most of the cases, brick red coloured blotches are seen on yellow background.

Figure 11.3: ***Tagetes tenuifolia***

Tagetes tenuifolia (Signet Marigold)

This species is an annual with branching habit. Plants grow 30-35 cm tall bushy with fine lacy foliage and covered with small single flowers of lemon-yellow and orange colour. Leaves are pinnately divided. These are grown for pot culture, edging or rock garden.

Tagetes lucida (Sweet Scented Marigold)

It is commonly called Pericon, Mexican mint marigold, Mexican tarragon, or Texas tarragon. *Tagetes lucida*, widely identified as a powerfully psychoactive strain of the marigold flower, was first documented by the Aztecs.The plants of this species are perennial. Leaves are sessile, small and lanceolate. Flowers are usually 2-3 rayed, produced in dense terminal corymbs.

Tagetes minuta

This species is an annual producing small white flowers in dense terminal clusters. Commercially oil is extracted from this species.

Tagetes lacera

The plant grows up to a height of 120-150 cm. It bears profuse flowers of yellow colour.

Tagetes lemmonii

It is a shrubby plant and grows up to 60-70 cm height. Leaves are slender and opposite. Leaflets are about 2-3 cm long. Flowers are showy having 2-3 cm diameter.

Figure 11.4: ***Tagetes lucida***

Figure 11.5: ***Tagetes minuta***

Breeding Methods

Intervarietal Hybridization (Types and Cultivars)

In marigold, many varieties and strains are available which vary in plant height, growth habit, flower shape and size. Mostly local varieties are being cultivated by the farmers.

African Marigold (*Tagetes erecta*)

Pusa Narangi Gainda

A cross of 'Cracker Jack` and `Golden Jubilee`,'Pusa Narangi Gainda' flowers in 125-136 days after sowing. Its plants are 73.30 cm tall, vigorous and uniform, foliage dark green and flowers orange coloured, carnation type, double, compact, 7.8 cm across disc florets present. It provides 34.9 t/ha of fresh flowers. It is the most suitable for garland making, religious offerings and carotenoid extraction.

Figure 11.6: Pusa Narangi Gainda

Pusa Basanti Gainda

It is a cross between 'Golden yellow' and `Sun Giant`. It is a late maturing variety which takes 135-145 days. The flowering duration is long (40-45 days), plants 60 cm tall, vigorous and uniform, foliage dark green. Flowers sulfur yellow coloured, carnation type, double, compact, 6.9 cm across, disc florets present but invisible. The row of whorls present in the capitulum is 7-10. The number of flowers produced by a single plant is 58-60. It is ideal for loose flower production and also for growing in pots and beds in the garden.

Figure 11.7: Pusa Basanti Gainda

The other important varieties/hybrids of important species are listed below:

Tall F_1 Hybrids

F_1 Gold Coin Series and F_1 Climax Series

Semi Tall F_1 Hybrids

This is also known as hedge type because of uniform and compact growth and grows up to 50 cm height. These hybrids have double flowers with lemon and golden yellow colour.

Dwarf F_1 Hybrids

Inca Series–Flowers are large, fully double and compact

Space age Series–Dwarf, uniform and early flowering

Galore Series–Long flowering duration and uniform.

Carnation Flowered

The plants grow to a height of 75 cm, flower up to 10 cm across with wide range of colours *viz.*, orange, golden yellow, deep orange or lemon yellow flowers.

Carnation flowered dwarf- grows up to 40 cm and having large flowers.

Chrysanthemum Flowered

Tall double chrysanthemum flowered–plants 60-80 cm tall, large flowers in various shades of yellow and orange.

Dwarf double chrysanthemum flowered–It has several strains *viz.*,

I. Luxor Series–Improved Cupid type, and

II. Rexor Series–Improved Spun Type

French Marigold (*Tagetes patula*)

Varieties

Red Brocade, Rusty Red, Butter Scotch, Valencia, Sussana.

Dwarf Double

Plants are 20-30 cm in height having colours- yellow, orange, reddish brown, mahogany red, golden yellow, bicolour.

Dwarf Double-Scabious Flowered

Flowers with cristed centre, wide range of colours- Golden yellow, Golden yellow with red, brownish red with orange centre, etc.

Dwarf Double petite

Very dwarf plants having 15-20 cm height. Bears numerous attractive flowers with golden yellow, orange, yellow and maroon colour.

French Dwarf Single

Compact flowers, single golden yellow, yellow, mahogany red with yellow centre, yellow with brown

Tagetes tenuifolia: Golden Gem, Lulu, Pumila.

Interspecific Hybridization

Interspecific hybridization was used in breeding improved cultivars (Kalloo, 1992) and to generate new crop species (Lukaszewski and Gustafson, 1987). The interspecific hybrids produced between African marigold and French marigold is intermediate in character. They exhibit the floriferousness of African marigold and the branching character of French marigold. The interspecific hybrids are triploids, as a result they are sterile and seed formation is not observed. These are favoured by private seed companies as they prevent illegitimate seed production.

Genetics and Breeding

The basic chromosome number of marigold is 12. *Tagetes erecta, T. tenuifolia, T. elongate, T. Lemmonii, T. jalicensis, T. lucida*, etc. are diploids (2n) having a chromosome number of 24.

Tagetes patula, T. minuta, T. remotiflora, T. mendocina, T. biflora, etc. are tetraploids (4n) having a chromosome number of 48. *Tagetes patula* is an allotetraploid between *Tagetes erecta* and *Tagetes tenuifolia* (Towner, 1961). Unlike other plants, where size of the plants and flowers increase with increase in ploidy level, in marigold exceptionally, it decreases. Galica-Fuentes (1995) observed wide diversity of infloroscences in marigold germplasm, especially in *Tagetes erecta.*

Male Sterility

In marigold, male sterility is governed by recessive gene and is maintained by crossing the heterozygous plants with sterile ones. Two kinds of male sterility are found in marigold, one is apetalous and double flowered (Gold-smith, 1968). Apetalous sterility is preferred over full double flowers because latter type is prone to break down and it gives rise to a few disc florets at later stages which may be due to either age of the plants or environmental factors.

Mode of Inheritance

Flower Type

The crossing between cupid and chrysanthemum types resulted in intermediate flowers. The F_2 population segregated in a ratio of 9 intermediate: 3 cupid: 4 chrysanthemum and showed that cupid and chrysanthemum types were governed by two genes. Carnation was dominant over chrysanthemum in a cross between these two types. The F_2 population segregated in a ratio of 15 carnation: 1 chrysanthemum which could be due to duplicate epistasis. Based upon these studies, the following genetic makeup of different types was proposed (Singh, 1970).

Carnation (double flower)	:	RR TuTu CC
Carnation (Single flower)	:	rr TuTu CC
Cupid	:	rr TuTu CuCu
Chrysanthemum	:	rr tutu cc
	:	RR tutu cc
Intermediate	:	RR tutu cucu

Flower Colour

The flower colour is governed by two or more genes (Punnet, 1937). The flower colour of interspecific hybrid between *T. erecta* and *T. jaliscensis* revealed that disc corolla anthocyanin from *T. jaliscensis* segregated as simple dominant over the absence in *T. erecta*. Here, the gene symbol R (Red) represents dominant allele for presence of anthocyanin in lobes of disc corolla (Towner, 1962). Different colors are available in marigold ranging from orpiment orange to yellow. Recently white marigold was also developed. Orange colour was dominant over yellow colour.

Flower Form

Flatness of the floret is governed by a single gene. Flat form is monogenic dominant over tubular form. Towner (1962) also observed that flat floret was dominant over tubular ray floret in a cross between *T. erecta* and *T. jaliscensis*. The dominant and recessive alleles of *T. jaliscensis* and *T. erecta* were assigned as Tu and tu, respectively.

Double Flower *vs* Single Flower

Double flowering habit is monogenic dominant over single flowering habit. Towner (1961, 1962) observed that double flowering habit was monogenic dominant over single flowering in F_2 crosses of *T. erecta* × *T. tenuifolia*. Singh (1970) also observed dominance of double flowering over single flowering in diallel crosses including 7 parents of marigold. The monogenic nature of dominance was confirmed in the test crosses.

Number of Flowers

Singh and Swarup (1973 a) studied inheritance of a number of flowers in six crosses of marigold and observed the larger absolute magnitude of the total epistatic gene effects of these crosses than the mean effects, thus indicating major contribution of the epistatic effect to variation in these crosses.

Days Required for First Flowering

From a diallel cross between seven parents, two crosses showed significant additive gene effects whereas in three crosses, dominance was significant. The epistatic gene effects in all crosses were more pronounced than additive and dominance gene effects (Singh and Swarup, 1973 a).

Leaf Colour

Kantor (1938) and Towner (1962) reported that the duplicate recessive genes were responsible for virescent character. Little *et al.* (1940) in *T. erecta* crosses found that virescent character was monogenic recessive and partially lethal.

Genetic Parameters

Genotypic and phenotypic coefficients of variation, heritability and genetic advance constitute the important genetic parameters frequently applied in plant breeding. Coefficients of variation allow a meaningful comparison of the variation of several traits of plants belonging to the same population, as well as a comparison of the variation of the same trait as expressed by different populations (of the same or different crops). Heritability is used to describe the properties of inheritance of quantitative traits. Plant breeders are interested in heritability because characters with higher values can be improved more rapidly with less intensive evaluation than those with low heritability (Nyquist, 1991). Heritability tells us about the additive genetic variance and phenotypic variance. The main purpose of estimating the heritability and the genetic parameters which compose the heritability estimate is to compare the expected gains from selection based on alternative selection strategies (Holland *et al.*, 2003).

Singh and Sen (2000) conducted an experiment on 12 cultivars of African and French marigold and recorded significant differences in all the characters studied. The coefficient of variation was minimum for days to flowering (PCV = 15.20, GCV = 14.89) and maximum for duration of flowering (PCV = 64.67, GCV = 63.84) in French marigold but in African marigold it was minimum for plant height at 60 days after transplanting (PCV = 8.14, GCV = 7.74). High heritability along with high genetic advance was observed for flower yield/plant, flower yield/ha and number of flowers/plant in French marigold whereas in African marigold, high values were observed for dry weight of flowers, flower yield/plot and flower yield/ha. Patnaik and Mohanty (2002) collected 13 genotypes of marigold for evaluation and observed that genotype of *Tagetes erecta* depicted significant differences in all the characters except number of primary branches. The maximum genotypic and phenotypic levels for flower yield and high heritability along with high genetic gain were recorded for characters like plant height, weight of individual flower and yield of flowers/plant, which may prove effective criteria for selection.

Sreekala *et al.* (2002) studied variability for carotenoids and yield components among 10 African marigold genotypes and concluded that except flower diameter, all characters *viz.*, plant height, plant spread, number of branches, days to flower, number of flowers/plant, flower weight, flower diameter, flowering duration, flower yield and carotenoids content showed higher phenotypic and genotypic variance. Singh and Singh (2004) assessed variability and heritability studies in three species of marigold and concluded that the coefficient of variation was minimum for duration of flowering (PCV = 18.90, GCV = 18.39) and maximum for number of flowers/plant (PCV = 331.20, GCV = 331.22). Heritability estimates for all the characters were high. High heritability with high genetic advance was observed for number of flowers/plant (h^2 = 99.99 per cent; GA = 1469.18)

Combining Ability Effects

In the progression of hybridization, identification of genetically superior parents is a pre-requisite. Combining ability provides an empirical summary of complex observations and a reasonable basis for forecasting the performance of yet untested crosses. Sprague and Tatum (1942) developed the concept of combining ability and provided an early systematic exposition in the context of evaluation of inbred line for crossing. Concept of combining ability was further apportioned and genetic variance was divided into two components *viz.*, variance due to general combining ability and variance due to specific combining ability. The general combining ability (gca) is the average or overall performance of a line in a series of crosses, whereas specific combining ability (sca) is the deviation in the performance of a particular cross than what would be expected based on the GCA effect of the lines involved.

Reddy *et al.* (1989) observed that in African marigold, GCA and SCA variances were statistically significant indicating that both additive and non additive component of variances were important for inheritance of plant height, flower size, flower weight, number of flowers/plant and yield/plot whereas, additive component of variance was important for number of branches/plant and early flowering. Greater magnitude of gca variance than that of sca variance, however, revealed preponderance of additive gene effects for all the traits, except for flower size for which both additive and non additive gene effects were important. Mohanty *et al.* (2003) analyzed the combining ability in marigold based on 6 × 4 line × tester analysis for 9 important characters. Variance due to general combining ability (gca) highly exceeded variance due to specific combining ability (sca) for all the characters and was also significant. The varieties *viz.*, Sutton's Double Orange, Bhubaneswar Local Orange, New Tech Orange and Early Yellow Single were good combiners for flower yield and component traits.

Correlation and Path Analysis

Correlation is a measure of association between observed performance of individuals for a pair of quantitative traits. Correlation studies help in knowing suitability of various traits for selection. Knowledge of the mechanisms underlying correlations among different traits is fundamental for understanding the degree of integration of the phenotype and to resolve constraints imposed on evolutionary processes (Lynch and Walsh, 1998). The correlations may be caused by (i) a functional relationship; (ii) pleiotropy and/or linkage; and/or (iii) variation in environmental conditions. Wright (1921) developed the method of path analysis as a means of interpreting the correlation between two variables in terms of hypothetical paths of causation between them. The purpose of path analysis is the quantification of the relative contributions of causal sources of variance and covariance once a certain network of interrelated variables has been accepted (Lynch and Walsh, 1998). Correlation studies coupled with path analysis will serve as a better criterion for selection of traits influencing yield.

Sreekala *et al.* (2002 a) reported very high genotypic and phenotypic correlation among total carotenoid yield, morphological and yield characters. The direct influence of total carotenoid yield was maximum from flower yield/plant, followed by flower diameter and flower weight. Singh and Singh (2005) observed positive correlation between days taken to flowering, flower diameter and average fresh weight of flower. Significant negative correlation was observed between days to flowering, flower duration and number of flowers/plant and between flower duration and flower diameter. Highly significant positive correlation between flower diameter, average fresh weight of flower and weight of seeds/peduncle and a negative correlation between number of flowers/plant and fresh weight of flower and weight of seeds/peduncle were also observed by them. Reena *et al.* (2005) studied variability and correlation in African marigold and concluded that characters such as seed yield/plant, buds/plant and number of flowers/plant showed highly significant correlation with flower yield/plant, while the number of flowers, number of buds and flower yield/plant showed highly significant correlation to seed yield.

Bandyopadhyay *et al.* (1997) analyzed correlation and path analysis in seed production of marigold as affected by micronutrient application and assessed that the number of seeds/flower and Zn uptake by the flower with the greatest positive effects on seed yield. They also reported that leaf area and average flower weight had the greatest negative effect on seed yield.

Heterosis

Heterosis usually refers to superior performance of first generation off-spring over its parents. The phenomenon of heterosis was recognized during last 200 years even before Mendel's contribution was discovered. Although, workable methods of utilizing hybrid vigour in crop production were known (Johnson, 1891 and Shamel, 1905), yet its genetic concepts were not recognized until the investigation of Shull (1908), East (1908) and Jones (1917) were completed. Falconer and Mackay (1996) demonstrated that heterosis could be expressed as $H_{mp} = y^2d$, where H_{mp} is termed as mid parent heterosis, d is the level of dominance and y is the difference in allele frequency of the parents. Heterosis was further elaborated on the basis of physiological, biochemical and molecular level. According to physiological basis of heterosis, metabolic efficiency may substantially relate to heterozygosity and heterosis. Hageman *et al.* (1967) proposed metabolic concept determining hybrid response.

According to Stuber (1999) inbreds tend to have unbalanced metabolic systems which differ among individual lines. It may be overcome in a heterozygous condition by proper selection of complementary inbred parents, resulting better balanced metabolic system in the resulting hybrids. Srivastava (1981) reported that heterotic hybrids were endowed with superior systems of chloroplasts and mitochondria and such superiority was provided by genomic and intergenomic interactions. General biochemistry shows that the metabolic processes in living organisms are regulated by both regulatory proteins (and their coding genes) and the mechanisms of regulation of gene activity. Studies, however, suggested that they are significantly related to manifestation of heterosis (Tsaftaris *et al.*, 1999). The molecular level shows that the quantitative expression for regulatory proteins and their coding genes is largely obtained from the analysis of RNA amount polymorphism (RAP) and protein amount polymorphism (PAP). Tsaftaris and Polidoros (1993) suggested that quantitative expression of certain genes is responsible for heterosis manifestation in hybrids. Another mechanism involved in regulating expression output of genes is methylation of cytosine residues in their DNA (Cedar, 1984). By extrapolating these observations, Stuber (1999) hypothesized that degree of DNA methylation may be regulating heterosis, such as that there is a gradual accumulation of methylation during selfing, which is then released/repatterned in hybrid. Although, a number of theories of heterosis are proposed to explain the phenomenon of heterosis, yet to date, none of these concepts either individually or jointly have shown the potential to predict heterosis, so that plant breeder can choose the parents with confidence to develop heterotic hybrids.

Production of F_1 hybrids in ornamental crops was carried out before Shull's classical theory of hybrid vigour which was reported in 1971 for the first time in plant breeding. The interspecific hybridization by crossing *Tagetes erecta* × *Tagetes tenuifolia* was done by Towner (1961) and it was observed that the natural and synthetic allotetraploids were intermediate between *Tagetes erecta* and *Tagetes tenuifolia* for nearly every character. F_1 of *Tagetes erecta* and *Tagetes tenuifolia* showed a striking morphological similarity to *Tagetes patula*. Interspecific crossing between *Tagetes jaliscensis* × *Tagetes erecta* was also studied by Towner (1962). Triploid hybrid was produced by interspecific crossing between *Tagetes erecta* and *Tagetes patula* which had relatively larger double flowers (Jalil *et al.*, 1974). Singh and Swarup (1971) evaluated 7 parents of African marigold in a diallel cross excluding reciprocals. The maximum heterosis was observed for total number of flowers. The crosses Alaska × Hawaii, Alaska × Cupid Orange Mum and Katrain Local × Cupid Orange Mum showed the highest heterosis (per cent) on flower size, flower weight, early flowering and total number of flowers. Relichova and Petrova (1988) observed the greatest heterosis effect in the number of heads/plant and head diameter, when 8 maternal and 5 paternal lines of African marigold were crossed in a diallel set.

Reddy *et al.* (1989) conducted study in African marigold using seven parental lines in diallel set (including reciprocals). The crosses African Double Yellow × Spun Gold, Giant Double African Orange × Coimbatore Local and Giant Double African Orange × African Double Yellow were recommended for exploitation of heterosis for commercial use due to high specific combining ability effects and percent heterosis for a number of characters. Kumar *et al.* (1989) evaluated 27 hybrids from crossing of lines and 3 testers of marigold. They found that hybrids involving lines L_2 and L_9 gave heterobeltiosis up to 64 per cent for flower diameter, 97 per cent for flower height, 58 per cent for number of flowers and 748 per cent for flower yield. Through diallel crosses, 15 F_1 hybrids of marigold were developed by involving 6 parents (excluding reciprocals). One hybrid raised from the cross of Giant Double African Orange and Crackerjack was promising for commercial purpose (Barad *et al.*, 1993). In a line × tester (3 × 11) crossing programme in *Tagetes erecta*, 33 F_1's along with parents were evaluated (Gupta *et al.*, 2001). Significant heterosis was observed in all traits except for stalk length and it was the highest in flower yield (310.2 per cent and 535.7 per cent for winter and summer crops, respectively).

Mohanty *et al.* (2003) conducted a study in marigold with crosses of 6 lines and 4 testers into line × tester design and found relative heterosis (260 per cent) and heterobeltiosis (211 per cent) for flower yield, respectively in Sutton's Double Orange × Early Yellow Single and also found heterosis and heterobeltiosis in other crosses for several characters, while negative heterosis for days to bud initiation, which is a desirable character for earliness. In the same year, they also concluded that the mean sum of squares due to parents versus crosses was highly significant for all the characters indicating presence of significant heterosis. Sreekala and Raghava (2003) conducted experiments in line × tester design with 7 testers and 3 male sterile lines and developed 21 hybrids. They reported for the first time that exploitation of heterosis should be useful in breeding for increased carotenoid esters and lutein and for this character positive heterobeltiosis was recorded.

Velmurugan *et al.* (2003) studied 11 lines of marigold, out of which two lines recorded higher mean of lines. General combining ability (gca) recorded significant positive and negative gca effects ranging from -0.4619 (L_3) to 0.2672 (L_2). Specific combining ability (sca) varied from -0.51136 to -0.486.

Genetic Diversity Using Molecular Markers

Many of the limitations of morphological and biochemical markers can be overcome by using molecular markers. An important application of DNA based markers is the development of unique genetic fingerprints, which help in characterization, hybridity confirmation and cultivar identification. The other major advantage of molecular markers is the abundance of polymorphic loci, which enables estimation of genetic distance among elite breeding lines and populations with high precision. Currently, there are three broad categories of DNA based molecular marker systems used for genetic diversity analysis in crop plants. These are: a) DNA hybridization based markers such as RFLP (Restriction Fragment Length Polymorphism); b) Polymerase Chain Reaction (PCR)-based sequence arbitrary methods which utilize short sequences of nucleotides as primers, like Randomly Amplified Polymorphic DNA (RAPD), Arbitrary Primed Polymerase Chain Reaction (AP-PCR), DNA Amplification Fingerprinting (DAF) and Amplified Fragment Length polymorphism (AFLP); c) PCR based sequence dependent methods like sequence tagged sites (STS), Sequence characterized Amplified Regions (SCARs) and Simple Sequence Repeats (SSRs) or Microsatellites (Staub *et al.*, 2000).

RAPDs are well suited for assessment of genetic diversity because of their simplicity, speed and relatively low cost compared to other molecular markers (Williams *et al.*, 1990; Staub *et al.*, 1996). In addition, RAPD assessment often detects more polymorphic loci than either isozymes or RFLPs (Rafalski and Tingey, 1993; Waugh and Powell, 1992).

Genetic Analysis Using RAPD Markers

Random Amplified Polymorphic DNA analysis provides a quick and efficient method for resolving genetic relationship. Most of the genetic diversity studies in crop plants are carried out using morphological markers only. Now-a-days Polymerase Chain Reaction (PCR) based molecular markers have developed into powerful tools to analyze genetic relationships and genetic diversity. RAPD technique is one among them. The genetic variation as detected by RAPD analysis opens up the avenue for the proper identification and selection of the genotypes that could be used for varietal identification and planning for future crop improvement programme.

Nine genotypes of marigold, selected from two species, *Tagetes erecta* L. and *T. patula* L. were characterized through electrophoresis of protein and using RAPD markers. The SDS-PAGE electrophoresis of total seed soluble proteins revealed 16 bands with Rm value ranging 0.07 to 0.67. A polymorphism of 68 per cent was recorded among nine genotypes of marigold. The highest dissimilarity was recorded between MG-1 and MG-5, 6, 7, 8 and 9, whereas MG-5, 6, 7, 8 and 9 were similar. Twenty five random RAPD primers were used for polymorphism based on DNA fingerprinting analysis of nine genotypes of marigold. Out of these, 15 primers gave amplification by producing total 113 bands with an average of 4.5 bands/primers. The dendrogram based on protein electrophoresis grouped the nine genotypes into two clusters species wise, whereas RAPD analysis showed clear cut genotypes and species difference, which confirm the reliability of RAPD markers over protein electrophoresis (Mor *et al.*, 2008).

Genetic Analysis Using ISSR Markers

To overcome the limitations like low reproducibility of RAPD, high cost of AFLP and the need to know the flanking sequences to develop species specific primers for SSR polymorphism, ISSR-PCR technique is very useful (Neyer *et al.*, 1993; Zietkiewiez *et al.*, 1994; Gupta *et al.*, 1994; Wu *et al.*, 1994). This technique is useful in areas of genetic diversity, phylogenic studies, gene tagging, genome mapping and evolutionary biology in a wide range of crop species. Inter simple sequence repeat (ISSR) technique is PCR based method, which involves amplification of DNA segment present at an amplifiable distance in between two identical micro satellite repeat regions oriented in opposite dissection. ISSR segregates mostly as dominant markers following simple Mendelian inheritance (Gupta *et al.*, 1994; Tsumura *et al.*, 1996; Ratnaparthe *et al.*, 1998; Wang *et al.*, 1998). However, they also segregate as co-dominant markers in some cases thus enabling distinction between homozygotes and heterozygotes (Wu *et al.*, 1994; Akagi *et al.*, 1996; Wang *et al.*, 1998; Sankar and Moore, 2001). He *et al.* (2009) identified a SCAR marker linked to a recessive male sterile gene (Tems) and showed that this SCAR marker, therefore, permits the efficient marker assisted selection of male sterile individuals in breeding programmes of marigold and will greatly facilitate the breeding of F_1 cultivars.

A marigold flower cDNA library was constructed and two cDNAs (*hmg1 and hmg2*) encoding 3 hydroxy-3methylglutaryl coenzyme A reductase (HMGR) were isolated and sequenced. The two cDNA have identical 3′- and 5′- untranslated regions. Southern blot analysis revealed the presence of three or more genomic sequences of marigold showing homology to the hmg1 cDNA, while Northern blot analysis detected at least two different size patterns of HMGR mRNAs in flowers, leaves and roots (Villar-Martinez *et al.*, 1999).

References

Akagi, H., Yokozeki, Y., Inagaki, A., Nakamura, A. and Fujimura, T., 1996. A co–dominant DNA marker closely linked to the rice nuclear restorer gene, Rf–1, identified with inter–SSR fingerprinting. *Genome,* 39: 1205–1209.

Bandyopadhyay, P., Das, D.K. and Chattopadhyay, T.K., 1997. Correlation and path analysis in seed production of marigold as affected by the micronutrient application. *Hort.l J.,* 10 (2): 73–78.

Barad, A.V., Sachan, S.C.P. and Kukadiya, M.U., 1993. Heterosis in African marigold (*Tegetes erecta* L.) *In*: *Golden Jubilee Symposium Horticulture Research: Changing Scenario*, held at Bangalore, p. 91.

Cedar, H., 1984. DNA methylation and gene expression *In*: A. Razin *et al.* (Eds.), *DNA methylation. Biochemistry and Biological Significance.* Springer Verlag, New York, pp. 147–164.

East, E.M. 1908. Inbreeding in corn. *Rept. Comm. Agric. Exp. Stn.* for 1907, 419–428.

Falconer, D.S. and Mackey, T.F.C., 1996. Introduction to quantitative genetics. 4th ed. Longman, Essex, England.

Gupta, M., Chyi, Y.S., Ramero–Severson, J. and Owen, J.L., 1994. Amplification of DNA markers from evolutionary diverse genomes using single primers of simple sequence repeats. *Theor. Appl. Genet.,* 89: 998–1006.

Gupta, Y.C., Raghava, S.P.S. and Misra, R.L., 2001. Heterobeltiosis in African marigold (*Tagetes ereeta,* L.). *Indian J. Genet.*, 61, 1, 65–68.

Hageman, R.H., Leng, E.R. and Dudley, J.W. 1967. A biochemical approach to corn breeding. *Advanced Agron.,* 19: 45–86.

He, Y.H., Ning, G.G., Sun, Y.L. and Bao, M.Z., 2009. Identification of a SCAR marker linked to a recessive male sterile gene (*Tems*) and its application in beeding of marigold (*Tagetes erecta*). *Plant Breed.* 128: 92–96.

Holland, J.B., Nyquist, W.E., and Cervantes–Martines, C.T., 2003. Estimating and interpreting heritability for plant breeding: An update. *Plant Breed. Rev.,* 22 (9): 112.

Jalil, R., Khoshoo, T.N. and Pal, M. 1974. Origin, nature and limit of polyploidy in marigolds. *Current Sci.,* 43: 777–779.

Johnson, H.W., 1891. *How Crops Grow*. Orange Judd and Co., New York.

Jones, D.F. (1917). Dominance and linked factors as a means of accounting for heterosis. *Genet.,* 2: 466–479.

Kalloo, G., 1992. Utilization of wild species. (In): Kalloo,G. and Chowdhury,J.B. (Ed.). Distant hybridization of crop plants. Springer–Verlag, New York. pp. 149–167.

Kantor, J. H. 1938. Master's thesis, Bucknell Univ. Lewisburg, pa., USA.

Kumar, S., Shanmgavelu, K.G. and Irulapan, I., 1989. Hybrid vigour in marigold for economic characters. *S. Indian Hort.,* 38, (3): 173–174.

Little, T.M., Kantor, J. H. and Robinson, Jr. B.A. 1940. *J. Hered.,* 31: 73–78.

Lukarzewski, A.J. and Gustafson, J.P., 1987. Cytogenetics of triticale. *Plant Breed. Rev.,* 5: 41–93.

Lynch, M. and Walsh, J.B., 1998. Genetics and Analysis of Quantitative Traits.Sinaure Assocs. Inc., Sunderland, Massachusetts, USA, 980p.

Mohanty, A., Mohanty, C.R. and Mohapatra, R.C., 2002. Heterosis studies in African marigold. *J. Ornamental. Hort.,* 6 (1): 55–57.

Mohanty, A., Mohanty, C.R. and Mohapatra, R.C., 2003. Combining ability for yield and its components in marigold. *J.Ornamental Hort.,* 6 (1) 34–38.

Mor, V. S., Deswal, D.P., Mann, A., Dahiya, B.S. and Beniwal, B.S., 2008. Characterization of marigold (*Tagetes* sp p.) genotypes using SDS–PAGE and RAPD markers. *Seed Sci Technol.,* 36: 757–766.

Neyer, M., Novy, R.G. and Vorsa, N., 1993. Evidence for RAPD Heteroduplex formation in craneberry: imolications for pedigree and genetic–relatedness studies and a source of co–dominant RAPD markers. *Theor Appl. Genet.,* 92: 840–849.

Nyquist, W.E.1991. Estimation of heritability and prediction of selection response in plant populations. *Critic. Rev. Plant Sci.,* 10: 235–322.

Rafalski, J.A. and Tingey, S.V., 1993. Genetic diagnosis in plant breeding: RAPDs, microsatellites and machines, *Trends Genet.,* 9: 275–280.

Ratnaparkhe, M.B., Tekeoglu, M. and Muchlbauer, F.J., 1998. Inter–simple–sequence Repeat (ISSR) polymorphisms are useful for finding markers associated with disease resistance gene clusters. *Theor Appl. Genet.,* 97: 515–519.

Reddy, N.T., Muthuswamy, S., Itulappan, I. and Abdul Khader, M., 1989. Heterosis and combining ability for yield and yield components in African marigold (*Tagetes erecta* L.). *S. Indian Hort.,* 36 (1): 51–56.

Reena, M., Beniwal, B.S., Bhatia, S.K. and Deswal, D.P., 2005. Variability and correltion studies in African marigold (*Tagetes erecta* L.). *Res. Crops,* 6 (2): 322–327.

Relichova, J. and Petrova, I., 1988. Combining ability and heterosis in *Tagetes erecta* L. *Sbornik– UVTIZ, Genetika–Slechteni,* 24 (1):75–84.

Sankar, A.A. and Moore, G.A., 2001. Evaluation of Inter–simple sequence repeat analysis for mapping in citrus and extension of genetic linkage map. *Theor. Appl. Genet.,* 102: 206–214.

Shamel, A.D. 1905. The effect of inbreeding in plants. U.S. Deptt. Agric. Yearbook for 1905. pp. 377–392.

Shull, G.H. 1908. The composition of a field of maize. *Rept. American Breeders' Association,* 4:296–301.

Singh, B. and Swarup, V. 1971. Heterosis and combining ability in African marigold. *Indian J. Genet. Plant Breed.,* 31 (9): 407–415.

Singh, B. and Swarup V. 1973a. Inheritance of qualitative characters in African marigold. *Indian J. Genet.,* 33 : 172–175.

Singh, D. and Sen N.L., 2000. Genetic variability, heritability and genetic advance in marigold. *J. Ornamental Hort.,* 3 (2): 75–78.

Singh, D. and Singh, A.K., 2005.Correlation and path coefficient analysis in marigold (*Tagetes* spp.). *Progressive Hort.,* 37, (2): 385–388.

Singh, D. and Singh, A.K., 2004. Studies on genetic variability, heritability and genetic advance in *Tagetes* spp. *Abst*: National Symposium on Recent Trends and Future Strategies, ISHS, New Delhi and UAS, Karnataka, p. 14.

Sprague, G.F. and Tatum, L.A. 1942. General versus specific combining ability in single cross of corn. *J. American Soc. Agron.,* 34:923–932.

Sreekala, C. and Raghava, S.P.S., 2003. Exploitation of heterosis for carotenoid content in African marigold (*Tagetes erecta* L.) and its correlation with esterase polymorphism. *Theor Appl. Genet.,* 106 (4): 771–776.

Sreekala, C., Raghava, S. P.S., Misra, R.L. and Voleti, S.R., 2002. Assessment of variability for carotenoids and yield components in African marigold. *J.Ornamental Hort.,* 5 (2): 5–7.

Sreekala, C., Raghava, S.P.S., Mishra, R.L. and Maini, S.B., 2002a. Path analysis for total carotenoid yield in African marigold. *J. Ornamental Hort.,* 5 (2):8–10.

Srivastava, H.K., 1981. Intergenomic interaction, heterosis and improvement of crop yield. *Advanced Agro.,* 34: 117–195.

Staub, J., Serquen F. and Gupta, M., 1996. Genetic markers, map construction and their application in plant breeding. *Hort. Sci.,* 31: 729–741.

Staub, J.E., Danin–Paleg, Y., Horejsi T.F., Reis N. and Katzir, N., 2000. Comparative analysis of cultivated melon groups (*Cucumis melo* L.) using RAPD and SSR markers. *Euphytica,* 115: 225–241.

Stuber, C.W., 1999. Biochemistry, molecular biology and physiology of heterosis. In: J.G. Coors and S. Pandey (*eds.*), *Genet. Exploit. Heterosis Crop*, pp. 173–183.

Towner, J.W. 1961. Cytogenetics studies on the origin of *Tagetes patula* I. meiosis and morphology of diploid and allotetraploid *T.erecta* × *T. tenuifolia*. *American J. Bot.,* 48:743–751.

Towner, J.W. 1962. Cytogenetics of *Tagetes jaliscensis* × *T. erecta*. *American J. Bot.,* 48:1064–1067

Tsaftaris, A.S. and Polidoros, A., 1993. Studying the expression of genes in maize parental inbreds and heterosis and non–heterotic hybrids *In*: Bianci (*ed.*), *Proceedings XII Eucarpia Maize and Sorghum Conference*, Bergamo, Italy.

Tsaftaris, A.S., Kafka, M., Polidoros, A. and Tani, E., 1999. Epigenetic change in maize DNA and heterosis *In*: J.G. Coors and S. Pandey (*eds.*), *Genet. Exploit. Heterosis Crop*, pp. 195–203.

Tsumura, Y., Ohba, K. and Strauss, S.H., 1996. Diversity and inheritance of inter–simple sequence repeat polymorphisms in Douglasfir (*Pseudotsuga menziesii*) and Sugi (*Cryptomeria Japinica*). *Theor. Appl. Genet.*, 92: 40–45.

Velmurugan, K., Vijayakumar, M. and Jawaharlal, M., 2003. Heterosis and combining ability effects for xanthophyll content of marigold (*Tagetes erecta* L.). *S. Indian Hort.*, 51, (1–6): 237–240.

Villar–Martinez, A.A. Del, Ma, R. and Paredes–Lopez, O., 1999. Molecular characterization of cDNAs encoding 3 hydroxy–3methylglutaryl coenzyme A reductase in marigold (*Tagetes erecta*). *J. Plant Physiol.*, 155: 205–211.

Wang, G., Mahalingam, R. and Knap, H.T., 1998. (C–A) and G–A) anchored simple sequence repeats (ASSRs) generated polymorphism in soybean, Glycine Max (L.). *Merr. Theor. Appl. Genet.*, 96: 1086–1096.

Waugh, R. and Powell, W., 1992. Using RAPD markers for crop improvement. *Trends Biotech.*, 10: 186–191.

William, J.G.K., Kubelic, A.R., Livac, K.J., Rafalski, J.A. and Tingey, S.V., 1990. DNA polymorphism amplified by arbitrary primers is useful as genetic markers. *Nucleic Acid Res.*, 18: 6531–6535.

Wright, S. 1921. Correlation and causation. *J. Agril. Res.*, 20: 557–585.

Wu, K., Jones, R., Dannaeberger, L. and Scolnik, P.A., 1994. Detection of micro satellite polymorphisms without cloning. *Nucleic Acids Res.*, 22: 3257–3258.

Zietkiewick, E., Rafalski, A. and Labuda, D., 1994. Genome fingerprinting by simple sequence repeat (SSR)–anchored polymerase chain reaction amplification. *Genomic.*, 20: 176–183.

2011, Biodiversity in Horticultural Crops Vol. 3 *Pages* **291–299**
Editor: **Professor K.V. Peter**
Published by: **DAYA PUBLISHING HOUSE, NEW DELHI**

Chapter 12

Zinnia

R. Crofton Sloan
North Mississippi Research and Extension Center,
Mississippi State University, Verona, Mississippi, USA
E-mail: rcsloan@ra.msstate.edu

Zinnias are grown around the world as cut flowers and as annual flowers in the gardens. *Zinnia violaceae* is the predominant ornamental species used by gardeners because of its large, showy, and colorful ray flowers. Zinnia seeds were first sent from Mexico to Europe at the end of the 18th century and are popular among gardeners since then. Flowers may be double with many ray flowers, or single with many disk flowers. Zinnias are very reliable summer annuals which bloom during the summer heat. They are easy to start from seed and are the most often sold that way, but bedding plants are also available. *Z. angustifolia* is a bedding plant considered to be well adapted to low water requirement gardens (Henson, *et al.*, 2006), but is less popular than *Z. violaceae* because the ray flowers are less showy and the number of colors is restricted. Since 1933, there are 43 zinnia cultivars selected as All-America Selections (AAS) winners. Among the annual flowers only petunia and marigold have more winning cultivars. (All America Selections, 2009).

Taxonomy and Origin

Zinnias are members of the Heliantheae tribe within the Asteraceae family. There are numerous popular ornamental plants in the tribe Heliantheae:–*Coreopsis, Cosmos, Echinaceae, Heliopsis, Rudbeckia,* and *Zinnia.* The subtribe Zinnia is composed of *Zinnia, Heliopsis, Philactis,* and *Sanvitalia* (Bremer, 1994). The first zinnia to be described was put in the genus *Chrysogonum* in 1753 by Linnaeus and named *C. peruvianum* (Torres, 1961). Linnaeus in 1759 changed the name of this specimen to *Zinnia peruviana.* Linnaeus named the genus *Zinnia* after German botanist Johann Gottfried Zinn. The natural range of all species of zinnia (except *Z. peruviana*) is restricted to North America. *Z. peruviana* grows in both North and South America (Torres, 1963).

Sessé and Moçiño discovered what they named *Z. elegans* near Tixtlae, Guerrero, Mexico in 1789, but their finding was not published until 1890. Seeds send by Sessé and Moçiño to Madrid, Spain were probably identified as *Z. elegans*. Cavanilles first planted the seed in 1791 in Madrid and named the plants *Z. violaceae*. Seed must have been sent to Vienna from Spain and planted by Jacquin in 1792 who described and named them *Z. elegans*, following the lead of Sessé and Moçiño (Kirkbride and Wiersema, 2007). *Z. violaceae* seed was planted in other parts of Europe by 1796 (Torres, 1963). *Z. haageana* was grown in Europe as an ornamental plant by 1861, and *Z. peruviana* was introduced to England in 1753 (Torres, 1963). 'Zennia', probably *Z. violaceae*, was first described in a U.S. publication in 1793 in the *Practical Farmer* (Adams, 2004). In the U.S., zinnias were described in old seed catalogus as having fifteen different colors and ranging in size from the dwarf cultivars, 9-36 cm tall to tall cultivars, 60-90 cm tall (Adams, 2004).

The taxonomy of *Zinnia* was debated and modified many times since the first description of the genus. The following description is based on Stimart and Boyle (2006) and Torres (1963). This classification of *Zinnia* recognizes seventeen species which arose in North America.

Family Asteraceae

Tribe Helianthaea

Genus *Zinnia*

Subgenus *Diplothrix*. Perennial, cespitose (matted, turf like) shrubs or subshrubs. Grows in Southwestern U.S. and Northern Mexico. Grows in dry environments

Z. juniperifolia (DC) A. Gray. Shrubs or subshrubs. Up to 90 cm tall. Much branched at base. Coahuila, Zacatecas, San Luis Potosá, Tamaulipas at elevations of 1350-2580 m.a.m.s.l. Long peduncles and bright orange ray flowers.
synonym: *Diplothrix juniperifolia* DC

Z. citrea Torres. Up to 20 cm tall. Coahuila and San Luis Potosí at elevations of 1080-2400 m.a.m.s.l. Morphologically similar to *Z. acerosa*.

Z. acerosa (DC) A. Gray. Subshrub up to 16 cm tall. Found in Sonora, Chihuahua, Zacatecas, Nuevo Leon, San Luis Potosí, Arizona, New Mexico, Texas at elevations of 780-1815 m.a.m.s.l.
Synonym: *Z. pumila* A. Gray, *Diplothrix acerosa* DC.

Z. oligantha I.M. Johnston. Up to 40 cm tall. Chihuahua and Coahuila at elevations of 1200-1500 m.a.m.s.l. Restricted distribution.

Z. grandiflora Nutt. Subshrub, 8-22 cm. Kansas, New Mexico, Oklahoma, Texas, Sonora to Coahuila, Durango to Zacatecas at elevations of 630-2100 m.a.m.s.l. Wide distribution. Northern most range of *Zinnia*. Grows in dry environments. Roadside weed forming dense mats.

Z. anomala A. Gray. Subshrub, low growing plant. Up to 11 cm tall. Zacatecas, Coahuila, Nuevo Leon, western Texas at elevations of 210-1500 m.a.m.s.l. Inconspicuous flowers.

Subgenus *Zinnia* (Annual or perennial, erect or procumbent herbs).

Section Zinnia (erect annual, showy flowers)

Z. peruviana (L.) L. Annual, 30-50 (-100) cm. Southeastern Arizona, Mexico, Central America, West Indies, Columbia, Ecuador to Peru and Argentina at elevations up to 3000

m.a.m.s.l. Grassy or rocky slopes, wooded bluffs, grasslands. Naturalized in southeastern U.S. Only *Zinnia* with native range in South America. History of nomenclature is long and confusing; many names. Flowers less showy than *Z. violaceae*.
Synonyms: *Z. multiflora* L., *Z. pauciflora* L., *Z. revoluta* Cav., *Z. tenniflora* Jacq., *Z. leptopoda* DC., *Z. verticillata* Andr., *Z. hybrid* Hort., *Z. intermedia* Engelm., *Z. pumila* A. Gray.

Z. violacea Cav. Annual, 90-100 (-200) cm. Sinaloa and Durango to Guerrero at elevations up to 1560 m. Naturalized in Florida, Central America, South America, and Australia. Widely cultivated around the world. Wild plants usually have orange-red to purple ray flowers.
Synonym: *Z. elegans* Jacq., *Z. australis* F. Bailey.

Z. haageana Regel. Annual, up to 75 cm. Jalisco to Guerrero at elevations from 900-2010 m. Grasslands, grassy or rocky hills. Cultivated as an ornamental. Easy to grow.
Synonyms: *Z. multiflora* HBK., *Z. angustifolia* DC., *Z. roezlii* Hort., *Z. mexicana* Hort., *Z. aurea* Hort.

Section *Mendezia* Annual or perennial, erect or procumbent herbs. Low growing plants. Restricted distribution in southern Mexican Plateau, Sierra Madre Occidental, and West coast of Mexico

Z. littoralis Robbins and Greenm. Perennial. Grows up to 14 cm tall. Found in Sinoloa from sea level to 100 m. Very limited distribution; found only near Mazatlán, Sinaloa

Z. greggii Robbins and Greenm. Perennial. Grows up to 30 cm tall. Found in Sinaloa to Michoacán at elevation from 1000-1560 m.
Synonym: *Z. angustifolia* var. *greggii*

Z. maritima HBK. Annual. Grows up to 50 cm tall. Found in Nayarit to Guerrero from sea level to 1200 m. Plants along the west coast are low growing while inland plants are more erect.
Synonym: *Z. palmeri*

Z. angustifolia HBK. Erect annual, 20-40 cm. Found in Sonora and Chihuahua to Michoacán at elevations from 60-2100 m. Rocky or grassy hills. Incorrectly called *Z. linearis* for years.
Synonym: *Z. linearis*.

Z. purpusii Brandegee. Erect annual. Grows up to 41 cm tall.

Z. bicolor DC. Annual, low growing, branched, up to 40 cm. Found in Nayarit to Guanajuato and San Luis Potosí at elevations from1000-1500 m. a.m.s.l.
Synonym: *Z. maritima*.

Z. tenella B.L. Rob. Erect annual. Found in Durango, Zacatecas and Jalisco at elevations 1500-2000 m.a.m.s.l. Often a red spot at base of yellow ray flowers.

Z. leucoglossa Blake. Grows to 12–20 cm tall. Found in Sinaloa near sea level.

There is much literature concerning changes in the taxonomy of *Zinnia*. Recently, two new species are named in subgenus Diplothrix: *Z. acerosa* (DC.) A. Gray var. *huanajuatensis* and *Z. zamudiana* (Calderon de Rzedowski and Rzedowski, 1996). Other new species combinations are *Z. angustifolia* HBK var. *littoralis* and *Z. maritima* HBK var. *palmeri* (Turner 1990).

The most common zinnia in cultivation is *Z. violaceae*, grown around the world. This species is often called *Z. elegans* in the horticultural industry where that is the preferred species name since its introduction to the Western world. In 1802, Curtis's Botanical Magazine recognized *Z. elegans* over *Z. violaceae* because they thought that Jacquin used *Z. elegans* before Cavanilles used *Z. violaceae* (Sims, 1801). *Z. elegans* was commonly used until 1984 when McVaugh used *Z. violaceae* to describe *Zinnia* in Mexico (Kirkbride and Wiersema, 2007). There is a proposal to change the name from *Z. violaceae* back to *Z. elegans* in order to go along with current and commercial usage since *Z. violaceae* is not used by most people (Kirkbride and Wiersema, 2007). *Z. violaceae* additionally is known by the common names of 'Youth and Old Age' (Bailey, 1917) and 'Old Maids'.

In 1917, Bailey divided commercial *Z. elegans* into 3 classes:

1. Tall, 60-90 cm. The Giant or Mammoth strain was developed by C. Lorenz and introduced in 1886.
2. Medium–in horticulture trade since1917; known as *Z. pumila*, *Z. nana*, *Z. compacta*
3. Dwarf–'Tom Thumb', 'Lilliputian'

Z. haageana ranked behind only *Z. violaceae* in importance to horticulture trade in 1917 before the popularity of *Z. angustifolia*. The single form of *Z haageana* was introduced in 1861 and the double form in 1871. The first hybrids between *Z. violaceae* and *Z. haageana* were introduced to trade in 1876 and named *Z. darwinii* (Bailey, 1917). There are commercially available cultivars of *Z. acerosa*, ('Apache Plumes'), and *Z. haageana*, ('Old Mexico' and 'Persian Carpet') (Bailey, 1976). Currently, however, *Z. violaceae* and *Z. angustifolia* are the most important commercial species.

Morphology

A simplified description of the morphology of the genus *Zinnia* was given by McVaugh (1984): ray flowers fertile; disk flowers all, or mostly, fertile; leaves mostly entire; involucres evidently graduated; phyllaries mostly with dark or discolored band at tip; disk achenes strongly flattened, and often awned.

Zinnias, primarily cultivars of *Z. violaceae*, are grown in gardens for their wide range of bright colored flowers. The colors of the ray flowers of *Z. violaceae* are very diverse and include all colors except blue. Diversity of color in ray florets is due to variations in biosynthesis of carotenoids and flavonoids in the epidermal cells and is dependent on the epidermal cell pH. The anthocyanidins pelargonidin and cyanidin are responsible for pink, lavender, rose, maroon, and violet colors when yellow caroenoids are absent in *Z. violaceae* and for orange, scarlet, and red colors when carotenoids are present. When anthocyanidin is not present in *Z. violaceae* and carotenoids are present, the ray color is yellow, and the ray color is white when anthocyanidin and carotenoids are both absent (Boyle and Stimart, 1989). The range of colors in *Z. angustifolia* is not as great as in *Z. violaceae* and is limited to orange, ivory, and white. Colors in *Z. angustifolia* are due to flavonoids and not carotenoids. Ray floret color in *Z. angustifolia* is due to ivory flavonoids plus two anthocyanidins and the absence of carotenoids and the anthocyanidins pelargonidin and cyanidin (Boyle and Stimart, 1989).

Z. violaceae cultivars are considered double when a large number of ray florets and a low number of disk florets are on the capitulum (Stimart and Boyle, 2006; Miyajima, 1998). Ray flowers which open day after day are more likely to result in double-flowered plants (Miyajima, 1998).

In greenhouse trials, *Z. violaceae* was shown to be facultive short day (SD) plants. Floral initiation and development occur faster under SD (12 hours or less) and are delayed under long day (LD)

Figure 12.1: Profusion Cherry (*Z. marylandica*)

Figure 12.1: Star White (*Z. angustifolia*)

Figure 12.1: Zinnia Dreamland Coral

conditions (14 hours or greater); however floral initiation occurs under all photoperiods. Plants grown under LD conditions had taller plants, larger leaf size, larger flower diameter, and more nodes than plants grown under SD conditions (Boyle and Stimart, 1983).

Breeding Efforts

Important selection criteria for zinnia breeding include flower color, rapid and uniform seed germination, strong stems, adequate branching, upright growth habit, disease resistance, uniform flowering, mature plant vigor, and double flower habit for *Z. violaceae* (Stimart and Boyle, 2006). *Z. violaceae* has many desirable garden qualities, but is susceptible to powdery mildew (*Erysiphe cichoracearum*), alternaria blight (*Alternaria zinniae*), and bacterial leaf and flower spot (*Xanthomonas campestris* pv. *zinniae*) (Spooner *et al.*, 1991)

Initial efforts which were made to cross breed *Z. violaceae*, with its superior colors and forms, with *Z. angustifolia*, with its superior disease resistance, were not successful (Boyle and Stimart, 1982). Later efforts produced *Z. marylandica* which is a hybrid of *Z. violaceae* (male parent) and *Z. angustifolia* (female parent). It is resistant to powdery mildew, alternaria blight, and bacterial leaf and flower spot (Spooner *et al.*, 1991). Sakata Seed Co. offers three cultivars of *Z. marylandica*: 'Profusion Orange', 'Profusion Cherry', and 'Profusion White' which was an AAS Gold Medal winner in 2001.

Z. elegans is not grown much as a bedding plant due to production difficulties. Narrow-leaf zinnia cultivars would be preferred by bedding plant growers, so attempts at breeding for narrow leaf cultivars were conducted by Duffy Cowen and Ewart (1988). Double-flowered *Z. violaceae* does not set many seed due to high percentage of female ray florets and low percentage of disk florets (which are complete male and female). Single-flowered zinnias have a higher percentage of disk florets. One thing that prevents seed production in double-flowered zinnia is the newly opened ray florets covering the pistils of previously opened ray florets which prevent pollination (Miyajima, 1995). Miyajima (1998) conducted tests to determine which seed characteristics could be identified which would result in large percentages of double-flowered *Z. violaceae*. Thin seed from ray florets produced higher percentages of double-flowered zinnia compared with wide seed from ray florets and seed from disk florets. Zinnia germplasm in the U.S. is conserved by the USDA National Plant Germplasm System.

Table 12.1: Zimmia Cultivars

Z. violaceae Cultivar or Series	*Height (inches)*	*Attributes*
Zinnita	6-8	Globe shaped plants
Short Stuff	8-10	Early bloom
Dreamland	8-12	Early bloom
Magellan	12-14	Large flowers
Distance	12-18	Upright
Swizzle	10-12	Bicolor
Pulcino	12-15	Compact
Zowie	22-28	Cut flower
Ruffles	24-30	Cut flower
Sun	24-30	Fade resistant
Uproar	24-30	Cut flower, vivid color
Lilliput	30	Garden and cut, strong stems
Envy	30	Dahlia flowered
Oklahoma	30-40	Small blooms, cut flower
Benary's Giant	40-50	Vigorous, standard cut flower
Envy	30	Cut flower, chartreuse
Cactus Jewels	24-36	Cut flower
State Fair	48	Mildew and Alternaria resistance

Species	*Cultivar Series*	*Height (in)*	*Attributes*
Z. angustifolia	Crystal	8-10	Tolerant of powdery mildew
Z. angustifolia	Star	14	Disease, heat tolerant
Z. marilandica (*Z. interspecific*)	Zahara	12-18	Disease resistant
Z. marilandica (*Z. interspecific*)	Profusion	12-18	Disease, humidity tolerant
Z. marilandica (*Z. interspecific*)	Profusion Knee High	18-20	Cut flower
Z. maritima	Solcito	10-12	

Garden Usage

Numerous cultivars of *Z. violaceae* are available for home gardeners. Zinnias are popular summer annual flowers because they germinate readily from seed and grow rapidly to bloom. Zinnias grow and have repeat bloom during summer heat. Many gardeners prefer compact plants which do not fall over due to the weight of a heavy flower on a long stem. Some of the popular series commercially available are listed in Table 12.1.

Cut Flower

Trials were conducted in Mississippi to evaluate the effects of cultivar and planting date on zinnia cut flower production under field conditions. While there was little difference in stem length due to planting date or cultivar, there was a trend for increased stem and bloom diameter as planting dates progressed from May to the July. 'Benary's Giant Deep Red' and 'Red Sun' were consistently the top producers of flower stems across the May, June, and July planting dates. The median number of stems produced by the zinnia cultivars in this study from the May, June, and July planting dates were 21.6, 10.8, and 14.5 stems/plant, respectively, for plants spaced 30 cm apart in the row. The potential stem yield for a single 33m row of the zinnia cultivars included in this trial was 2160, 1080, and 1450 stems for the production life from May, June, and July plantings, or 4690 stems for the three plantings combined (Sloan and Harkness, 2008).

References

Adams, D.W., 2004. *Restoring American Gardens: An Encyclopedia of Heirloom Ornamental Plants.* Timber Press. Portland, Cambridge.

All-America Selections, 2009. AAS winners 1933–present. 1 July 2009. < http://www.all–americaselections.org/AAS_Winners.asp?Sort1=Year_Won&Sort2=DESC >.

Bailey, L.H., 1917. *Cyclopedia of Horticulture*, Vol. 6. The Macmillan Co. London.

Bailey, L.H., 1976. *Hortus Third*. Macmillan Publishing Co. New York, London.

Boyle, T.H. and Stimart, D.P., 1982. Interspecific hybrids of *Zinnia elegans* Jacq. and *Z. angustifolia* HBK: Embryology, morphology and powdery mildew resistance. *Euphytica,* 31: 857–867.

Boyle, T.H. and Stimart, D.P., 1983. Developmental responses of Zinnia to photoperiod. *J. Amer. Soc. Hort. Sci.*, 108(6): 1053–1059.

Boyle, T.H. and Stimart, D.P., 1989. Anatomical and biochemical factors determining ray floret color of *Zinnia angustifolia, Z. elegans,* and their interspecific hybrids. *J. Am. Soc. Hort. Sci.,* 114(3): 499–505.

Bremer, K., 1994. *Asteraceae: Cladistics and Classification*. Timber Press, Portland, OR.

Calderon de Rzedowski, G. and Rzedowski, J., 1996. Dos novedades en *Zinnia* subgenera *Diplothrix* (Compositae: Heliantheae) del centro de México. *Acta Botania Mexicana*, 36: 77–83.

Duffy Cowen, R.K. and Ewart, L.C., 1988. Inheritance of a narrow leaf shape in *Zinnia elegans. J. Amer. Soc. Hort. Sci.*, 113(4): 62–65.

Henson, D.Y., Newman, S.E. and Hartley, D.E., 2006. Performance of selected herbaceous annual ornamentals grown at decreasing levels of irrigation. *HortScience*, 41(6): 1481–1486.

Kirkbride, J.H. and Wiersema, L.H., 2007. Proposal to conserve the name *Zinnia elegans* against *Z. violacea* (Compositae). *Taxon.*, 56(3): 958–959.

McVaugh, R., 1984. Zinnia. In: *Flora Novo–Galiciana: A Descriptive Account of the Vascular Plants of Western Mexico, Vol. 12: Compositae,* (Ed.) W.R. Anderson. The University of Michigan Press. Ann Arbor, MI.

Miyajima, D., 1995. Causes of low double-flowered seed production in breeding zinnia. *J. Amer. Soc. Hort. Sci.*, 120(5): 759–764.

Miyajima, D., 1998. Improvement of ornamental value by seed selection in double-flowered zinnias. *HortScience*, 33(4): 696–698.

Sims, J., 1801. *Zinnia elegans*. Violet-colored zinnia. *Curtis's Botanical Magazine*. 15: plate 527.

Sloan, R.C. and Harkness, S.S., 2008. Evaluation of zinnia cultivars for field grown cut flower production. *J. Applied Hort.*, 10(1).

Spooner, D.M., Stimart, D.P. and Boyle, T.H., 1991. *Zinnia marylandica* (Asteraceae: Heliantheae), a new disease-resistant ornamental hybrid. *Brittonia*, 43(1): 7–10.

Stimart, D. and Boyle, T., 2006. Zinnia, *Zinnia elegans, Z. angustifolia*. In: *Flower Breeding and Genetics; Issues, Challenges, and Opportunities for the 21st Century*, (Ed.) N.O. Anderson. Springer, Netherlands. 12: 337–357.

Torres, A.M., 1961. Taxonomy of the genus Zinnia (Compositae). *Ph.D. Diss.*, Indiana University.

Torres, A.M., 1963. Taxonomy of Zinnia. *Brittonia*, 15: 1–25.

Turner, B.L., 1990. New species, names, and combinations in Mexican Asteraceae. *Phytologia*, 68(6): 14–19.

2011, Biodiversity in Horticultural Crops Vol. 3 *Pages* ***300–310***
Editor: **Professor K.V. Peter**
Published by: **DAYA PUBLISHING HOUSE, NEW DELHI**

Chapter 13

Byadgi Chillies

H.D. Mohan Kumar, M. Shivaprased, M.H. Tatanaga and R.K. Mesta

Horticulture Research Station,
Devihosur, Haveri, University of Horticultural Sciences, Bagalkot
E-mail: ars_devihosur@rediffmail.com

Chilli (*Capsicum annuum* L.) is one of the most important commercial vegetable cum spice crops of Karnataka and is being grown on an area of 1.9 lakh hectares with a production of 1.26 lakh tones annually. Many traditional and improved varieties are grown in different parts of the state both for green chilli and dry chilli purpose. Byadgi Kaddi and Byadgi Dabbi are two most important and very popular export quality chilli cultivars extensively grown for dry chilli purpose in Haveri, Dharwad, Gadag, parts of Belgaum, Koppal, Shimoga, Chickmagalur, Davanagere and Chitradurga districts under rain fed situation. Now its cultivation is spreading to non-traditional areas like Bellary, Raichur and Gulbarga districts where it is being cultivated under irrigated conditions. For green chilli purpose the crop is raised mostly under irrigated situation throughout the state. Green chilli occupies nearly 30-35 per cent of the total chilli growing area in the state. Haveri district stands first followed by Hassan, Belgaum, Kolar and Mysore districts in green chilli cultivation. Sankeshwar, Kadaravalli local, Chincholi local, Gouribidanur are some of the local green chilli cultivars. GPC-82, KDC-1, G-4, Pusa Jwala and Samruddhi are improved popular varieties. Besides, most of the farmers grow private hybrids for green chilli purpose. Consumer preference for green chillies differs from place to place. Accordingly cultivation of varieties/hybrids also varys from place to place. Locally people prefer medium sized, light green, moderate to less pungent chillies. Where as for export to other states and outside countries medium sized, glossy, dark green with thick pericarp, moderate to highly pungent chillies, which can withstand long distance transport, are preferred.

Byadgi chilli grown in Karnataka is well known for quality throughout the world. They possess a unique characteristic flavor and aroma, high colour, oleoresin and very less or no pungency, which

Figure 13.1: Byadgi Kaddi

Figure 13.2: Chilli Drying

Figure 13.3: Chilli Grading

Figure 13.4: Chilli Drying

Figure 13.5: Byadgi Dabbi

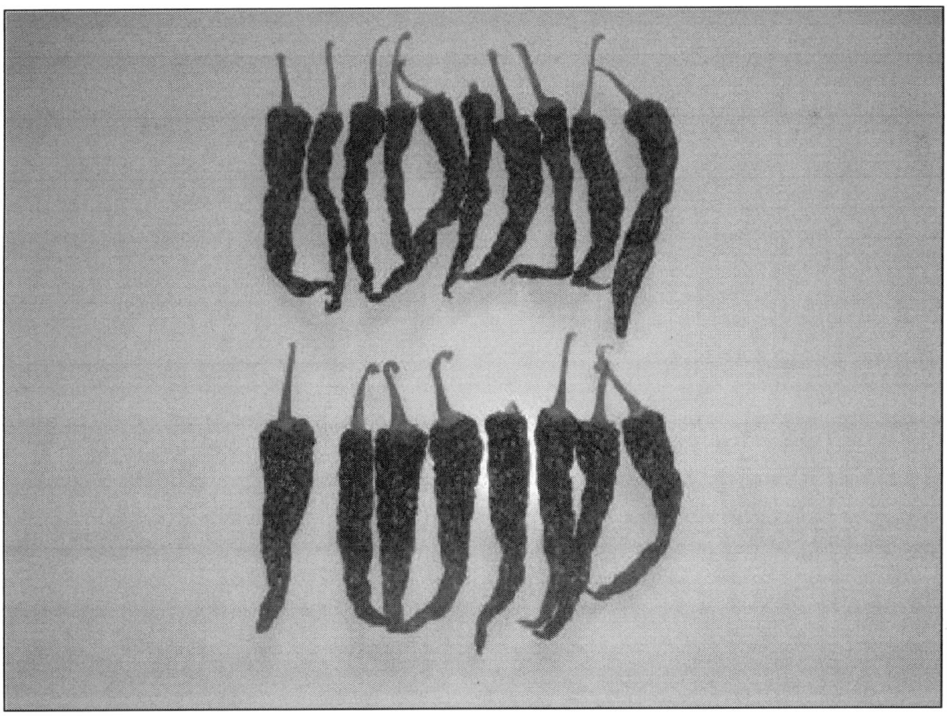

Figure 13.6: Byadgi Dabbi

Figure 13.7: Byadgi Kaddi

Figure 13.8: Variability in Chilli Byadgi

Figure 13.9: Byadgi Dabbi

Figure 13.10: Byadgi Kaddi

makes it the best variety for oleoresin extraction and chilli powder industries. Quality wise, it is comparable only to Paprika varieties grown in European countries. There is lot of demand for these chillies both in domestic and international market. Nearly 75 per cent of Byadgi chillies sold in the market is exported in either direct or indirect forms mostly to European, American and middle East countries. Thus India earns nearly Rs. 800 crore worth foreign exchange every year. Hence, there is lot of scope for improving the production and productivity of these chillies to expand the export market.

Byadgi varieties are highly branching types, grow to a height of 1 meter with a spread of 1 meter. Leaves are small, light green in colour. Fruits light green at immature stage, turn to deep red at maturity and on drying develop characteristic wrinkles on the surface which is unique to Byadgi chillies. Kaddi fruits are very thin, 15–18 cm long while dabbi fruits are 12–15 cm long broader at the shoulder and narrow down towards the tip. Byadgi chillies fetch premium price in the market for their characteristic aroma, high colour and oleoresin with very less or negligible pungency. These cultivars are being named after a small town called "Byadgi" in Haveri district where it is believed to have been identified and cultivated by the farmers since several decades. Unfortunately, records indicating origin and/or the exact period of commencement of Byadgi chilli cultivation is not available. Byadgi chilli market is one of the biggest markets in Asia. Traders from within and outside the country visit the market every year to buy Byadgi chilli. Farmers from far off places bring their produces to this market as they are sure of getting better price here than elsewhere.

Continuous selection by the farmers, generation after generation, has resulted in the congregation of unique genetic combination which produces quality chillies with the highest colour, very less pungency, captive flavor and aroma lending the Indian food its exquisite taste. The colour value in these varieties is around 160-180 ASTA units, next only to Paprika varieties grown in European countries. Byadgi varieties of Karnataka and Tomato chilli (Warangal Chappatta) grown in Warangal, East and West Godavari districts of Andra Pradesh are the only varieties of India, which have export potential for oleoresin industries within and outside India. Off late, with the introduction and cultivation of exotic varieties and hybrids in the traditional rainfed chilli growing areas, hitherto exclusively occupied by Byadgi cultivars, there has been continuous deterioration of original qualities of Byadgi cultivars mainly due to natural out crossing with the introduced exotic varieties and hybrids and also due to accumulation of spontaneous heritable changes (mutations) in the population over generations. In a survey conducted recently by our research station, it is the common opinion of the Farmers, Merchants, Industrialists (Oleoresin, Chilli powder and Pickle industries) and consumers alike, that the quality (mainly the colour, aroma and taste) and yield of Byadgi cultivars are going down every year. Unfortunately, so far no individual or organization has made any serious efforts to undertake purification of these varieties and supply pure and quality seeds to the farmers. Since these cultivars have not been officially released, they do not come under the existing quality seed production chain and thus pure and quality seed material is not available to the farmers, for realizing higher yields. Byadgi chillies are valuable gifts to us from our forefathers. If, this unique combination is lost, it is very difficult to regain the original combination. They have to be conserved in their original state and further improved for resistance to biotic stresses making them more productive agronomically.

Considering the importance of Byadgi chillies in agriculture, trade and commerce and also the problems faced by the farmers, an attempt was made to collect and assess the variability/diversity in cultivators field and also to purify Byadgi chillies. An extensive tour was undertaken during 2004 *kharif* season to visit the farmers fields in traditional Byadgi chilli growing areas of Karnataka. On

Table 13.1: Genetic Parameters in Respect of Productivity and Quality Traits in Byadgi Kaddi Chilli Accessions

Sl.No.	Characters*	Mean	Range	σ2 e	σ2 p	σ2 g	PCV (%)	GCV (%)	h2 (%)	GA	GA as % Mean
1.	DFF	65.188	60-69.5	3.91	7.51	3.60	4.21	2.91	48	2.71	4.15
2.	PH	68.735	49.38-85.87	43.28	79.20	35.92	12.95	8.72	45.4	8.31	12.09
3.	PB	2.613	2.2-3.2	0.60	0.083	0.023	11.05	5.86	28.2	0.17	6.51
4.	SB	5.776	5.0-6.9	0.21	0.29	0.072	9.33	4.67	25	0.28	4.86
5.	FL	14.056	11.93-18.04	1.37	1.95	0.577	9.95	5.4	29.5	0.85	6.05
6.	FD	0.599	0.53-0.77	0.0024	0.0033	0.0009	9.62	5.03	27.4	0.03	5.01
7.	STL	3.548	2.25-5.0	0.091	0.441	0.350	18.74	16.68	79.2	1.09	30.80
8.	FPP	52.887	21.4-82.8	40.38	166.50	126.12	24.4	21.23	75.7	20.13	38.07
9.	SNPF	59.234	32.6-79.2	31.09	102.06	70.96	17.06	14.22	695	14.47	24.43
10.	SWPF	0.337	0.22-0.63	0.0026	0.0059	0.0032	22.82	16.97	55.3	0.09	27.28
11.	TW	5.762	4.4-7.55	0.205	0.288	0.082	9.31	4.98	28.6	0.32	5.55
12.	AA	104.86	30.02-173.52	2.61	934.31	391.70	29.15	29.11	99.7	62.79	59.88
13.	COLV	136.893	17.41-242.06	11.71	1294.27	1282.55	26.28	26.16	99.1	73.44	53.64
14.	CAPV	0.553	0.155-1.2	0.0008	0.0718	0.070	48.41	48.12	98.8	0.55	99.45
15.	RS	15.813	9.71-21.11	0.383	7.22	6.83	16.99	16.54	94.7	5.24	90.15
16.	NRS	9.174	1.06-18.79	0.12	12.28	12.15	38.2	38.01	99	7.15	77.98
17.	TS	25.365	18.3-39.50	0.83	14.13	13.30	14.82	14.38	94.1	7.29	28.75
18.	PHE	3.054	0.79-4.28	0.015	0.519	0.504	23.59	23.24	97.1	1.44	47.21
19.	DYPP	49.205	15-89	41.81	269.17	227.36	33.34	30.64	84.5	28.55	58.03
20.	DYPH	13.667	4.16-24.72	3.22	20.76	17.54	33.34	30.64	84.49	7.97	58.29

*DFF: Days to 50 per cent flowing; pH: Plant height; PB: No. of primary branches per plant; SB: No. of secondary branches per plant; FL: Fruit length; FD: Fruit diameter; STL: Stalk length; FPP: No. of fruits per plant; SNPF: Seed No. per fruit; SWPF: Seed weight per fruit; TW: 1000-Seed weight; AA: Ascorbic acid; COLV: Colour value; CAPV: Capsaicin value; RS: Reducing sugar; NRS: Non-reducing sugar; TS: Total sugars; PHE: Free phenols; DYPP: Dry fruit yield per plant; DYPH: Dry fruit yield per ha.

Table 13.2: Genetic Parameters in Respect of Productivity and Quality Traits in Byadgi Dabbi Chilli Cccessions

Sl.No.	Characters*	Mean	Range	σ2 e	σ2 p	σ2 g	PCV (%)	GCV (%)	h2 (%)	GA	GA as % Mean
1.	DFF	68.62	62.5-70.0	1.39	2.25	0.86	3.69	3.19	74.6	3.8	5.54
2.	PH	70.16	60.71-93.03	25.35	52.05	26.70	10.28	7.36	51.3	7.62	10.87
3.	PB	2.58	1.65-3.3	0.09	0.15	0.05	15.1	8.96	35.2	0.28	10.85
4.	SB	5.51	4.7-7.1	0.30	0.40	0.10	11.57	5.83	25.4	0.33	5.99
5.	FL	12.33	8.41-14.48	0.98	1.83	0.84	10.99	7.47	46.2	1.29	10.47
6.	FD	1.37	0.98-2.8	0.02	0.16	0.13	29.11	26.79	84.7	0.7	50.87
7.	STL	3.55	1.8-5.1	0.14	0.57	0.43	21.27	18.47	75.4	1.17	32.96
8.	FPP	29.03	14.5-46	20.78	53.39	32.61	24.92	19.97	64.2	9.57	32.97
9.	SNPF	96.76	67.136.8	157.81	264.97	107.16	16.82	10.7	40.4	13.56	14.01
10.	SWPF	6.63	0.45-0.98	0.007	0.01	0.009	20.01	15.02	56.3	0.15	23.48
11.	TW	6.68	5.5-8.35	0.398	0.53	0.13	10.9	5.43	24.8	0.37	5.54
12.	AA	75.51	18.125.5	4.99	482.59	477.59	29.09	28.94	99	44.79	59.32
13.	COLV	201.87	134.91-280	12.55	811.33	798.78	14.11	14	98.4	57.77	28.62
14.	CAPV	0.40	0.09-1.01	0.0002	0.04	0.04	51.85	51.72	99.5	0.43	105.91
15.	RS	31.12	19.65-39	2.34	16.00	13.65	12.85	11.87	85.3	7.03	22.59
16.	NRS	27.14	17.15-41.9	1.67	28.51	26.83	19.68	19.09	94.1	10.35	38.14
17.	TS	59.76	51.4-67.35	3.71	16.53	12.82	6.8	5.99	77.5	6.49	10.86
18.	PHE	3.03	0.80-4.23	0.02	0.49	0.47	23.09	22.59	95.7	1.38	45.55
19.	DYPP	56.6	22-104.5	72.46	359.03	286.27	33.48	29.91	79.8	31.16	55.05
20.	DYPH	15.72	6.11-29.32	5.59	27.70	22.10	33.48	29.91	79.82	8.67	55.18

*DFF: Days to 50 per cent flowing; pH: Plant height; PB: No. of primary branches per plant; SB: No. of secondary branches per plant; FL: Fruit length; FD: Fruit diameter; STL: Stalk length; FPP: No. of fruits per plant; SNPF: Seed No. per fruit; SWPF: Seed weight per fruit; TW: 1000-Seed weight; AA: Ascorbic acid; COLV: Colour value; CAPV: Capsaicin value; RS: Reducing sugar; NRS: Non-reducing sugar; TS: Total sugars; PHE: Free phenols; DYPP: Dry fruit yield per plant; DYPH: Dry fruit yield per ha.

careful verification of the characters of Byadgi chilli plant and fruits in the fields, the samples (approx. 500 g. dry fruits) were drawn from the middle of the plot, brought to HRS, shade dried and then seeds were separated and kept ready for sowing. Ninety such B. kaddi and B. dabbi collections were made covering all the villages in the traditional Byadgi chilli growing areas. A few samples were also drawn from non-traditional growing areas covering Bellary, Raichur and Gulbarga districts. Normally the farmers and seed merchants in non-traditional growing areas buy fresh fruits/seeds every year from Kundgol, Saunshi, Gudigeri and Annigeri villages in traditional chilli growing areas.

A field experiment was conducted at HRS, Devihosur during 2005 *kharif* season to assess the extent of variation in Byadgi chilli cultivars. Analysis of variance revealed significant variation for all the twenty productivity and quality characters studied with considerable amount of range in both Byadgi kaddi and dabbi accessions.Estimates of components of variability, heritability and genetic advance for the characters under study are presented in Table 13.1 and Table 13.2. Almost all the characters exhibited superior mean values with considerable range indicating existence of variability in the population suggesting the possibility of improving these characters through suitable selection criteria. Phenotypic co-efficient of variability for all the characters is slightly higher than genotypic coefficient of variability in both kaddi and dabbi accessions suggesting that the apparent variation is not only due to genotypes but also due to the influence of environment. Selection for such traits could be sometimes misleading. However, environmental variances observed for all the characters are very less suggesting very little influence of the environment. Knowledge about the magnitude of heritability of a trait is very essential for the breeder. All the traits in both kaddi and dabbi accessions exhibited high amount of heritability. Selection based on these traits would be effective as they are likely to be controlled by additive genes. Except for total sugars, all the traits recorded higher per cent of genetic advance over mean in both kaddi and dabbi accessions. Higher mean values coupled with high heritability and genetic advance clearly indicate the possibility of further improvement by employing suitable selection criteria.

The genetic diversity among 90 accessions each of Byadgi kaddi and dabbi was measured by employing D2 statistics. Out of 19 characters studied in Byadgi kaddi, number of seeds/fruit contributed maximum to diversity (44.87 per cent) followed by number of fruits/plant (39.40 per cent), reducing sugar, no. of secondary branches/plant, fruit length, total sugars, clolur value. Where as in Byadgi dabbi number of seeds/fruit contributed maximum to diversity (45.44 per cent) followed by no. fruits/plant (26.74 per cent), reducing sugar, fruit length, total sugars, no. of secondary branches/plant, ascorbic acid and colour value. Based on D2 values Byadgi kaddi accessions were grouped into 18 clusters while Byadgi dabbi accessions were grouped into 14 clusters. Grouping of accessions in different clusters indicated no relationship between genetic divergence and geographical location. Based on the results a few accessions in kaddi and a few in dabbi are identified for recombination breeding programme.

Genetic purification and crop improvement of Byadgi chillies are in progress at Horticulture Research Station, Devihosur, Haveri. Purified varieties will be released for general cultivation by the farmers in a couple of years.

References

Hosmani, M.M., 1993. *Chilli crop (Capsicum annuum)*. Bharath Photo Offset Works, Dharwad, Karnataka.

Mohan Kumar, H.D., Shivaprasad, M., Tatagar, M.H. and Astaputre, S.A., 2005. *Preliminary Survey Report (Unpublished)*. Horticulture Research Station, Devihosur, Karnataka.

Salimath, P.M., Mohan Kumar, H.D. and Basavaraj, N., 2008. Byadgi chillies: Dollar earning high quality chillies from Karnataka. *Spice India*, 21(7): 6–8.

Sandeep, 2007. Genetic variability, correlation, morphological and molecular diversity in Byadgi Kaddi and Byadgi Dabbi Chilli (*Capsicum annuum* L.) accessions. *M.Sc. (Agri.) Thesis*, University of Agricultural Sciences, Dharwad, Karnataka, India.

2011, Biodiversity in Horticultural Crops Vol. 3 *Pages* **311–323**
Editor: **Professor K.V. Peter**
Published by: **DAYA PUBLISHING HOUSE, NEW DELHI**

Chapter 14

Bambara Groundnut

S.L. Venter[1], E van den Heever[1] and C.J. Swanevelder[2]*
[1]*Agricultural Research Council,*
Vegetable and Ornamental Plant Institute (ARC-VOPI),
Private Bag X293, Pretoria, South Africa, 0001
[2]*Agricultural Research Council*
Grains Institute, Post Bag X1251, Potchefstroom, South Africa, 2520
**E-mail: evdheever@arc.agric.za*

A major future challenge is to support livelihoods of people within the global context and with the prospect of climate change. The effect on the potential production of sufficient food within environments in which temperature and rainfall are likely to change, and the resulting potential pressure on genetic diversity, as well as the potential increase in disease pressure and drought incidence in areas with high agricultural potential, need to be carefully monitored. Furthermore, care should be taken that the food basket diversity is not negatively influenced by these factors.

Figure 14.1: Bambara Groundnut

Within this context, much more emphasis is placed on the major crops over the past decades. The risks associated with the dependence on only a few crops for the world's demand for food, feed, fuel, shelter, industry, medicine, improvement of the environment, as well as the reduced diversity of species on which the worlds food and nutrition security and environmental safety relay, will have immense implications for nutritional security and hidden hunger of the world (Haq, 2007). This risk will be of increasing concern in future, since the options for rural people will become more limited because of the increasing narrowing of the base of food basket. There are, however, many other species, currently of diverse importance, which remain to be tapped.

There is increasing awareness throughout the world that additional plants should be developed into crop plants to avoid future problems in terms of food shortages (Venter *et al.*, 2007). This requirement stems from the rapidly rising demand for food, particularly in developing areas, and the fact that little new land will become available for food production. On the contrary, more and more high potential agricultural land is being lost annually to urban sprawl, particularly in developing countries. Coupled with this is the fact that most of today's common crops have already been pushed close to their maximum yields. Hope for the future may well lie in a number of unexploited edible plant species which have largely been ignored in the past.

In South Africa, hunger and malnutrition are still prevalent in many rural and urban areas. It is estimated that at least three million people under the age of 15 are suffering from malnutrition (Venter *et al.*, 2007). Various studies conducted in South Africa emphasized the role of indigenous plants in providing many of the basic needs of rural people, *i.e.*, shelter, fuel, medicine, as dietary supplements and as source of income. The high volumes of plant material are used for fuel (wood), fencing and building materials. Medicinal plants are an important element in the traditional health care system, in both rural and urban areas (Cunningham *et al.*, 1992). Indigenous plants are also harvested and used for crafts, which have the advantage of utilizing renewable resources, generating valuable income and preserving traditional skills. Certain indigenous plants produce good quality oils which are sought after in the perfume industry (Cunningham *et al.*, 1992). Flower pickers in the Western Cape area of South Africa, started a multi-million Rand industry by scouring the Cape Mountains for indigenous flowers from the fymbos and selling them. This tradition is still in existence and these flowers are now also cultivated commercially (Coetzee and Littlejohn, 1994).

In South Africa, a country with national food self-sufficiency and food exports, malnutrition still prevails in many rural areas. This is due to a number of reasons, not the least of which is the harsh climatic conditions encountered in these areas (Allemann *et al.*, 1996). It is estimated that there are at least three million people under the age of 15 suffering from malnutrition in South Africa (FRD, 1992). Diseases in African countries are often associated with malnutrition and as a result, those who do survive these diseases often never recover completely. Malnutrition contributes to high infant, maternal and childhood mortality rates found in many African countries (Sofowora, 1996). The use of undomesticated food plants in regions with low agricultural potential or during periods of drought adds to food security in these areas and provides dietary supplements to the predominantly starchy staple foods (Cunningham *et al.*, 1992).

One way of increasing food production in arid and semi-arid regions is through use of indigenous crops known to grow in these areas without substantial agronomic care. These indigenous plants may be better adapted to the often harsh climatic conditions, and have a few insect and disease problems in comparison to the introduced crops (Chweya, 1989; Cunningham *et al.*, 1992; Attere, 1990; Mwashayeni, 1994 and Venter *et al.*, 2007). Indigenous plants could be developed and used as food crops to broaden the food base of African countries (Chweya, 1989). The number of plant species in southern Africa exceeds 20,000, and more than 1,000 of these are edible (Wehmeyer, 1986). In fact, the National Botanical Institute of South Africa has records of more than 1,700 plants referred to as being edible by the local population.

Water is a scarce commodity in South Africa. In general, the subtropical countries of southern Africa are prone to droughts, thus, planning for future crop development should take cognisance of this fact (Tyson, 1990). There is a great need, therefore, to develop crops from appropriate indigenous plants which are eminently suited for cultivation in large areas of southern Africa which have low agricultural potential for conventional crops (Cunningham, *et al.*, 1992, Venter, *et al.*, 2007). Most indigenous vegetables are, unfortunately, associated with poverty and are, therefore, shunned by many, or considered inferior to the exotic species. As a result, these crops have not been fully domesticated and the management requirements remain virtually unknown (Mwashayeni, 1994). These plants should be developed to the stage that they can be cultivated as specialist crops in their own right and no longer perceived as food only in times of famine, or to be harvested from the wild. Both scientific data and traditional knowledge suggest that plant products from indigenous plants are not inferior in taste and nutritional quality in species more widely known and used (Venter *et al.*, 2007).

The use and harvest of indigenous crops grown in the wild in many areas of South Africa are well known in survival strategies of the rural poor. Many under-utilized indigenous species have important multiple uses and consumption of these foods enable people, specifically women and growing children, to obtain nutrients for a balanced diet, cash to supplement family incomes and generally improve health. They are also valuable sources of food and nutrition during periods of emergency, *e.g.*, drought, food failure, during war and seasonal effects on production of crops.

One crop that has the potential to address some of the challenges mentioned above is the Bambara groundnut. Bambara groundnut, *Vigna subterranea*, is one of the a few indigenous tropical African crops (Figure 14.2). Even so, in many countries its place as a major crop has now been taken over by the true groundnut (*Arachus hypogaea*), except on the poorer soils and in areas with uncertain rainfall, where it is still important as a subsidiary crop. Nevertheless, considerable interest is taken in the crop in recent years, especially in view of its high yield and resistance to disease.

The rural poor usually rely on an energy-rich carbohydrate food as staple in their diet and to balance the diet, a protein-rich food like Bambara groundnut, should be added. Kwashiorkor, which is prevalent in many areas is a sign of protein deficiency in the diet. Increasing the vegetable protein supply in these malnourished areas is less difficult and lesser expensive than boosting the supply of animal protein (Venter *et al.*, 2007). Bambara groundnuts are a good source of protein, as well as being drought tolerant and able to grow in poor soils, making them eminently suitable for production in areas with low agricultural potential. They are regarded important and cultivated in many areas of South Africa. The use of this crop as a food crop could lead to the conservation of our natural plant and physical resources, as well as contributing to more sustainable agriculture. More resistant crops will be available for production in problematic areas and this will help to facilitate food security. With the high nutritional status they have the potential to assist in addressing some malnutrition problems and fill a valuable niche in this regard (Venter *et al.*, 2007).

The ARC-VOPI became involved in this line of research after it was identified as a research priority by a number of people involved in the rural development field. A number of plants were identified as priority species in rural areas, including Bambara groundnut. Good adaptation of these species to unfavourable conditions is one of the major reasons for the growing interest in adapting some of these crops as possible substitutes for, or to be grown in conjunction with, commercial crops,

Figure 14.2: A Good Stand of Bambara Groundnuts

particularly by small scale farmers under low input conditions. With the increasing interest in indigenous cuisine, and the popularity of so-called "ethnic" restaurants in South Africa (Anderson, 1995), these crops could be developed into specialist horticultural crops in their own right to serve this growing market.

General Information

The Bambara groundnut, *V. subterranea* (L.) Verdc. (Verdcourt, 1980) is an indigenous tropical African crop (Figure 14.4). Bambara groundnut is mainly grown on small farms by female farmers for home consumption and only surpluses are sold on local markets. According to Coudert (1984), about half of the world production is produced in West Africa. It is one of Africa's most popular grain legumes after groundnut (*Arachus hypogaea* L.) and cowpea (*Vigna unguiculata* L.) On the poorer soils and in areas with uncertain rainfall, it is an important subsidiary crop. The name most commonly used, Bambara groundnut, appears to be derived from the district of Bambara, near Timbuktu in Northern Nigeria. Bambara groundnut is known by many different names over the world - every language and dialect has its own name for this nutritious food crop, which gives an indication of the popularity of the crop. Considerable interest was taken in the crop in recent years, especially in view of its high yield and resistance to disease, even under harsh climatic conditions (Venter *et al.*, 1997). Bambara groundnuts are cultivated extensively in Africa at a subsistence level. They are reported to be the third most important leguminous crop south of the Sahara, being superseded only by cowpeas and groundnuts. Most of the production is, however, consumed domestically and Bambara groundnuts do not usually enter international trade (Schippers, 2000; Venter *et al.*, 2007). As indicated above, the Bambara groundnut has lost importance in many parts of Africa because of the expansion of groundnut production, but in recent years there is renewed interest in the crop for cultivation in the arid savannah zones, because of its resistance to drought conditions and its ability to yield a reasonable crop when grown on poor soils (Venter *et al.*, 2007).

Names of Bambara groundnut

Scientific Names

Vigna subterranea (L.) Wilcz. is the correct scientific name for this crop. The synonym, *Voandzeia subterranea* is often seen in literature, but it is no longer valid. This species is a member of the Fabaceae and subfamily Papilionoideae, or pod plant family, the same as other types of beans (Hepper, 1970).

Common Names

The plant is known under a wide variety of common names throughout the world. These can vary from district to district even within a country. Some of the most common names in South Africa are as follows: Bambara groundnut or Njugo bean (English), Dopboontjie or Jugoboon (Afrikaans), Izidlubu (Zulu), Nduhu-mvenda or Phonda (Venda), Tindhluwa or Ndhluwa (Shangaan), Ditloo-marapo (Sutu) and Njugu-mawe (Swahili) (Hepper, 1970; Holm and Marloth, 1940).

Varieties

A large number of varieties are recognized by indigenous people and are distinguished mainly by the colour of shelled seed. (Holm and Marloth, 1940; Tardieu, 1957; Dalziel, 1937).

1. *Black variety*–The skin of the kernel is black and the hilum white. It is an early maturing variety and consequently the most commonly grown.
2. *Spotted variety*–usually yellowish-brown with dark specks. Some seeds have larger spots which almost become dark patches. The true variety, however, is finely speckled.
3. *Yellowish brown variety*–is able to withstand continuous moist conditions better than any other variety.
4. *Red variety*–this is the most prolific bearing variety. It is, however, liable to rot sooner in the ground than other varieties because it requires a longer growing period.
5. *Cream variety*–it is not very popular because of its small kernels and its unprolificacy.

Centre of Origin and Distribution

According to Burkill (1906) in Holm and Marloth (1940), Bambara, a district on the Upper Niger near Timbuktu, has given its name to this African groundnut, but as it is widely cultivated throughout tropical Africa, Bambara has no pre-eminent claim to the plant (Holm and Marloth, 1940).

That Africa is the native home of the species has, however, been proven beyond doubt. Its cultivation extends from the Sahara and over the major portion of tropical Africa right down to KwaZulu-Natal. The Bambara groundnut is widely cultivated throughout tropical Africa, as well as in Madagascar,

Mauritius, India, Sri-Lanka, New Caledonia, Northern Australia, Tropical Central America, Surinam, Brazil and South America (Venter *et al.*, 2007).

Botanical Description

The crop is an annual herb with variable growth habits, often spreading or trailing. The plant produces a mainly fibrous root system, with many nodules formed by the associated *Rhizobium* bacteria. Stems are branched and hairy, with short internodes, and roots are often produced at nodes in contact with the soil. These stems carry trifoliate leaves on long erect petioles which are thickened at the base. The three leaflets are between 2.5 and 7.5cm long and 1 to 3cm wide (Holm and Marloth, 1940).

The self-pollinated flowers can be yellow, red or pink, and are borne on axillary racemes arising from the stem nodes. The gynophore elongates after fertilisation to push the ovary into the soil where the fruits are formed. The fruits are pods which are hard, wrinkled, 15 to 25 mm in diameter, developing below the soil surface with 1 to 2 seeds/pod. The large seeds are round, smooth, hard, up to 15 mm across and the colour varies from white to red, black or mottled (Figure 14.3). The approximate weight of 100 seeds is between 60 and 70g.

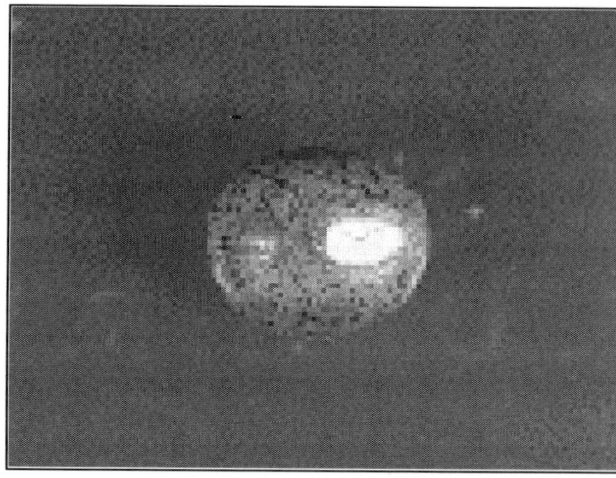

Figure 14.3: A Bambara Seed

Climatic and Soil Requirements

Although a very adaptable crop, the Bambara groundnut prefers a sunny, hot climate with average day temperatures between 20 and 28°C, with a frost-free period of 100-120 days (Kay, 1979; Holm and Marloth, 1940; Hepper, 1970). The crop requires an evenly distributed moderate rainfall from sowing until flowering. Optimum yields are obtained with an annual rainfall of 900 and 1 200 mm (Kay, 1979). Satisfactory yields can be obtained in areas with a pronounced dry season, since the crop is relatively drought resistant. It is also tolerant to periods of heavy rainfall, except during flowering stage. The semi-arid regions of South Africa with rainfall of less than 500mm/annum (especially when evenly distributed throughout the growing season) can still be suitable for production.

Although adapted to a wide range of soils, the Bambara groundnut thrives on light sandy, well-drained loams of medium fertility with a pH of 5.0-6.5. Soils with a high organic content will give increased yields compared to those cultivated on soils poor in organic matter. This plant can be grown on poor soils unsuitable for groundnuts and will give higher yields than groundnuts on low fertile soils. Calcareous soils are detrimental to the crop and are, thus, not suitable (Kay, 1979). It is recommended that nitrogen rich soils or heavy fertile soils be avoided, since excessive vegetative growth inhibits pod production (Venter *et al.*, 2007).

Most cultivars are adapted to short days, meaning that they start flowering when the day length shortens towards end of summer. The Bambara groundnut produces satisfactory yields at elevations of up to 1 600 m.a.m.s.l (Kay, 1979).

Cultivation

Propagation

Seeds, normally shelled, are used to propagate Bambara groundnut, although whole pods are sometimes used. Bambara groundnut seeds are usually planted directly into the soil by hand, but for large scale production, the seed can be planted by a maize/bean planter, using plates as specified for beans or groundnuts. The use of shelled seed that treated with Thiram is recommended (Kay, 1979).

Sowing Methods and Spacing

The Bambara groundnut may be planted as a pure stand, often as the first crop in a rotation, followed by root crops like sweet potato, cassava or potatoes, but is more than often grown as an intercrop with cereals (usually pearl millet), root crops or other legumes. The soil should be ploughed 30 cm deep and 1-4 seeds are sown in a hole about 5-7.5 cm deep (Kay, 1979).

Seeds are usually planted 10-30 cm apart in single rows 40-50cm apart on the flat, or in double rows 20 cm apart on beds or flat-topped ridges 90 cm apart. Between 40 and 50 kg of seed are required to achieve a plant population of 80 000 plants.ha^{-1}. In South Africa, the average seed-rate is 50 to 60kg.ha^{-1} (Kay, 1979). The seedbed must be ploughed at least 300mm deep to enable plants to bury its seedpods after fertilisation of the flowers. The final seedbed must be uniform, fine, slightly firmed, free from clods, weeds and weed remains (Venter *et al.*, 2007).

Planting Dates

In South Africa,sowing takes place in mid-October up to early February, although the better results are obtained when planted about mid-November (Venter *et al.*, 2007). Early and late plantings produce lower yields. In food insecure areas where Bambara groundnuts are grown as part of a balanced diet, and also in vulnerable areas, it is, however, advantageous that the growing season can be extended by early and late plantings, resulting in food still being available in those areas and for longer periods, even though yields are sub-optimum.

Fertilization and Irrigation

As indicated, Bambara groundnuts grow in poor soils without fertilizer. For optimum yields, however, a balanced N:P:K fertiliser should be applied prior to planting according to the requirements as determined by a soil analysis. Nitrogen and potassium can be applied as top dressings to stimulate growth approximately three weeks after planting, if required. As this plant is a legume, it is possible to inoculate with an appropriate strain of *Rhizobium* bacteria to enable the plant to utilise atmospheric nitrogen. If this is done, all nitrogenous fertilisers should be avoided.

Details on the effect of irrigation are not yet available, but the ARC performed a preliminary study during 1998 and 1999 summer seasons at ARC Roodeplaat (25°35'S, 28°21'E, altitude 1164m) to determine effect of irrigation on crop yield. This was, however, a preliminary trial and further research is recommended. The experiment was set out under a rain shelter as follows: plot size was 5.4 m^2 consisting of four rows with seed planted at 20cm in-row and 50cm between rows spacing. Three levels of irrigation were maintained (100 per cent, 60 per cent and 40 per cent) and the trial was replicated three times. The soil was fertilized with chicken manure prior to planting according to the soil analysis results at a rate of 10.9kg/5.4m^2 to supply 30g/m^2 nitrogen. The plants were irrigated using an overhead irrigation boom system fitted with different nozzle sizes to provide low, medium and high water level treatments. The 40 per cent and 60 per cent levels were calculated as percentages

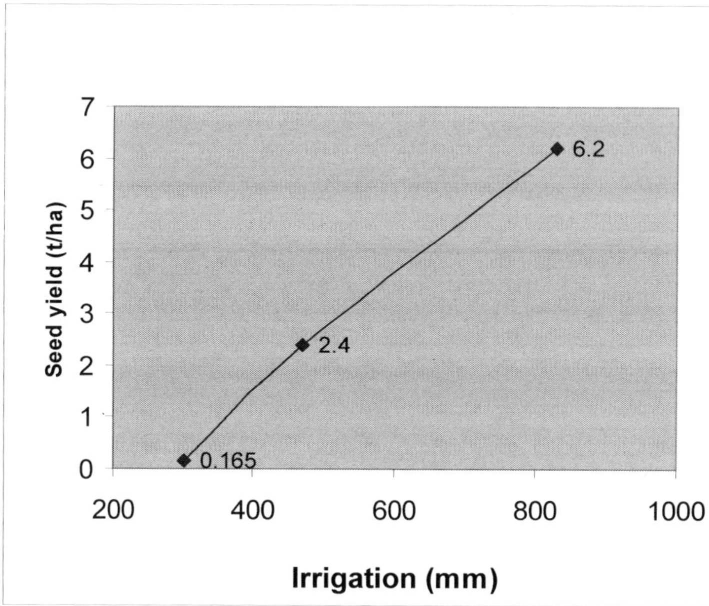

Figure 14.4: Seed Yield (t/ha) at Different Irrigation Levels during 1998 Summer Season

of the high water level percentage (100 per cent). The soil water content was measured weekly using a neutron water meter inserted *via* an access tube in the middle of each plot. Weekly total crop water use of each irrigation level was calculated, from the water balance equation, as the change in total soil profile water content during that week, allowing for the amount of irrigation applied. The plots consisted of four rows each, from which the two middle rows were harvested. The average number of plants harvested/plot totalled thirteen. The seed yields under the different irrigation regimes are indicated in Figure 14.4.

Figure 14.4 shows that a variation in irrigation levels had a large effect on yield. The highest seed yield of 6.5 t/ha was obtained at an irrigation level of 830mm, followed by a seed yield of 2.4 t/ha at 470mm irrigation level. The lowest yield of 0.165 t/ha was obtained at an irrigation level of 302mm for the entire growing season. These results indicated that increased water application in *V. subteranea* cultivation would increase seed yield. However, it is also clear that a relatively good yield could still be obtained at lower irrigation levels.

To conclude, the seed yield gave a set pattern for the different irrigation levels for the 1998 and 1999 summer seasons. The results suggest that seed yield in *V. subteranea* production is highly dependant on water availability. The results obtained under high water application surpassed the yield that farmers normally get, which was in the region of 0.5-1t/ha under dry land production. From the results obtained, it can be concluded that seed production in *V. subteranea* can be substantially increased if grown under irrigation.

Under dry land conditions in areas with sufficient rainfall, irrigation is rarely required, but the producer should be guided by the rainfall requirements of the crops as set out above.

Weeding

Weeding is essential and is usually carried out 15 days after germination, and as the clusters join. This is essential to reduce weed competition with the crop and is normally carried out by hand as there are no herbicides registered for use on this crop. The lower part of the plants should be earthed up when they are 15 to 25cm high to encourage development of pods underground.

Growth Period, Harvesting and Yield

Bunch types take approximately 90 to 120 days to mature, while spreading types are mature at around 120 to 150 days after planting. For eating purposes, the seed should not be harvested when fully mature, but should be soft still. When harvesting, the entire plant is lifted by uprooting by hand or with hoes. For seed production, plants are uprooted with a hoe, ploughed out or pulled by hand and allowed to dry in the field in windrows for one or two days. Harvesting should take place during dry weather. The pods may be shelled as soon as harvested, or they may be allowed to dry in the sun for one to four weeks. Threshing is usually carried out manually, cracking the pods by various means (Venter *et al.*, 2007).

Yields are very variable, ranging from 800 to 1 500kg.ha^{-1} of mature pods, representing between 300 and 600kg.ha^{-1} of dried seed. Average yields of seed in Africa are reported 650-850 kg.ha^{-1}. Yields are very dependant on the variety and environment interaction, and a variety that yields very well in one area may not yield at all when planted in a different area (Venter *et al.*, 2007).

Insect Pests and Diseases

V. subterranea is remarkably free from diseases during growth, although an unidentified leaf virus infection is reported that results in reduced yields in Nigeria. The following fungal diseases are reported to infect the plant, but none appear to be serious: *Sphaerotheca voandzeiae, Cercospora voandzeiae, Ascochyta phaseolorum, Phaeolus manihotis, Meliola vignae-gracilis, Cercospora canescens, Colletrotrichum capsici, Corticium rolfsii Curzi (Sclerotium rolfsii Sacc), Corticium solani, Elsinoe* sp., *Erysiphe polygoni, Leptosphaerulina trifolii, Phyllosticha voandzeiae* and *Phynchrytrium dolichi* (Hepper, 1970).

The important diseases appear to be a seedling blight caused by *Corticum* spp., powdery mildew, caused by *Erysiphe polygoni*, and *Fusarium* wilt (Hepper, 1970).

Stored seeds are very liable to attack from beetles and one species of moth is reported. The following insect species are recorded on *Voandzeia* plants and stored seed: Coleoptera (beetles, bruchids, true weevils), Hemiptera (bugs, leafhoppers) and Lepidoptera (larvae of moths and butterflies) (Hepper, 1970).

Viruses that can affect growth and yield of Bambara groundnut include Alfalfa mosaic (3 strains), bean chlorotic ring spot (4 strains), bean mosaic, bean necrosis and white clover mosaic. Most viruses are transmitted mechanically or by insect vectors. There is no chemical control for virus diseases, and disease monitoring and management are, therefore, important to reduce the potential spread of viruses (Hepper, 1970).

Nutritional Qualities and Uses

The Bambara groundnut is a complete food (Begemann, 1988). The fat content varies 5.3 to 7.8 per cent, carbohydrate proportion ranges 54.5 to 69.3 per cent. It is also an important crop for its high protein content (Nx6.25), ranging from17.0 to 24.6 per cent. The protein quality is high and rich in lysine, but limiting in isolycine. The biological value and protein digestibility are 56 and 82 per cent, respectively as quoted by Begemann (1988), from numerous authors. A review article, Linnemann (1987), quoted the following detail from the literature:

1. Amino acid composition (per cent) of the protein fraction of 18 per cent : aspartic acid 10.9 - 11.9, threonine 3.2 -3.6, serine 5.1 - 5.4, glutamic acid 16.7 - 17.0, proline 4.1, glycine 3.6 - 3.8, alanine 3.6 - 4.2, valine 4.5 - 4.8, cystine 1.2 - 1.4, methionine 1.2 - 1.3, isoleucine 3.5 - 4.0, leucine - 7.4 - 7.7, tyrosine 3.2 - 3.4, phenylalanine 5.4 - 5.8, lysine 6.8 - 7.7, histidine 3.1 - 3.5 and arginine 6.5 - 7.7.
2. The main fatty acids (per cent) of the oil extracted from Bambara groundnut seed: palmetic (C16:0) 18 - 24, stearic (C18:0) 5 - 9, oleic (C18:1) 18 - 24, linoleic (C18:2) 35 - 40, linolenic (C18:3) 1 - 2. Furthermore, no phytohaemagglutinins, cyanogenetic glucosides or alkaloids were found, but it has, however, trypsin inhibitory activity which may vary per 100g seed from 6.75 to 15.44 units/mg.

The immature semi-ripe seeds are more palatable than the hard mature seed. The immature seeds are normally eaten fresh, boiled or grilled and the mature seeds are ground into flour, and then mixed with oil or butter to form porridge and sometimes they are roasted in oil. Young whole pods can be washed and boiled and used in soups. Seed may be roasted and flour may be prepared from roasted or unroasted seed which can be used for livestock feeding after soaking in water.

The roasted ground meal is used as a coffee substitute. The seeds of Bambara groundnuts are often mixed with other foods. They are often added to meat stew, rice, spinach, other vegetables, maize and sorghum. Cakes or balls may be made from the flour mixed with maize. There exist numerous superstitions or taboos regarding the Bambara groundnuts. Planting in virgin soil or before the maize crop had germinated was not permitted. It was prohibited that Bambara groundnuts be planted for periods varying from one to five seasons when a member of the ruling house died, or when drought was predicted by a weather prophet. If no rain had fallen for a considerable time Bambara groundnuts were often pulled up. Males were never allowed to pass through Bambara groundnut fields. The belief was that bad yields would result. Lands belonging to males were planted with Bambara groundnuts only every alternate year, otherwise they might die or some other evil might befall their families. The lands of women could be planted annually. When Bambara groundnuts were not mounded, young people were not allowed to eat the beans. It was considered that permanent or temporary sterility would result if this law was not observed. To what extent these superstitions and taboos are observed today are not known. Many of the beliefs had acceptable reasons for their existence.

Apart from the above, it is alleged that vomiting is checked by chewing and swallowing the raw beans. Pregnant women who suffer from nausea use the above prescription for relief.

Bambara groundnut is eaten in various forms, either immature or fully ripe. The immature seeds are normally eaten fresh, boiled, or grilled, while the mature ones are often pounded into flour and then mixed with oil or butter to form porridge, or sometimes they are roasted in oil. Occasionally the young whole pods are washed and boiled, or used in soups (Kay,1979). Commercial canning of the Bambara groundnut has been successfully developed in Zimbabwe and Ghana.

Marketing

Bambara groundnuts are cultivated extensively in Africa at a subsistence level. They are the third most important leguminous crop south of the Sahara, being superseded only by cowpeas and groundnuts. Total production in Africa is about 330,000 tonnes/annum. Most of the production is consumed domestically and Bambara groundnuts do not usually enter international trade (Kay, 1979).

This crop has lost importance in many parts of Africa because of the expansion of groundnut production. In recent years there is renewed interest in the crop for cultivation in the arid savanna zones because of its resistance to drought conditions and ability to yield a reasonable crop when grown on poor soils. Little work is done on the agronomy of the crop or its improvements and there is an urgent need to develop improved cultivars.

The Bambara groundnut is relatively unknown in many countries and their people. The reason for this may be the lack of stock for commercialization of the product and this can be attributed to difficulties experienced in harvesting the crop, caused by the poor attachment of the pod to the plant. It is very difficult, even almost impossible, to harvest it mechanically and, therefore, production is restricted to limited acreage and tonnage. Even with hand harvesting a significant part of the yield becomes detached in the process and has to be retrieved by hand from the soil. In recent years, more attention is given to this nutritious crop.

The South African market for Bambara groundnuts is growing steadily and all the products sold locally are imported from Zimbabwe. The current political turmoil in Zimbabwe has led to the local traders looking for a local supplier. In 2002 they were willing to approve a contract for 500 tonnes of Bambara groundnut/year. This corresponds to an area of roughly 200 to 300ha at present yields. The price of this commodity is linked to that of sugar beans. During an informal survey between 1998-2002 in informal markets in rural areas in South Africa, it was observed that Bambara groundnuts were sold/cup, earning an amount at the time of $ 0.5/cup.

Conclusion

Bambara groundnut has the potential to address some of the nutritional problems experienced in the world today, specifically those related to protein deficiency. There is an urgent need to develop innovative strategies to address the increased demand for nutritious food, and strategies enabling farmers to overcome the adverse effects and risks caused by climate change. We therefore, need to carry out research to ensure that the value of different underutilised crops is fully exploited and develop new crops adapted to adverse conditions. This enhances our ability to mitigate the adverse effects that climate change may have in future. Seed production and seed supply systems are important aspects which need a lot of attention, as well as the development of sustainable food systems and an increase in the diversity of the foodbasket. In the light of the high percentage of malnutrition in South Africa and the serious effect it has on the health, growth and development, of especially children, this is one crop that can be included in such food systems. A strategy also needs to be developed to create formal and local markets for the products of Bambara groundnuts. The development of underutilized crops clearly has the potential to not only address malnutrition problems and increase the diversity of the food basket, but also to enhance the competitiveness of the agricultural/farming sector by adding additional crops, better adapted to harsh climatic conditions and poor soils. It will assist in increasing food supply to vulnerable areas, raise productivity of agricultural land, and produce and contribute to natural resource management and conservation.

References

Allemann, J., Van den Heever, E. and Viljoen, J., 1996. Evaluation of Amaranthus as a possible vegetable crop. *Appl. Plant Sci.,* 10: 1–4.

Anderson, C., 1995. Eat ethnic in Johannesburg. *Air Tales* 1: 28–30.

Anonymous, 1993. Nurturing a cornucopia of potential. *Washington Post,* September 20.

Anonymous, 1994. Under-utilized plants: widening Africa's food base. *IPGRI Newsletter for Sub–Saharan Africa,* 4 (2).

Attere, A.F., 1990. Conservation and use of indigenous vegetable resources in the SADCC region. In: *Vegetable Research and Development in SADCC Countries: Proceedings of a Workshop,* held in Arusha, Tanzania, 9–13 July.

Begemann, F., 1988. Ecogeographic differentiation of Bambarra groundnut (*Vigna subterranea*) in the collection of the International Institute of Tropical Agriculture (I.I.T.A). *Dr. Agr. Dissertation.* Techischen Universität Munchen.

Chweya, J.A., 1989. The role of indigenous vegetation in human food production. *A Paper Presented at the Indigenous Vegetation Experience Sharing Symposium*. Organised by Kengo, 20–22 September, JKCAT (Thika), Kenya.

Burkill, T.H., 1906. Bambara groundnut *Voanzeia subterranea,* Thou. *Bulletin of Miscellaneous Information of the Royal Botanic Gardens,* Kew, London, UK, 20: 68–70.

Coudert, J., 1984. Market openings in West Africa for cowpeas and Bambara groundnuts. *International Trade Forum,* 20(1):14–15, 28–29.

Cunningham, A.B., De Jager, P.J. and Hansen, L.C.B., 1992. The indigenous plant use programme. Foundation for Research Development, Pretoria, South Africa.

FRD, 1992. *Rural Poverty Alert: Challenging Science and Technology*. Foundation for Research Development, Pretoria, South Africa.

Grubben, G.J.H. and Almekinders, C.J.M., 1996. Developing the potential of local vegetables using experiences from Africa and South-East Asia. *Paper Delivered at Cameroon Meeting of NRI.*

Haq, N., 2007. Under-utilized fruits: A resource for sustainable livelihoods. In: *Breeding of Neglected and Under-utilized Crops, Spices and Herbs,* (Eds.) S. Ochatt and S.M. Jain. Science Publishers, Enfield, Jersey, Plymouth, pp. 1–28.

Hepper, F.N., 1970. Bambara groundnut (C. *Voandzeia subterranean*). *F.C.A.,* 23: 1–6.

Kay, D.E., 1979. *Crop and Product Digest, 3–Food Legumes*. Tropical Products Institute, London.

Mwashayeni, E., 1994. Indigenous vegetables: Challenge to researcher, extensionist and consumer. *ILEIA Newsletter,* July, 31.

Schippers, R.R., 2000. *African Indigenous Vegetables: An Overview of the Cultivated Species.* Natural Resources Institute/ACP–EU Technical Center for Agricultural and Rural Cooperation Chatham UK.

Sofowora, A., 1996. Research on medicinal plants and traditional medicine in Africa. *J. Alternative Complementary Medicine,* 2: 365–372.

Tardieu, M., 1958. Secondary crops in the action zone of the Agronomic Research Centre, Bambey. Bulletin Agronomique Minestère de la France d'Outre Mer 17: 5–54. French. In: *Bambara Groundnut (Vigna subteranea) Literature: A Revised and Updated Bibliography,* Compiled by Anita R. Linnemann. Wageningen Agriculture University, 1992.

Tyson, P.D., 1990. Modelling climatic change in southern Africa: a review of available methods. *Sth. Afr. J. Sci.*, 86: 318–330.

Venter, S.L., Jansen van Rensburg, W.S., Van den Heever, E., Vorster, H.J. and Allemann, J., 2007. Indigenous crops with potential but under-utilized in South Africa. In: *Breeding of Neglected and Under-utilized Crops, Spices and Herbs,* (Eds.) S. Ochatt and S.M. Jain. Science Publishers, Enfield, Jersey, Plymouth, pp. 391–427.

Verdcourt, B., 1980. The correct name for the Bambara groundnut. *Kew Bull.*, 35(3): 474.

Wehmeyer, A.S., 1986. Why so little research on the ennoblement of indigenous edible wild plants? *Acta Hort.*, 194: 47–53.

Index